ENGINEERING for FOOD SAFETY and SANITATION

ENGINEERING for FOOD SAFETY and SANITATION

A Guide to the Sanitary Design of Food Plants and Food Plant Equipment

THOMAS J. IMHOLTE

Best of Luck
Hope You Like The Book

TIFS

Technical Institute of Food Safety

Independently Produced by Thompson & Company and
Taly Design Group, Inc., Minneapolis, Minnesota

Managing Editor: Amy Dwyer
Editors: Ann Kavanaugh
Denny McClelland
Editorial Assistants: Ruth Lent
Michaelene Kelly
Proofreading: Chris Larsen
Book Design and Production: Henry Taly
Cecilia Wong
Catherine Knaeble
Adam Goldstein
Lisbeth Conigliaro
Marci Howe
Cover Design: Henry Taly
Photos by General Mills
Galen Erickson, Photographer
Technical Illustrations: Illustrations provided by author
Sandra Kapitan
Gary Burton
CAD-CAM illustrations courtesy
of General Mills

Printed in the United States of America

ISBN 0-918351-00-6

To the Food Industry
In the Interest of Their Continued Success
In Producing Safe, Wholesome Food Products

CONTENTS

ILLUSTRATION and PHOTOGRAPH CREDITS ix

PREFACE xi

ACKNOWLEDGEMENTS xiii

1 INTRODUCTION to FOOD SANITATION 1

FOOD MANUFACTURING INDUSTRY 2

PRODUCT RISKS 2

RECOGNIZING RISKS IN THE MANUFACTURING ENVIRONMENT 9

PERSONAL HYGIENE AND PERSONNEL PRACTICES 10

SANITATION PROGRAMS 11

CONCLUSION 12

2 PLANT LAYOUT 13

SELECTING THE FOOD PLANT LOCATION 14

CHOOSING THE BUILDING TYPE AND SIZE 14

FACILITY PLANNING 16

EQUIPMENT LAYOUT 21

EQUIPMENT SUPPORTS 29

CATWALKS AND ACCESS PLATFORMS 32

SANITATION SPECIFICATIONS 34

3 BUILDINGS 37

GENERAL CONSIDERATIONS IN BUILDING DESIGN 38

FOUNDATIONS AND FRAMING 41

FLOORS 46

WALLS 56

ROOFS 63

CEILINGS 67

BUILDING INSULATION 68

WINDOWS 68

ENTRANCE AND EXIT DOORS 69

LIGHTING 76

VENTILATION 79

PLUMBING 86

UTILITY PIPING 93

EMPLOYEE FACILITIES 101

4 EQUIPMENT 105

GENERAL EQUIPMENT DESIGN CONSIDERATIONS 106

EQUIPMENT CONSTRUCTION MATERIALS 113

CONSTRUCTION FEATURES 118

PROCESS CONTROLS 130

CONVEYING SYSTEMS 132

BULK SOLID STORAGE BINS AND USE BINS 143

LIQUID INGREDIENT STORAGE AND TRANSFER SYSTEMS 147

DUST COLLECTION SYSTEMS 151

5 ELECTRICAL SYSTEMS 155

ESTABLISHING DESIGN REQUIREMENTS 156
ELECTRICAL EQUIPMENT 156
ELECTRICAL HARDWARE 160

6 CONSTRUCTION 163

BUILDING MATERIAL AND EQUIPMENT INSPECTIONS 164
CONSTRUCTION PROGRESS INSPECTIONS 165
FIELD FABRICATION 168

7 MAINTENANCE 169

PREVENTIVE MAINTENANCE 170
BUILDING MAINTENANCE 170
EQUIPMENT MAINTENANCE 174
TEMPORARY REPAIRS 178
EQUIPMENT CLEANOUT 178
MAINTENANCE SHOP AND STORAGE AREAS 180
MAINTENANCE DURING OPERATIONS 180
MISCELLANEOUS MATERIALS 181
NEW INSTALLATIONS 181
APPROVED LUBRICANTS 181

8 CLEANING 183

WET CLEANING 184
CLEAN-OUT-OF-PLACE (C.O.P.) SYSTEMS 185
CLEAN-IN-PLACE (C.I.P.) SYSTEMS 187
CENTRAL CLEANING SYSTEMS 191
INDIVIDUAL HOT WATER HOSE STATION 196
DRY CLEANING 200

9 SOLID WASTE HANDLING SYSTEMS 203

LOCATION 204
BUILDINGS AND EQUIPMENT 205
OUTSIDE LOCATIONS 206

10 PEST CONTROL SYSTEMS 209

COMMON INSECTS 210
CONTROL SYSTEMS 213
RODENT CONTROL WITH ULTRA-SONICS 232

11 FOREIGN MATERIAL CONTROL 233

MAGNET APPLICATIONS 234
METAL DETECTORS 254
SCALPING AND SIFTING 261
ASPIRATORS 266

12 FOOD LAWS and REGULATIONS 267

INSPECTIONS 268
FEDERAL FOOD, DRUG AND COSMETIC ACT, AS AMENDED 268
THE FEDERAL MEAT INSPECTION ACT AND THE FEDERAL POULTRY INSPECTION ACT 276
OTHER FEDERAL, STATE, AND LOCAL LAWS AND REGULATIONS 276

REFERENCES 277

INDEX 279

ILLUSTRATION and PHOTOGRAPH CREDITS

We gratefully acknowledge the use of the following photographs and illustrations:

CHAPTER 1

FIG. 1-1, Dr. Edmund A. Zottola, University of Minnesota, p. 3; FIG. 1-3, Industrial Fumigant Company, p. 6.

CHAPTER 2

FIG. 2-3, Ralston Purina Company, p. 16; FIG. 2-6, Economics Laboratory, Inc., Klenzade Division, p. 19; FIG. 2-7, Economics Laboratory, Inc., Klenzade Division, p. 20.

CHAPTER 3

FIG. 3-1, Ralston Purina Company, p. 39; FIG. 3-2, Ralston Purina Company, p. 40; FIG. 3-15, Atlas Minerals and Chemicals, Inc., p. 50; FIG. 3-49, Ralston Purina Company, p. 81; FIG. 3-57, Watts Regulator Company, p. 90.

CHAPTER 4

FIG. 4-3, Food Engineering Corporation, p. 108; FIG. 4-48, Economics Laboratory, Inc., Klenzade Division, p. 148; FIG. 4-49, Owens Corning Fiberglas Corporation, p. 149.

CHAPTER 8

FIG. 8-6, Economics Laboratory, Inc., Klenzade Division, p. 189; FIG. 8-7, Economics Laboratory, Inc., Klenzade Division, p. 191; FIG. 8-12, Economics Laboratory, Inc., Klenzade Division, p. 196; FIG. 8-16, Strahman Valves, Inc., p. 199; FIG. 8-17, Hoffman Air & Filtration Systems, Division of Clarkson Industries, Inc., p. 200; FIG. 8-18, Hoffman Air & Filtration Systems, Division of Clarkson Industries, Inc, p. 201.

CHAPTER 10

FIG. 10-1, Dr. Philip Harein, University of Minnesota, p. 210; FIG. 10-2, David K. Mueller, RPE, Fumigation Service and Supply, Inc./Insects Limited Inc., p. 211; FIG. 10-3, David K. Mueller, RPE, Fumigation Service and Supply, Inc./Insects Limited Inc., p. 211; FIG. 10-4, David K. Mueller, p. 211; FIG. 10-5, David K. Mueller, p. 211; FIG. 10-6, Degesch American, Inc., p. 212; FIG. 10-7, Degesch American, Inc., p. 212; FIG. 10-8, Degesch American, Inc., p. 212; FIG. 10-9, Don Gilbert Industries, Inc., p. 213; FIG. 10-10, Don Gilbert Industries, Inc., p. 214; FIG. 10-11, Don Gilbert Industries, Inc., p. 215; FIG. 10-12, Micro-Gen Equipment Corporation, p. 216.

CHAPTER 11

FIG. 11-7, Eriez Magnetics, Eriez Manufacturing Company, p. 240; FIG. 11-8, Eriez Magnetics, Eriez Manufacturing Company, p. 240; FIG. 11-9, Eriez Magnetics, p. 241; FIG. 11-10, Eriez Magnetics, p. 241; FIG. 11-11, Eriez Magnetics, p. 242; FIG. 11-12, Eriez Magnetics, p. 242; FIG. 11-13, Eriez Magnetics, p. 242; FIG. 11-14, Eriez Magnetics, p. 243; FIG. 11-15, Eriez Magnetics, p. 243; FIG. 11-16, Eriez Magnetics, p. 243; FIG. 11-17, Eriez Magnetics, p. 243; FIG. 11-18, Eriez Magnetics, p. 244; FIG. 11-22, Eriez Magnetics, p. 245; FIG. 11-32, Eriez Magnetics, p. 252.

PREFACE

The food industry has made a great deal of progress over the years in producing safe, wholesome food products. This progress first began with the advances in public health that brought about pasteurization, mechanical refrigeration, and sanitary improvements in manufacturing processes. More recently, progress has come both from the advent of food additives designed to stabilize and preserve foods and from technological improvements in manufacturing. As sanitation practices have improved, better designs that allow for easier cleaning have become the keys to improving sanitation while lowering operating costs. These design efforts are becoming apparent in today's modern and efficient manufacturing facilities, facilities which are being kept cleaner than ever before. What is needed now is to expedite this progress through more communication about these recent design improvements and about design improvements still needed. This is the goal of *Engineering for Food Safety and Sanitation.*

Engineering for Food Safety and Sanitation is written primarily for design engineers, to provide them with clear, concise, and easily accessible information on sanitation and product safety requirements. It is meant to be used as a tool to guide the engineer in the design of manufacturing facilities and equipment. Though *Engineering for Food Safety and Sanitation* identifies major design problems and presents possible solutions, it does not mandate certain practices; rather, it suggests solutions and challenges the designer to improve upon those solutions offered.

Engineering for Food Safety and Sanitation is also designed to aid professional sanitation managers and the numerous other plant personnel who identify sanitation failures and solve ongoing problems. The pictures and illustrations are intended to improve communication between food plant personnel and design engineers. The drawings expand upon the information presented in the text by showing important sanitary design details and concepts; the photographs take this process one step further by clearly illustrating, for both engineers and non-engineers, many sanitary design features.

Engineering for Food Safety and Sanitation is not intended to replace those standards written for specific segments of the food industry such as the 3-A Dairy Standards, Baking Industry Sanitation Standards Committee (BISSC) bakery standards, and the select group of United States Department of Agriculture standards for the meat industry. Instead, it is meant to complement these fine works and be used as a supplementary guide. *Engineering for Food Safety and Sanitation* is also intended to be used as a guide for the many segments of the food manufacturing industry that are without standards.

The book begins with an overview of the fundamentals of food sanitation. Engineering personnel will find this introductory material of particular benefit as it is these fundamentals that establish the sanitary design requirements upon which sanitary design is based. In later chapters, specific design requirements that prevent product contamination and make the facility and the equipment easier to clean are presented. Other chapters deal with product safety issues that demonstrate the importance of including specially designed equipment that can cope with bacteria, insect, and foreign material concerns. The last chapter highlights the important role of the various government regulatory agencies.

While an attempt was made to present the most current and complete overview of modern food sanitation, this book by no means represents the final word on the subject. New innovations are always needed to meet the demands of the modern consumer for wholesome, safe food products. *Engineering for Food Safety and Sanitation* will have served its purpose well if the information provided leads to further innovations in the field.

ACKNOWLEDGEMENTS

A very special thanks to General Mills, the Package Foods Operations Division Engineering Department, and their manufacturing plants for their many contributions to this book. In particular, I want to thank Karl Potter, electrical engineer, for his help in preparing Chapter Five and Shirley Timmerman, secretary, for her assistance in preparing the manuscript.

A special thanks also to James Rutledge, Lauhoff Grain Company, Danville, Illinois, for reviewing the manuscript.

A special thank you to the following people or firms for their pictures or drawings:

Dr. Philip Harein, Professor and Extension Entomologist, University of Minnesota, Minneapolis, Minnesota

Dr. Edmund A. Zottola, Professor of Food Microbiology, University of Minnesota, Minneapolis, Minnesota

Mr. Don Gilbert, Don Gilbert Industries, Inc., Jonesboro, Arkansas

David K. Mueller, RPE, (artist for insect drawings), Fumigation Service and Supply, Inc., Indianapolis, Indiana

Industrial Fumigant Company, Olathe, Kansas

Ralston Purina Company, St. Louis, Missouri

Atlas Minerals and Chemicals, Inc., Mertztown, Pennsylvania

Economics Laboratory, Inc., Klenzade Division, St. Paul, Minnesota

Food Engineering Corporation, Minneapolis, Minnesota

Owens Corning Fiberglas Corporation, Toledo, Ohio

Hoffman Air and Filtration Systems, Division of Clarkson Industries, Inc., E. Syracuse, New York

Eriez Magnetics, Eriez Manufacturing Company, Erie, Pennsylvania

Edward A. Bonelli, and Associates, San Francisco, California

Micro-Gen Equipment Corporation, San Antonio, Texas

Watts Regulator Company, Lawrence, Massachusetts

Strahman Valves, Inc., Florham Park, New Jersey

Entoleter Company, New Haven, Connecticut

Kice Metal Products Co., Inc., Wichita, Kansas

Degesch American, Inc., Weyers Cave, Virginia

Taylor Instrument, Rochester, New York

Barkley and Dexter Laboratories, Inc., Fitchburg, Massachusetts

CHAPTER 1 INTRODUCTION to FOOD SANITATION

CHAPTER CONTENTS

FOOD MANUFACTURING INDUSTRY

PRODUCT RISKS

RECOGNIZING RISKS IN THE MANUFACTURING ENVIRONMENT

PERSONAL HYGIENE AND PERSONNEL PRACTICES

SANITATION PROGRAMS

CONCLUSION

Safe and sanitary food manufacturing is a science. It requires a good working knowledge of microbiology, entomology, toxicology, pest animal behavior, and food technology. The safety and sanitation manager develops and implements programs to ensure a safe, sanitary product. The design engineers are responsible for the overall structure of the plant, including building type, equipment fabrication, and access.

The importance of good engineering for food safety and sanitation is suggested by the title of this book. Without good engineering, food safety and sanitation programs can become significantly more costly. In addition, the effectiveness of these programs is reduced, and problems such as food poisoning, spoilage, insect infestations, and injurious foreign material contaminations are more likely to occur.

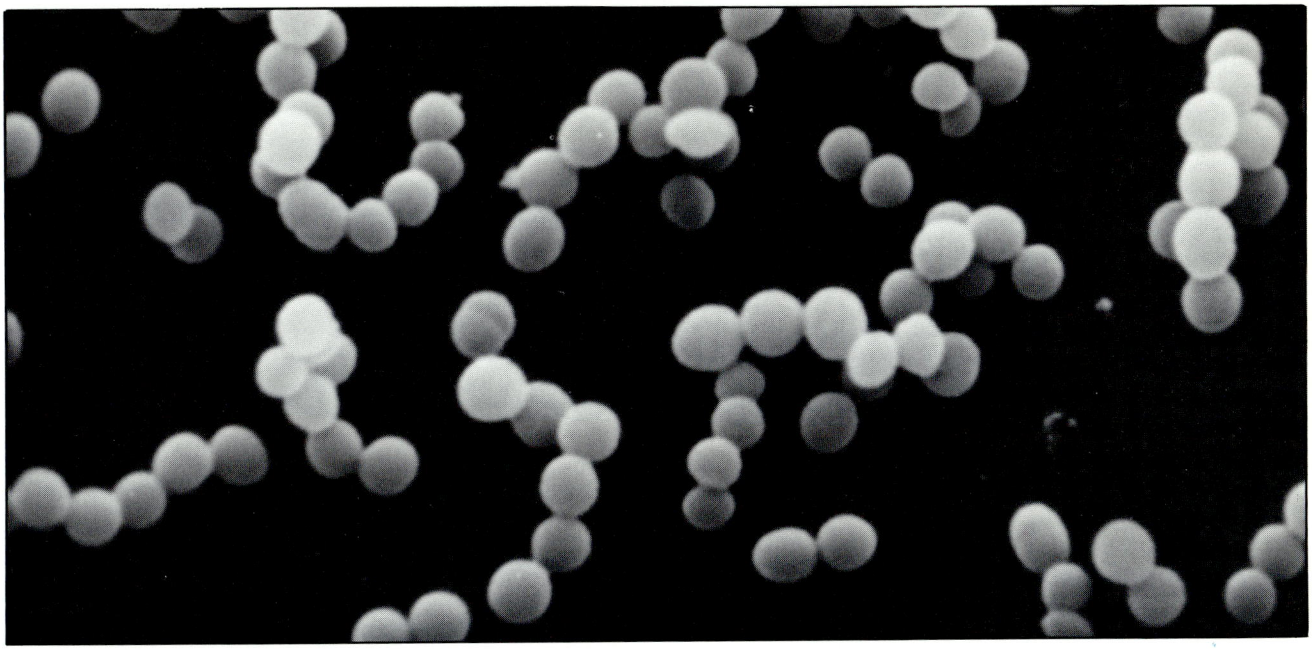

The principles of food safety and sanitation are sometimes perceived as common knowledge, which any sensible, responsible person should recognize. On the contrary, food safety and sanitation is a complex subject. The more that engineers, in particular, know about it, the better their judgments and decisions are likely to be.

This chapter presents an overview of food safety and sanitation issues. It begins with a summary of the different types of food manufacturing and the product risks associated with each. The product risk provides a good starting point, because it determines the nature of a food safety and sanitation program. The discussion of product risk touches on some fundamentals of microbiology, entomology, and pest animal behavior. Also covered are some common control techniques and the role of engineering in these techniques. The chapter addresses foreign material contamination from chemicals and physical objects. Because human beings can pose a significant contamination threat to food, good employee hygiene practices are set forth. These practices apply not only to manufacturing employees but to engineers, maintenance personnel, equipment servicemen, and construction crews. Finally, the chapter looks at how sanitation programs are planned for particular manufacturing environments and the product risks associated with these environments.

FOOD MANUFACTURING INDUSTRY

The food manufacturing industry can be divided into the following broad (and by no means all-inclusive) segments: baking and confectionary, grain processing, cereals and prepared dry mixes, pasta, canning, meat and poultry, dairy, and frozen foods.

Many of the food sanitation requirements are fundamental to the entire food manufacturing industry regardless of the specific segment. However, product safety risks and product sensitivities to certain types of food contamination vary from segment to segment. These risks and sensitivities can even vary within a particular product line. A food plant design engineer will find it advantageous to be aware of these specific sensitivities. Such awareness can result in significant design cost savings.

PRODUCT RISKS

Most product risks or product sensitivities fall into one or more of the following areas:

- Microbe-sensitive
- Insect-sensitive
- Rodent infestation
- Foreign material contaminations, both physical and chemical

MICROBE-SENSITIVE

It would be somewhat difficult to list all the products that are considered microbe-sensitive. Animal products or byproducts such as eggs, milk, or cheese fall into this category. Nonetheless, many high-protein products, low-acid vegetables and certain types of grains are also microbe-sensitive.

Bacteria. Bacteria are tiny living organisms which are visible only under a microscope. They measure approximately 1/25,000 of an inch in size. *(Figure 1-1)* When ideal growth conditions are present, they reproduce through an asexual method every 20-30 minutes. These conditions are temperatures of 80°-100°F, moisture levels of approximately 35%, and low acid levels. In the absence of one or more of these factors the reproduction rate is slowed.

The rapid growth rate of bacteria is of particular significance because they are considered most harmful in large numbers. Of most concern are the pathogens that cause disease. The presence of even a single pathogenic species is very dangerous because of the rapid reproductive capabilities of bacteria. Nonetheless, not all bacteria are necessarily bad. In fact, bacteria are used in the manufacture of such foods as yogurt and sour cream.

Food Poisoning. Certain bacteria cause food poisoning, which can result in serious illness and can even kill. The following table contains a partial list of the more common types of food poisoning caused by bacteria, along with the symptoms and the foods usually involved.

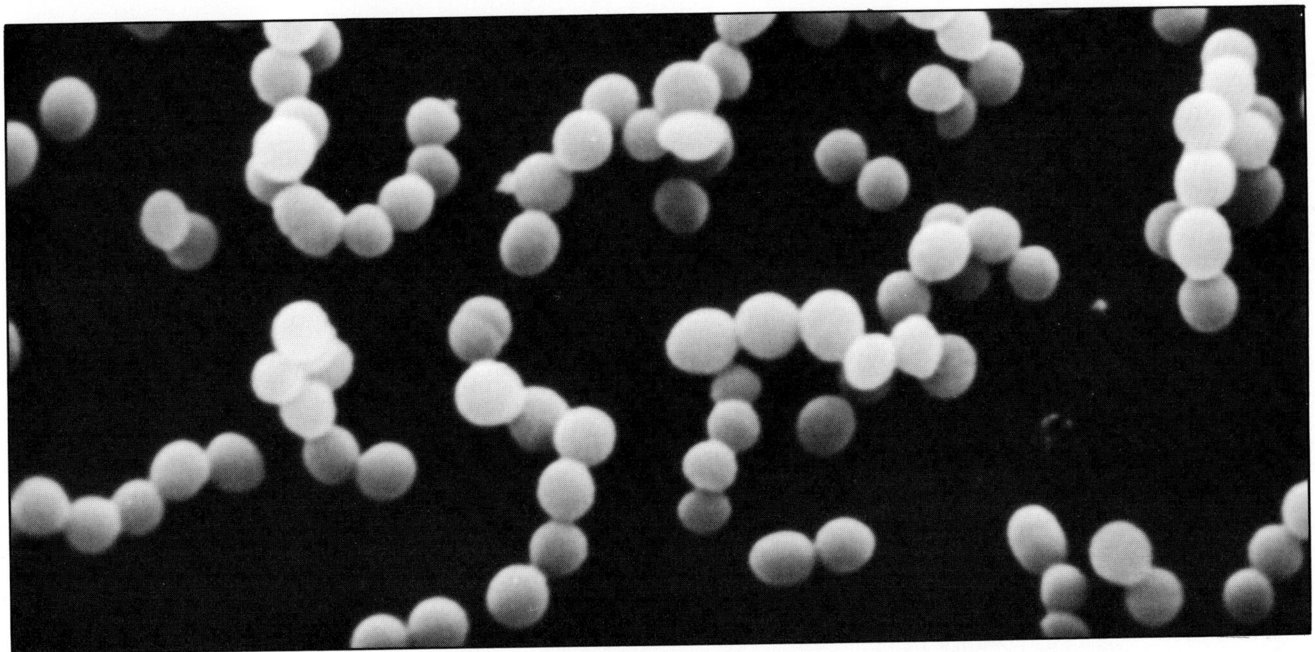

FIG. 1-1 Streptococcus lactis bacteria as seen using a scanning electron microscopy technique with a 5000X magnification.

TABLE 1-1 COMMON FOOD POISONINGS

Food Poison	Onset of Symptoms	Foods Involved	Symptoms
Salmonellosis	5 to 72 hours	Poultry, meats, fish, eggs	Stomach pain, diarrhea, frequent vomiting, chills, fever and prostration.
Staphylococcus (Staph)	1 to 6 hours	Bakery cream fillings, potato salad, meats	Stomach cramps, nausea diarrhea, vomiting, prostration, normally no fever.
Clostridium Perfringens	2 to 18 hours	Meats, reheated meats, meat pies cold cuts, stews meat dishes	Nausea, stomach cramps and diarrhea. Normally no vomiting. Lasts 8 to 12 hours.
Botulism	6 hours to 8 days	Home-canned, low acid vegetables	Can cause death through paralysis of respiratory system. Difficulty in swallowing and breathing.

- Salmonellosis. This type of food poisoning is commonly misdiagnosed as a serious case of influenza. Under some extreme conditions, it can cause death. Usually these conditions involve an infant, child, or elderly person whose bodily defense system has been weakened.

- Staphylococcus. Staphylococcus food poisoning is quite similar to salmonellosis, and it, too, is commonly misdiagnosed as the flu. However, staphylococcus only lasts from two to three days, while salmonellosis can last for seven to fourteen days, or longer.

- Clostridium Perfringens. This type of food poisoning is also similar to salmonellosis. However, the illness is not quite as severe, and it normally only lasts eight to twelve hours.

- Botulism. This is said to be the most deadly food poisoning known to man. Most statistics suggest a fatality rate of 20-25%. Death is caused by paralysis of the respiratory system. A bacteria called clostridium botulinum secretes a poison that causes botulism. This poison is not heat-stable and it can be destroyed by pasteurization temperatures.

  Botulism poisoning was more common during the home canning era. Many home canning processes did not complete pasteurization of the canned product; consequently, neither the bacteria nor the poison produced by the bacteria was destroyed. Modern canning technology includes temperature monitoring devices and failure alarms which virtually assure that effective pasteurization temperatures have been reached and held for sufficient time to complete pasteurization of the foods being canned. In addition, intense quality control sampling programs back up the modern canning process. Samples are taken regularly of all canned products considered sensitive to botulism contamination. These samples are thoroughly analyzed for the presence of clostridium botulinum.

Mold. Molds are plants classed as fungi. *(Figure 1-2)* Molds grow from single-celled reproductive microorganisms called spores. Spores can also reproduce through asexual methods. Although very tiny, mold spores are macroscopic, or faintly visible to the naked eye. They are easily airborne, and once they settle on food products and on food product surfaces, they develop rapidly.

Mold is found in a variety of foods and in many manufacturing environments, particularly where moisture is present. Moisture levels of only 15% are needed to start a culture. Mold is a serious problem, not only because it destroys food, but because several types of mold produce mycotoxins. Some of these mycotoxins are known to cause cancer in laboratory test animals.

Microbe Control. Microbiologically sensitive food manufacturing systems are generally more expensive to design because the control of bacteria and mold requires thorough cleaning and sanitizing of the entire surface of a product zone. The contact zones of equipment must be constructed of stainless steel and they must be entirely exposed to the cleaning process. An equipment design that provides for disassembly or for access doors assures thorough, easy cleaning.

Some sophisticated, fully automated clean-in-place systems can accomplish cleaning without any need for disassembly or for access doors. In such cases, the equipment design itself is critical. It must be completely free of pits, checks, cracks, and crevices where food particles can lodge and bacteria can grow. In addition, bacteria-sensitive products will require pasteurization and other sterilization equipment.

Mold control in damp, moist manufacturing environments generally calls for frequent air exchanges and, in extreme cases, dehumidifying equipment. Surfaces in these environments must be smooth, impervious to moisture, and easy to clean. Any wet manufacturing systems or parts of wet manufacturing systems require similar design requirements, regardless of their degree of microbe-sensitivity.

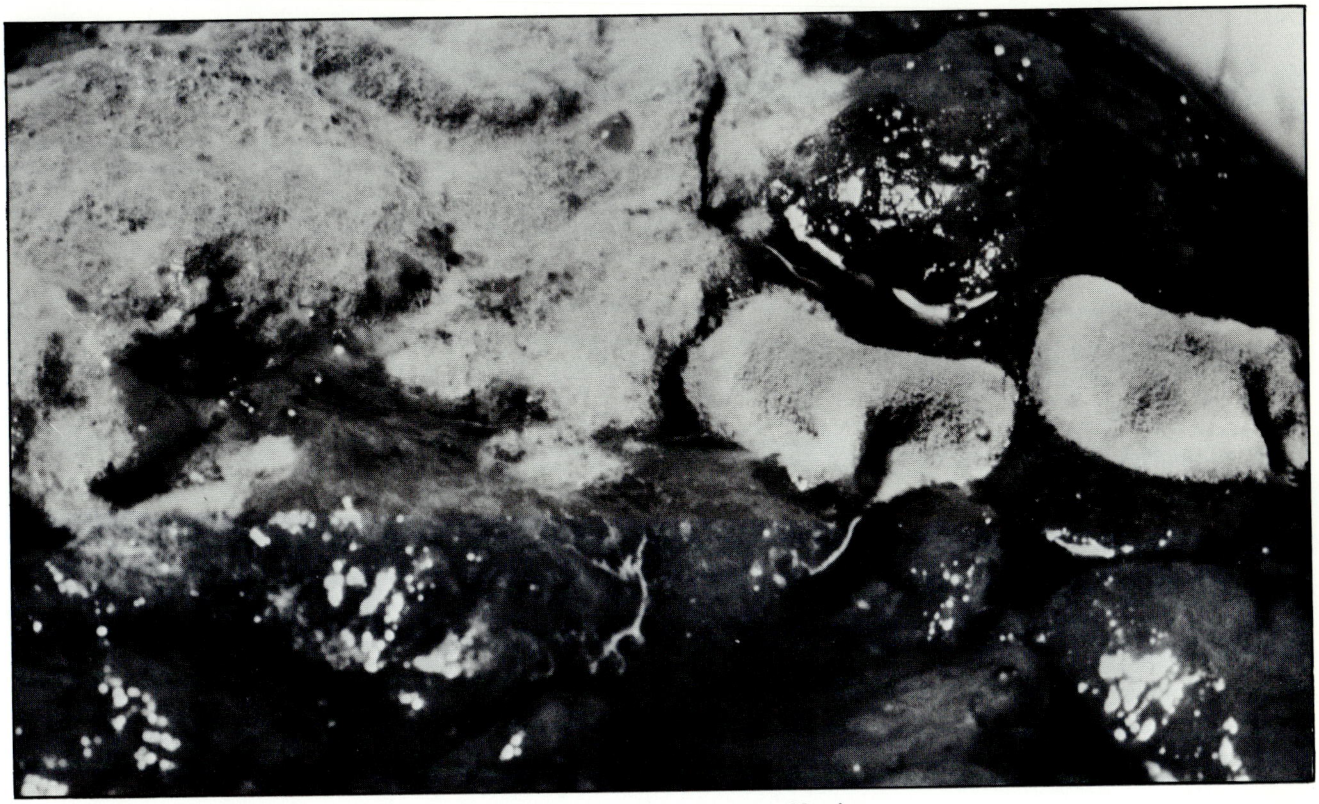

FIG. 1-2 Molds are plants classed as fungi. Some molds produce mycotoxins, or mold poisons.

INSECT-SENSITIVE

All segments of the food industry are bothered by insects to some extent. House and fruit flies are found almost everywhere. Dermestid groups, which consist of tiny beetles, are also commonly found. The grain processing segment in particular is troubled by the stored-product pests. These pests are tiny insects that feed on grain products in all stages of manufacture, from grain kernels to flour to the final end product.

Generally insects are regarded as only a minor nuisance. This is not necessarily true. Insects carry disease-causing bacteria, and the damage they cause to products costs manufacturers millions of dollars annually.

Additional millions are spent each year controlling insects. The engineer's sanitary design decisions have a tremendous impact on the magnitude of product losses and on control efforts. While engineers are certainly not required to become degreed entomologists, they should have some familiarity with the product's degree of insect sensitivity. They should also inquire about proposed control methods, as they could have a distinct effect on the selection of building materials. They can also determine some special equipment design needs or access needs into process equipment.

INSECT BIOLOGY

The development of insects from eggs to adults varies with the species of insect. Most insects, including stored product pests, develop through what is called a *complete metamorphosis.* This means they go through four distinct stages of development: egg, larva, pupa, and adult.

In the initial stages, wormlike larvae emerge from tiny eggs. In the pupa stage, the adult shape of the insect starts to develop within the larvae skins. When the insects reach the adult stage, they shed the skins and take on the shape and appearance common to all insects: six legs and three body segments. *(Figure 1-3)* Under good growth conditions, insects can move through these life stages in three to four weeks.

Another form of insect development is *gradual metamorphosis.* In this development process, the insect goes from the egg to an immature *nymph* stage. The nymph then develops into an adult. Cockroaches, termites, and grasshoppers are a few of the insects that develop in this manner.

INSECT CONTROL

Insects are controlled in the plant with good cleaning practices, with chemicals, and, in some cases, with heat. Electric insect light traps are also very effective in controlling flying insects. It is often said that 80% to 85% of insect control consists of good cleaning practices. Without such practices, insect control becomes an exercise in futility.

Some common methods used to control insects are spraying, space spraying/misting, equipment fumigations, space fumigations, general fumigations, and heat sterilization.

Spraying. Spraying consists of applying liquid insecticides with a one or two gallon compressed air sprayer. Chemical insecticides represent a serious contamination threat to food. Consequently, government regulations severely restrict their use in manufacturing areas. Only certain chemicals are approved. The quantities used and the method of application are also closely regulated. Many applications are restricted to crack and crevice treatment. This means that a fine pin stream nozzle must be used on the sprayer to direct the insecticide into the crack or crevice.

Space Spraying/Misting. Space spraying and misting use a pneumatic or mechanical misting device to apply pesticides. A thermo-fogging machine can also be used for this method. The type of liquid pesticide used for this treatment, and the quantity per given area are carefully specified. Product zone surfaces must either be protected from contact with insecticides or they must be washed before production begins.

Equipment Fumigations. Equipment Fumigations are more commonly used in the grain-related industry which includes baking, pasta, cereal and flour milling. Equipment

FIG. 1-3 A complete metamorphosis; the four-stage development process of most insects.

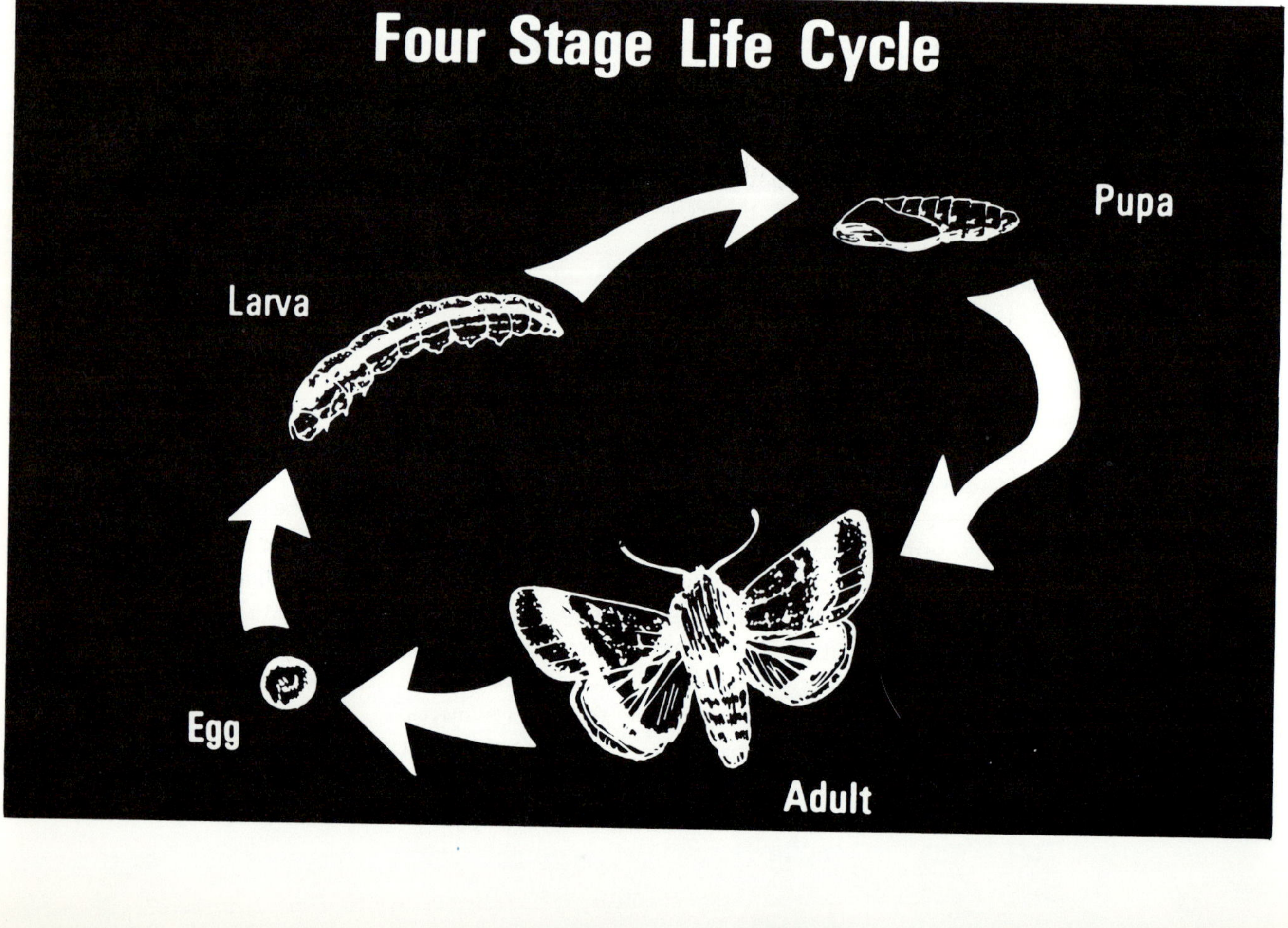

fumigations are accomplished by closing up the equipment and sealing any openings where the fumigants can escape. Then the fumigants are introduced into the equipment to be held for 16 to 24 hours. Until recently, liquid gas such as ethylene dibromide, ethylene dichloride and carbon tetrachloride were used for this purpose. One of these fumigants, ethylene dibromide, has since been removed from the market; similar measures are under serious consideration for the other two. If these chemicals are removed from the market, it is believed that substitute fumigants will be approved and used for fumigations of this type. If continued use of the chemicals is prohibited, then the need for better equipment design will increase. It has always been well known that insect infestations in equipment could be controlled through thorough equipment clean-outs. But the design of the equipment has sometimes caused cleaning problems. Equipment which was typically fumigated could not be thoroughly cleaned because its design did not allow good access or because it was full of small cracks and crevices where insects could harbor. There was simply no practical way to clean these cracks and crevices.

Space Fumigations. Space fumigations use a similar approach and some of the same fumigants as equipment fumigations. The difference is that space fumigations are used on small areas of the plant, not just on equipment. Solid form or gaseous fumigants are used, rooms or small areas of the plant are sealed, and fumigants are introduced into the area. After being held for approximately 24 to 48 hours, the fumigant is evacuated from the area through aeration. Before normal activities are resumed, the area is checked for the presence of gas with appropriate equipment.

General Fumigations. This type of fumigation usually involves large areas, such as the entire manufacturing or storage area. Once the fumigant has been introduced into the tightly sealed plant, concentrations are monitored to make sure they are sufficient to obtain kills. The fumigant is held for a predetermined period of time, and then evacuated. Finally, the structure is checked for the presence of any gas before normal plant activities are resumed. Fumigants commonly used are methyl bromide, phosphine and, sometimes, hydrogen cyanide. General fumigations are very costly to do. Sealing the structure to make it gas-tight accounts for the largest share of the cost.

Because some pesticide chemicals used for control are very corrosive, special materials must be used in the construction of manufacturing equipment. Some building materials, such as wood and cinder block, are not structurally suitable for fumigation. They do not hold gas well or their incompatibility with the fumigant causes odor problems. These issues can be resolved by consulting the plant sanitation and safety manager.

Heat Sterilization. Heat sterilization was used in the early days of flour milling to control stored product pests, but heat dried out wood housed equipment and caused it to crack. Since most equipment today is constructed of metal, drying and cracking are no longer a problem. Faced with increased pressure for removal of pesticides from the marketplace, many food plants are again giving consideration to heat sterilization. It is safe to work with and it is compatible with almost all food products. Only foods sensitive to higher temperatures need to be removed from the area to be heated. Heat sterilization and insect light traps for flying insect control will be discussed in more detail in Chapter Ten.

Elimination of objectionable design conditions such as cracks, crevices and ledges can do away with the need for fumigations or other chemical and perhaps even some mechanical pest controls. Insect-sensitive ingredients or finished food product can lodge or become trapped in these areas. Once this occurs, insects soon follow to set up homesteads and to reproduce.

RODENT INFESTATIONS

Short of a deadly food or chemical poisoning outbreak, nothing causes quicker regulatory action than rodent infestation in a food plant or storage area. Rats and mice are responsible for the spread of a number of diseases, either directly, through contamination of human food with their urine or feces, or indirectly, through fleas or mites that have been living on their bodies. Some of the more common diseases caused by rodents are: rat-bite fever, murine typhus fever, salmonellosis, trichinosis, and the Black Death plague.

Rodents are said to outnumber humans many times over. Like insects, they destroy millions of dollars worth of food every year.

Rodent Biology. Probably the greatest difficulty which rodent control encounters is the fantastic reproductive capability of these animals. A mouse reaches sexual maturity in 1-1/2 months. The gestation period of a female mouse is around 19 days, and she is capable of having as many as eight litters per year, with each litter containing five to six animals. Fortunately, the life span of a mouse is only one year, at most.

The reproduction characteristics of rats vary slightly. Sexual maturity is reached in 2-1/2 months, and the gestation period is 21 days. Rats have four to seven litters per year with six to eight animals or more per litter.

RODENT CONTROL

While the problem might seem overwhelming, given the sheer numbers, rodent control is not complicated. It simply requires the use of these basic control tools:

- Getting rid of exterior attractors (food and a place to live)
- Exterior baiting and trapping
- Rodent-proofing of buildings
- Getting rid of interior attractors (same as exterior attractors)
- Interior trapping
- Inspecting
- Ultra-sonic devices

Getting Rid of Exterior Attractors. Getting rid of exterior attractors means getting rid of any food or product spills on the ground and getting rid of places where rodents can harbor or live. Keeping the grounds well maintained makes it easier to observe any rodent activity. Elevating outside equipment 9 to 12 inches deprives them of a possible harbor. These measures are important not only for rodent control, but for the plant's community image.

Exterior Baiting and Trapping. Most food manufacturing facilities establish an exterior baiting and trapping program that consists of two lines of defense. The first line is placed on the perimeter of the property or approximately 50 yards from the buildings. This line consists of bait stations placed approximately 50 to 75 feet apart. Bait stations are locked metal boxes that have a hole in each end for rodents to enter. Bait, usually anticoagulant, is placed in the box. Anticoagulant bait interferes with the rodent's ability to form bloodclots; as a result, the rodent bleeds to death.

The second line of defense consists of bait stations at the foundation walls of the buildings; they are also placed at 50 to 75 foot intervals. A few mouse traps placed near building entrances further reduce any intrusions from mice. Wind-up traps or similar multiple catch traps are useful for this purpose. The whole idea of the exterior baiting and trapping program is to keep rodents outside of the buildings.

Rodent-proofing of Buildings. Rodent-proofing consists of sealing off every conceivable point where a rodent can get into the building. An opening no larger than 1/4 to 3/8 of an inch is all that a rodent needs to get into the plant. Moreover, rodents are excellent climbers and jumpers. With a running start, they can jump as high as three feet, and they can climb straight up vertical concrete walls. A full-grown rat can reach as high as 18 inches. During fall and spring migrational periods, rodents are on the move in search of food and a comfortable, warm place to live. The fall in particular is a good time to make a rodent-proofing inspection of the plant buildings.

Getting Rid of Interior Attractors. Interior attractors are essentially the same as exterior: food to eat and a place to live. All interior holes in floors, block walls and other parts of the plant should be filled. Idle equipment storage areas with rack storage will get the equipment off the floor. Finally, all food and waste containers should be covered.

Interior Trapping. The interior trapping program includes mouse traps placed about 30 feet apart against the walls. This is because rodents do not see very well, and they feel their way along the walls. Wind-up or multiple catch traps are preferred. Trap placement is documented on a floor plan of the area.

Bait stations are generally not recommended for use inside the plant, as there is always the danger of bait contaminating the foods in process. Also, with baits, there is no control over where the rodent will die. In some extreme cases, a bait station can be used to handle a specific problem; however, this is usually done on a closely supervised basis.

Inspecting. The inspection step is very important. Interior traps are checked on a weekly basis; however, a severe rodent infestation might require daily inspections. Exterior bait stations and traps are also checked, on a somewhat less frequent basis. The inspection involves looking for any signs of activity at the traps and recording such signs on a trap and bait station report. Activity signs might include tracks, rodent pellets and gnawing. The tightness of the structure and possible harbors for rodents are also noted. Recording is very important to the identification of an ongoing problem.

Ultra-sonic Devices. Ultra-sonic devices produce a high-frequency sound that hurts the ears of rodents and drives them off. This high-frequency sound is not harmful to man because it is beyond his hearing range. Ultra-sonic devices are controversial pieces of equipment, but there is evidence that they can be effective under some circumstances. These units will be discussed further in Chapter Ten.

Other tools that can be useful in bringing a rodent problem under control include glue boards and tracking powder.

Essentially, controlling rodents means changing the living environment of the rodent. Take away its food and its place to live and it will move on. The secret to *zero goal* rodent control is the implementation of these seven steps simultaneously. Should any one of them falter, the whole program could be jeopardized.

Engineering, of course, will find its challenge in the selection of building materials, landscaping features, and design details that will keep rodents from gaining access to the plant in the first place.

FOREIGN MATERIAL CONTROL

Physical Contamination. All segments of the food industry must contend with such contaminants as glass, wood, sharp metal objects, bones, string, paper, and stones. Of these, glass, wood and metal are particularly injurious, especially when they are sliver-shaped. Slivers can easily lodge in the throat or stomach and cause serious injury to the body. Physical contamination control is further complicated by the resemblance of these objectionable substances to food particles.

Physical contamination is best controlled by selecting materials and design details that preclude or minimize potential contamination in the first place. The HACCP system (Hazard Analysis and Critical Control Points) should be used to identify the potential sources of contamination through an analysis of the process flow. Potential sources of contamination can then be eliminated, wherever possible, or appropriate equipment such as magnets, metal detectors, and screens can be provided.

Chemical Contamination. This type of contamination also confronts all segments of the food industry. Equipment lubricants, cleaning solvents and detergents, paints, and pesticides are just a few possible sources of chemical contamination. Contamination does not always result from direct contact with the food product. It can also occur while the product is in process, in storage, or in transit. Many foods and food ingredients readily absorb various chemical odors, including lubricant odors.

The chemicals commonly found in food plants can range from nontoxic to highly toxic materials. The presence of even small quantities of some chemicals in food can be a major concern. Because of the swift, modern distribution methods used by food companies, a chemically contaminated food product can become widespread in a matter of days and jeopardize the health of many people.

One of the best ways to prevent chemical contamination is to implement an *approved only* chemical use program, including lubricants. Lubricants will be discussed in more detail in Chapter Seven.

RECOGNIZING RISKS IN THE MANUFACTURING ENVIRONMENT

As stated earlier in this chapter, the manufacturing environment can affect the product safety program by causing product contamination. The mere presence and proximity of building materials and fixtures represents a threat to the products in process. Such things as glass from an overhead light fixture, peeling paint, and pieces of foreign material falling from the overheads and insects harboring in building cracks, can easily end up in the product. Such factors, particularly in a large plant, can represent many concerns for the products in process.

Selection of the right materials and provisions for protecting the product in each manufacturing stage minimizes many of these concerns. A design engineer will find consultation with the plant food safety and sanitation manager of benefit in identifying and in dealing with risks from the manufacturing environment.

PERSONAL HYGIENE AND PERSONNEL PRACTICES

Personal hygiene and personnel practices should not be underestimated, as it is frequently people who are associated with serious outbreaks of different types of illnesses, including those involving food. People carry bacteria, which they easily transfer to food products in the course of performing their jobs.

The various regulatory agencies recognize bacterial contamination from humans as a major concern for food safety. For that reason, they have issued a number of regulations that clearly define personal hygiene and personnel practices for people working in a food plant. In addition, each food plant normally has rules and regulations of its own, which incorporate government standards and are generally more stringent.

Regulations for personal hygiene practices apply to all plant personnel, not just to those assigned to manufacturing areas. These regulations must be observed by engineers, draftsmen, servicemen, construction personnel, and plant visitors of all types. There have been many instances when equipment servicemen arriving at a plant to perform a service were not allowed entry because they could not meet plant personal hygiene standards.

Construction personnel are usually unaware of the personal hygiene and personnel practices required when they are working in a food plant. A specific effort should be made to review the plant regulations in detail with each construction employee arriving on the job for the first time. In particular, construction personnel should be made aware of rules concerning hair and open sores or wounds. The rules and regulations should be posted in the construction area.

Following is a list of plant rules and regulations for personal hygiene and personnel practices. They give an idea of what can be expected of all people who have business in the plant.

RULES AND REGULATIONS

1. Personnel with communicable diseases, infected wounds, or sores are not to work in areas where food might become contaminated.
2. Personnel must wear clean clothes and maintain a high degree of personal cleanliness.
3. Personnel must wash and sanitize hands prior to starting work and after each absence from the work station.
4. Fingernails must be kept clean.
5. Personal clothing must be stored in designated areas only.
6. Shirt pockets must be kept empty.
7. All insecure jewelry must be removed. Rings, such as wedding rings, may be worn, provided they can be sanitized.
8. Hair restraints are to be worn as indicated by hair standards.
9. Mustaches and sideburns are to conform to hair standards.
10. Employees may not wear beards. (This rule varies according to company policy.)
11. Cosmetics, personal medications, and similar items are to be stored and used in designated areas only.
12. Eating, drinking, or gum chewing is permitted in designated areas only.
13. Smoking is permitted in designated areas only.
14. Tobacco chewing in the plant is prohibited.
15. Spitting in the plant is prohibited.
16. Horseplay is prohibited.

HAIR STANDARDS

Both food plants and government regulatory agencies have special requirements for hair. Hair is of concern not only because its presence in food is highly objectionable, but also because it generally carries an exceptionally heavy load of bacteria.

Regulations regarding hair differ from plant to plant. The varying microbe sensitivities of food products and the difficulties of implementing a program account for these differences. Also, many hair restraints are uncomfortable, particularly those for facial hair. Following are representative food plant standards for hair.

To further protect the product from potential sources of foreign material contamination and to assure full compliance with good manufacturing practices, the following procedures will be enforced whenever ingredients or finished products are exposed:

1. All employees working in areas where ingredients or finished product are exposed must wear adequate head covering. Hats or caps may be used as hair restraints if hair is neatly trimmed on the sides and back, and if hair follows the contour of the head. Hair should taper as it extends down the back of the head. If hair does not meet these standards, an appropriate hair restraint selected and provided by the company is to be worn.

2. Sideburns are acceptable if they are neatly trimmed and if they do not extend below the bottom of the earlobe.

3. Beards are not to be worn.

4. Mustaches are acceptable if they are neatly trimmed and if they do not extend beyond the outer edges of the top lip.

5. The above regulations apply to all plant visitors.

SANITATION PROGRAMS

HACCP

A basic system called HACCP (Hazard Analysis and Critical Control Points) is commonly used in one form or another to set up a sanitation program. It involves a detailed analysis of the manufacturing flows beginning with the receipt of the raw materials or processed ingredients to the final warehousing of the end products. This analysis begins with the identification of product risks or product sensitivities as they relate to consumer safety and acceptance. Once the risks are identified, each piece of equipment in the flow is studied to determine its potential contribution to the identified risks. A sanitation control is then set up to prevent the determined potential hazard from developing.

For example, one of the potential risks identified for a pasta product might be insect infestation. A flour bin holding bulk flour might be identified as a potential hazard because insects can live and multiply in the cracks and crevices. The sanitation control selected to keep the hazard from developing would call for the bin to be cleaned and fumigated on a predetermined schedule.

The HACCP looks at each piece of equipment in the manufacturing process in a similar manner. The potential hazard is identified as it relates to the identified risks, and a sanitation control is set up as needed.

To assure that the sanitation controls are doing the job, the HACCP system uses control points. Control points are a specified spot in the manufacturing flow where the product can be sampled and measured to determine the presence of contaminants. Some control points are more critical than others in a typical food manufacturing flow, and they would be identified accordingly. Control points can also be specific pieces of equipment. Sifters to remove insects and foreign material, metal detectors to remove metal, and pasteurization equipment can all be used as control points.

In addition, the HACCP system identifies manufacturing environmental factors that can cause product safety problems. The sanitation program is set up to deal with these identified risks. Engineers can make use of finished HACCP flows in establishing the sanitary design, which has a tremendous impact on the final sanitation program.

MASTER SANITATION SCHEDULES

Master sanitation schedules are sometimes used to develop sanitation programs. Like HACCP, this system relies on the flow sheet to develop cleaning, sanitizing, and pest control schedules. Master sanitation schedules also depend on visual inspections or measurements at strategic control points to assure that the in-place sanitation program is working. This system effectively identifies and schedules the cleaning of floors, walls, overheads, and other parts of the building.

CONCLUSION

The value of understanding the science of food sanitation cannot be overemphasized. The more that is known about product risks and sensitivities, cleaning methods and materials, and pest control techniques, the more assurance there is that the sanitary design needs of the plant will be met. These needs can be met with a minimum capital commitment, and the return in lower ongoing maintenance costs will be tenfold.

Remember, food sanitation is a science. It involves the worlds of microbiology, entomology, toxicology, food technology and pest animal behavior.

CHAPTER 2 PLANT LAYOUT

CHAPTER CONTENTS

SELECTING THE FOOD PLANT LOCATION
CHOOSING THE BUILDING TYPE AND SIZE
FACILITY PLANNING
EQUIPMENT LAYOUT
EQUIPMENT SUPPORTS
CATWALKS AND ACCESS PLATFORMS
SANITATION SPECIFICATIONS

It is hard to conceive of another single factor that is of more importance than plant layout. The efficient production of a safe, wholesome food product begins with a good layout. Poor layout, on the other hand can undermine productivity and product safety by creating a variety of chronic problems for both.

A good plant layout generally means a good-looking plant. Appearance is important. Overall appearance has a positive effect on the work attitude of employees. A good work attitude, in turn, leads to higher productivity. Appearance also makes a good impression on visitors and on customers, which establishes good customer relationships. Finally, appearance will also influence sanitation inspectors. An experienced inspector responds favorably to a facility with a nice appearance.

In addition to improving appearance, a good layout

benefits the sanitation program in other ways. Layout planning that includes adequate space around, over, and under equipment allows for easier cleaning. Access to remote areas, such as overheads, permits thorough inspections which can head off many sanitation problems. Separation by partitions or by walls isolates many operations which might otherwise represent a product safety or product contamination threat.

This chapter focuses on sanitation problems that the plant layout must address. It will start with the location of the plant site. Other topics covered in this chapter include the selection of the building type or types, department planning, environmental controls, utility requirements, the placement of equipment, and the supporting and access requirements of the equipment.

SELECTING THE FOOD PLANT LOCATION

The selection of a plant location requires a great deal of thought. Certainly economic considerations, such as market area, raw material supply, energy availability, and transportation services are important. These must be balanced with product safety considerations, which in turn are affected by the proximity of the site to nearby industries and sanitary landfills. Smoke, dust, odors, and other contaminants are a concern to nearby food-manufacturing facilities. Many industries, including oil refineries, certain chemical plants, paper mills, steel mills, and coal yards, give off objectionable exhaust stack emissions and effluents despite diligent efforts to control them. Even the very best control systems are not completely efficient, nor are they free from malfunction.

Sanitary landfills, incinerators, junk yards, and service depots attract rodents and other vermin. During cold weather migrational periods, these vermin seek a warm and comfortable place to homestead.

Proximity to nearby residential areas should also be avoided. Although most zoning and building ordinances establish industrial boundaries, certain manufacturing circumstances or subsequent boundary changes may create problems for both the plant and the residents.

A good potable water supply and an adequate sewage disposal system are essential. Such things as municipal water supply additives that may be objectionable to your product quality or process should be investigated. If the municipal sewage handling facility is inadequate, then the site you select should be large enough to accommodate your own treatment facilities. If it cannot, another location should be chosen.

The problem of the plant's liquid waste or processed water is often overlooked. Frequently this waste water contains organic matter. The presence of certain organic materials, such as flour, sugar, starch, and fats, makes the liquid biodegradable, which increases the biochemical oxygen demand (B. O. D.). Many sewage plants object to an increase in B. O. D. and will add a compensating surcharge to normal sewage bills.

The plant location must be of adequate size to accommodate the immediate building plan, including needs for additional buildings for specific plant activities. The site size should also allow for the arrangement of buildings, access roads, parking lots, and utility lines to assure ease of operations. Finally, the site must be large enough to accommodate future expansion. All too often, the parcel of land purchased for the best of building plans becomes too small. Once a food manufacturing plant has outgrown itself, all kinds of compounding product safety problems can occur. Crowding limits the options for an efficient and sanitary plant layout. Good planning up front will prevent this from happening. Growth is difficult to measure, but product history and demographic data can be useful.

CHOOSING THE BUILDING TYPE AND SIZE

The controversy over single-story versus multiple-story buildings will probably go on forever. In fact, both have advantages and disadvantages which directly influence the manufacturing process.

Single-story buildings simplify the shifting of personnel from operation to operation. *(Figure 2-1)* Supervision of personnel and of the manufacturing operation is also enhanced. Process ingredients and other materials move more efficiently. Plant maintenance and sanitation benefits because there is less area to cover, and less time is required to move from job to job.

The major problem with single-story buildings is that they do not lend themselves to gravity flow of products in process; as a result, the use of additional mechanical and pneumatic elevating and conveying equipment becomes necessary. Most single-story structures require ceiling heights that are difficult to reach. These same ceiling areas become heavily congested with supporting service piping, duct work, and miscellaneous equipment. It is often difficult to get at these areas to perform cleaning and maintenance operations.

Multiple-story buildings have the opposite advantages and disadvantages. *(Figure 2-2)* These buildings allow for gravity-flow process techniques which eliminate a lot of mechanical and pneumatic elevating and conveying equipment. In addition, ceilings for each floor are lower, less congested, and more accessible. At the same time, shifting personnel from operation to operation and supervision become more difficult. Freight elevators must be used to move process ingredients and other materials

FIG. 2-1 Single-story plants have many labor-saving advantages.

FIG. 2-2 Multiple-story plants take advantage of gravity flow, eliminating some equipment needs.

from floor to floor. Maintenance operations generally require more travel time from job to job, and there is more area to clean.

Since the demands of the manufacturing process influence the building layout, it might be appropriate to compromise by selecting a single-story building with varying head rooms featuring mezzanine floors. This type of plan allows for gravity flow where it is necessary; at the same time, it offers the advantages provided by a single-story building of more efficient personnel shifting, supervision, and material movement. *(Figure 2-3)*

The size of the food plant will depend on what is to be manufactured and in what quantity. It will also be determined by the extent of related support functions that are to be included, such as a central billing department or a main product distribution center. Once these decisions have been made, layout planning can actually begin.

FACILITY PLANNING

When you design the building layout, you will want to make use of such planning tools as flowsheets, layout drawings, and sanitation specifications. You will also want to include those features of the building that play an important role in the total environment. These are such features as environmental control systems, pollution control systems, cleaning systems, and waste-handling systems.

FLOW SHEETS: THE KEYS TO FACILITY PLANNING

Flow sheets are the key documents in layout planning. Flow sheets should not be limited to the manufacturing process itself, but should also help establish the flow of materials and finished products from department to department.

FIG. 2-3 A compromise is a labor saving, single-story plant with varying head rooms to take advantage of gravity flow as needed.

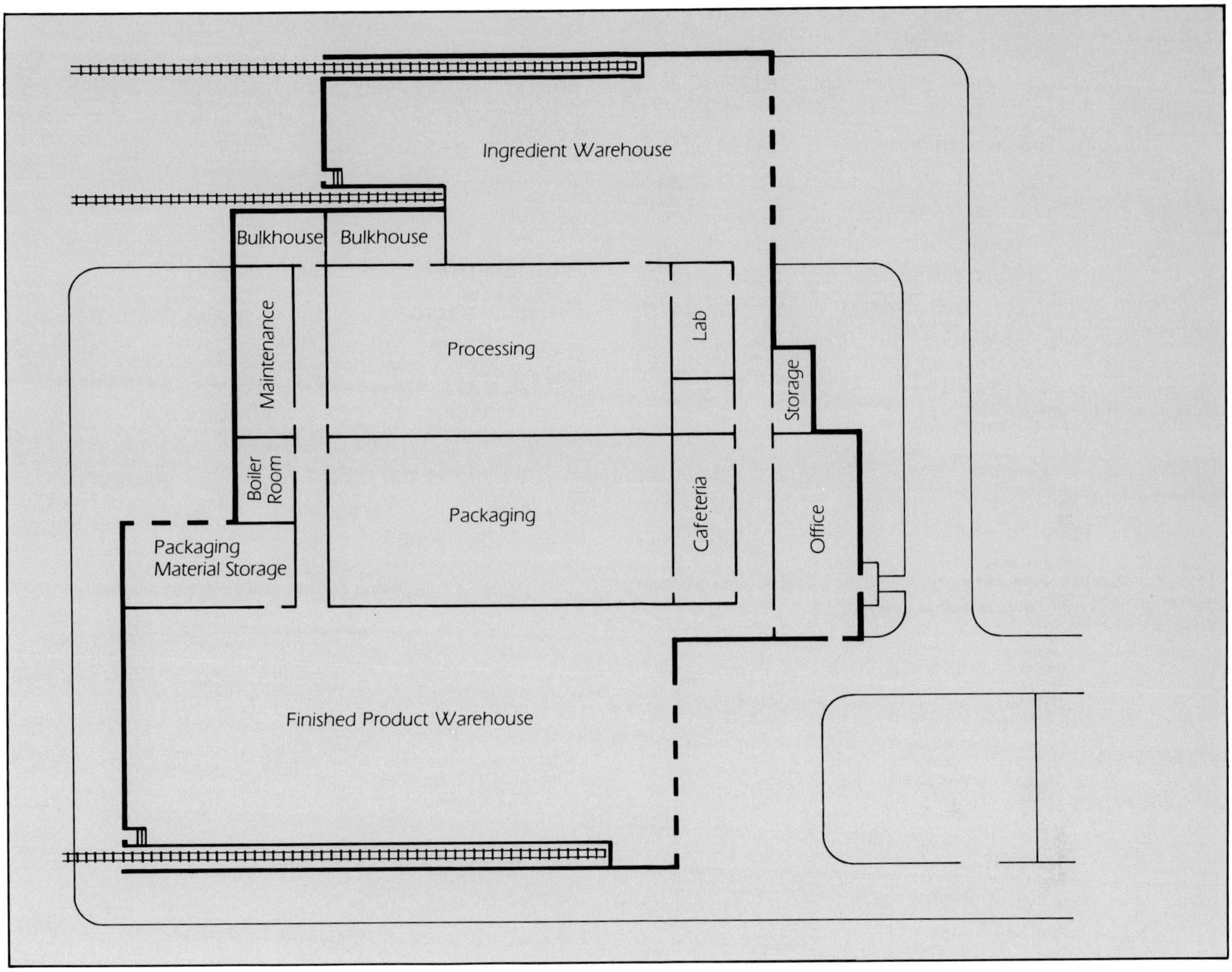

FIG. 2-4 Departmental flows can readily identify needed walls and partitions to improve product safety. They insure that the food is processed in a smooth flow from department to department over the shortest route.

Flow sheets can be used to group departments according to their function and to their relative importance.

Departmental Flows. Departmental flows are simple enough to be represented in a plan view. *(Figure 2-4)* The product safety requirements can be readily seen. Operations or departments which represent a potential hazard to the food product can be appropriately identified, and walls and partitions can be planned accordingly. The following should be separated by walls or partitions:

- Administrative areas
- Raw material or ingredient storage
- Processing
- Packaging and packaging material storage
- Finished product storage
- Maintenance operations
- Maintenance supply rooms and cleaning supply storage areas
- Pesticide storage
- Power and control rooms
- Cafeteria/locker rooms/rest rooms
- Offal/refuse handling areas

Special consideration should be given to separating wet manufacturing operations from dry where there is a need. Glass packaging materials should receive special handling and storage with the storage area remotely located from manufacturing areas and/or separated by walls or partitions.

Planning should include separate rooms or buildings for the power equipment, electrical switch gear, and the various panel modules from which the manufacturing systems will be operated. These rooms should be designed so they can be readily cleaned. Where it is necessary, electrical switch gear and control rooms should be pressurized with conditioned air. Processes which are considered to be dusty virtually dictate the need for this layout condition.

Process and Packaging Flows. Process and packaging flows might be the most important of all engineering drawings. They identify the equipment and space requirements, as well as the special sanitation design needs of a particular manufacturing process. *(Figure 2-5)* A process flow can tell you whether wet cleaning or dry cleaning will be required. This establishes the needed sanitation design criteria for both the equipment and surrounding area. Process flows also identify potential micro-biological and insect concerns.

PRELIMINARY LAYOUTS

Preliminary layouts utilizing templates should be used as the basis for the ultimate layout. The final layout should present a smooth, orderly flow of raw materials or ingredients through each manufacturing phase on to the storage of the finished product. It should take advantage of gravity flow and reflect the shortest distance between two points. The final layout should leave adequate space around the equipment to permit easy cleaning and

FIG. 2-5 Process or packaging flows are the major keys to product safety, establishing a foundation for the sanitation program and justification for the sanitary design criteria.

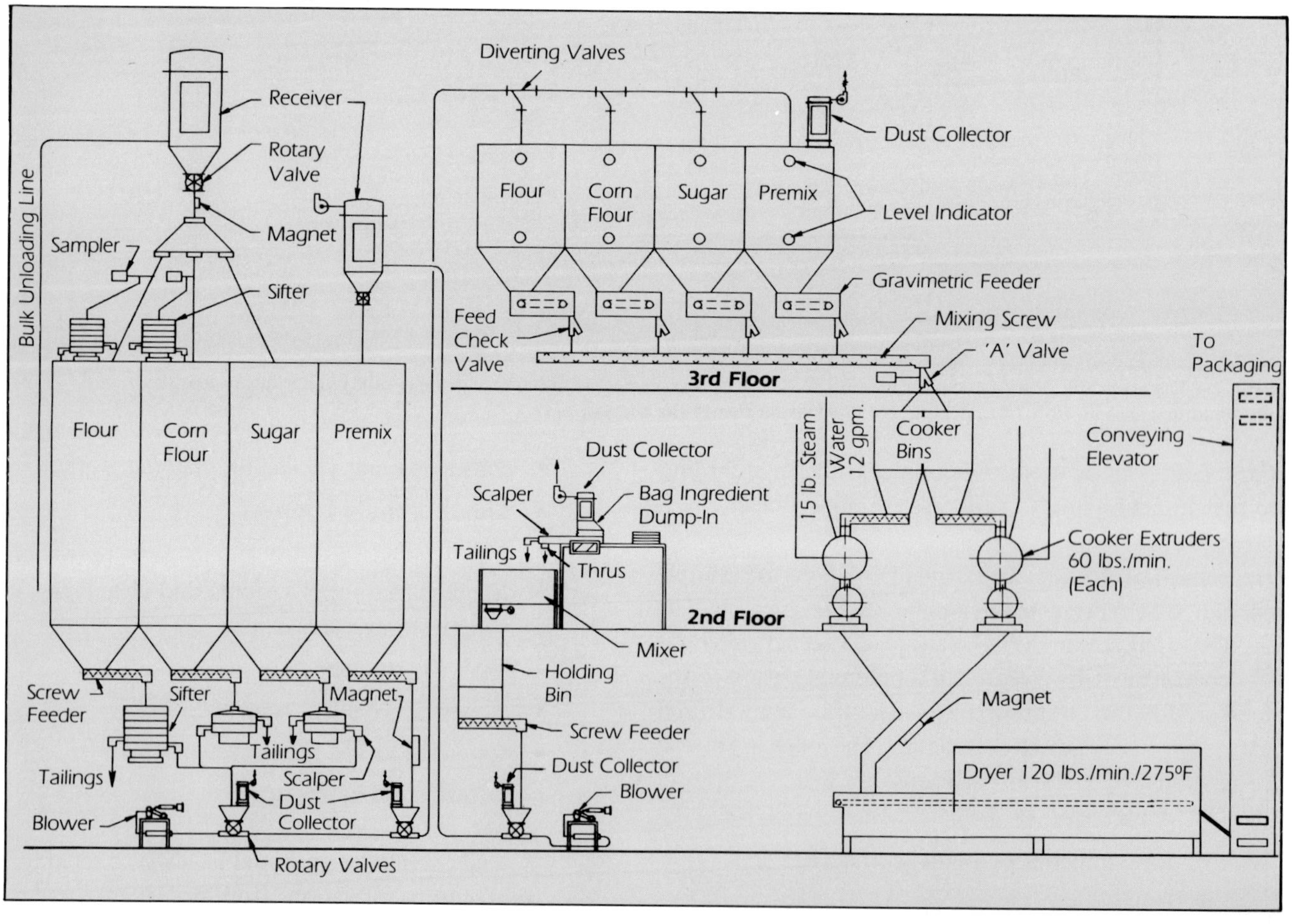

maintenance. Adhering to criteria of this type not only reduces equipment needs, but maintenance and cleaning tasks as well.

LIGHTING SYSTEMS

The importance of good lighting systems to the sanitation program cannot be overemphasized. In the opinion of many, the two most significant sanitation contributions to a well-planned facility are the floors and lighting. Both have a marked impact on that so very important consideration, appearance.

Standard lighting level recommendations as found in most illumination engineering handbooks are normally inadequate. The identification of sanitation failures and the performance of various sanitation cleaning tasks require higher lighting levels as suggested in Chapter Three. Further, certain types of lighting systems such as mercury vapor produce an ultra-violet light, and flying insects are attracted to ultra-violet light. On the other hand, high pressure sodium light is less attractive to the insect. This feature lends itself to the selection of high pressure sodium lighting systems for warehouses and outdoor security lighting applications.

CONTROLLING THE MANUFACTURING ENVIRONMENT

The manufacturing environment should be controlled where there is a need to do so. Many food processes release a great deal of moisture. Layout planning must include adequate air exchanges and wet exhaust systems to minimize the otherwise certain sanitation problems of mold and peeling paint that are often found in moist environments. Molds can grow in atmospheres containing as little as 15% moisture. Increasing air exchanges to reduce the moisture level is one of the best ways to control molds.

In some cases the hydroscopic nature of some food products might require the incoming air to be dehumidified. Some ingredients become sticky in the presence of higher humidity and hang-up in bins or spouts, developing into unwanted sanitation problems. Ingredient dust becomes sticky and forms crusts on equipment surfaces that are difficult to remove.

Specific areas should be heated or cooled, as the case may be, to enhance the total operation and to provide for the comfort of employees. Dust collection systems should also be provided where they are needed.

POLLUTION CONTROL

Every food plant will be confronted with a number of emission or effluent problems. The well-planned sewage handling system, as previously mentioned, along with the careful planning of certain process waste and storm sewer discharges, will eliminate most of the concern for effluent. Food process stack emissions are likely to be classified as a nuisance because they emit food particles, but nuisance or not, these fine emissions of food particles must be controlled. Odor control is also an important consideration. Pleasant as well as unpleasant odors must be controlled.

CLEANING SYSTEMS

Sophisticated, fully-automatic *clean-in-place* (C. I. P.) systems should be used where possible. *(Figure 2-6)* C. I. P. systems have the advantage of doing a better job of cleaning in much less time and can be easily justified on a *return on capital* basis.

FIG. 2-6 Clean-in-place (C. I. P.) cleaning systems reduce costs and improve cleaning quality.

Sophisticated cleaning systems also include wet central cleaning systems which pipe a supply of hot water, and in some cases cleaning chemicals, throughout the entire facility. *(Figure 2-7)* The volume and pressure of the water are provided to accommodate the cleaning need. Although high water volume generally cleans best, there are some jobs that can be nicely managed by high pressure/low volume systems.

Central vacuum systems for both wet and dry pickup

FIG. 2-7 Central cleaning systems also reduce cleaning costs and are an important part of good layout planning.

should also be part of the layout plan. Specially designated areas for wet cleaning of mobile or portable equipment and utensils should be provided and adequately furnished with the appropriate cleaning equipment as well.

WASTE HANDLING SYSTEMS

The plant layout must include provision for handling solid wastes such as product offal, plant refuse, and other debris, in addition to liquid wastes. These facilities should be located at the back of the plant and be completely isolated from the manufacturing process. Waste systems represent a potential contamination source. In addition, they can harm the overall image of the plant unless they are located out of sight.

PRODUCT SAFETY EQUIPMENT

The product safety equipment needed varies with the specific food processing employed. It includes pasteurization or sterilization equipment such as heat exchangers, cookers, blanchers, and ovens. It also includes equipment which can detect, trap, or remove physical foreign material objects such as metals, stones, or other solid objects. Equipment of this kind includes filters, liquid strainers, stone traps, magnets, metal detectors, scalpers, sifters, gravity tables, and aspirators. Processing and packaging flows should be used to determine if and where this equipment should be located.

Metal sifter or scalper screens or liquid strainer screens should be made from wire having a magnetic property. Magnetic metals are more easily detected by metal detectors, and magnets can be used to remove magnetic metals. If corrosion-resistant metal wire is needed, 400 stainless steel or plated mild steel wire should be considered.

Nuts, bolts, machine screws, and other fasteners should not be used in product zones. However, if they must be used, they should also have a magnetic property. Glass sight glasses and pressure gauge lenses should be changed to poly-carbonated plastic as glass is a serious product safety hazard.

CHEMICAL CONTAMINANTS

Another important product safety issue involves the use of certain chemicals in or around the processing equipment. These chemicals include boiler feed water chemicals, lubricants, solvents, paints, and caulking materials. The chemicals selected should be in accordance with available regulatory requirements, or at the very least, considered to be food grade, or safe for use around foods fit for human consumption. Good product safety regulations prohibit the use of caulking materials in product zones and require that all paints be free of heavy metals such as lead. Food grade epoxy paints are preferred for most applications. Another important consideration involves thermometers, monometers, and other liquid containing instruments that may contain mercury, as mercury is highly toxic and should not be used around food. The way to prevent chemical contamination of food products around a food plant is through the implementation of an approved chemical program.

PEST CONTROL SYSTEMS

The types of pests that are common to the plant location and common to the products that will be manufactured should be identified, so that planning can include the required pest control systems. Flying insect problems will require the use of electric insect light traps. Stored-product insect problems may suggest that heat sterilization of the entire plant be undertaken, in which case, the proposed heating systems will need sufficient capacity to provide the higher temperatures needed to kill insects. Rodent control programs should include plans to rodent-proof all buildings and the use of ultra-sonic equipment, if deemed necessary.

In addition, mechanical equipment can be added to the manufacturing systems to destroy and remove insect pests. This equipment includes infestation destroyer impact machines, aspirators, sifters, and scalpers.

EQUIPMENT LAYOUT

It would be ideal to lay out the equipment and then design the building, in other words, to lay out the equipment and build the building around it. Unfortunately, in most cases, this approach is simply not possible. Instead, templates should be used to plan the equipment layout. It is very important to use templates to obtain the best possible layout, one that provides for good access to all the equipment.

SPACE REQUIREMENTS

It is difficult to be specific about the space needed around equipment. About the only thing that most people are willing to be specific about is to take as much space as they can get! Some systematic planning proposals use published or other self-proclaimed standards to determine space requirements. Standards are considered risky, however, and space determination can be more accurately defined through individual calculation. This method of calculation considers the space occupied by the equipment itself, the space required by the operator, the space required for materials or supplies, and finally the space required for maintenance and cleaning. Although the calculation method is accurate, many engineers prefer the preliminary layout method using templates cut to scale.

Equipment layout with an emphasis on sanitation provides enough space around the equipment to allow for easier cleaning. In general, a clear distance of 36 inches all around the equipment is considered the bare minimum. *(Figure 2-8)* Actually, more space is preferred and should be the rule rather than the exception. This is particularly

FIG. 2-8 Cleaning and maintenance tasks require additional floor space around equipment.

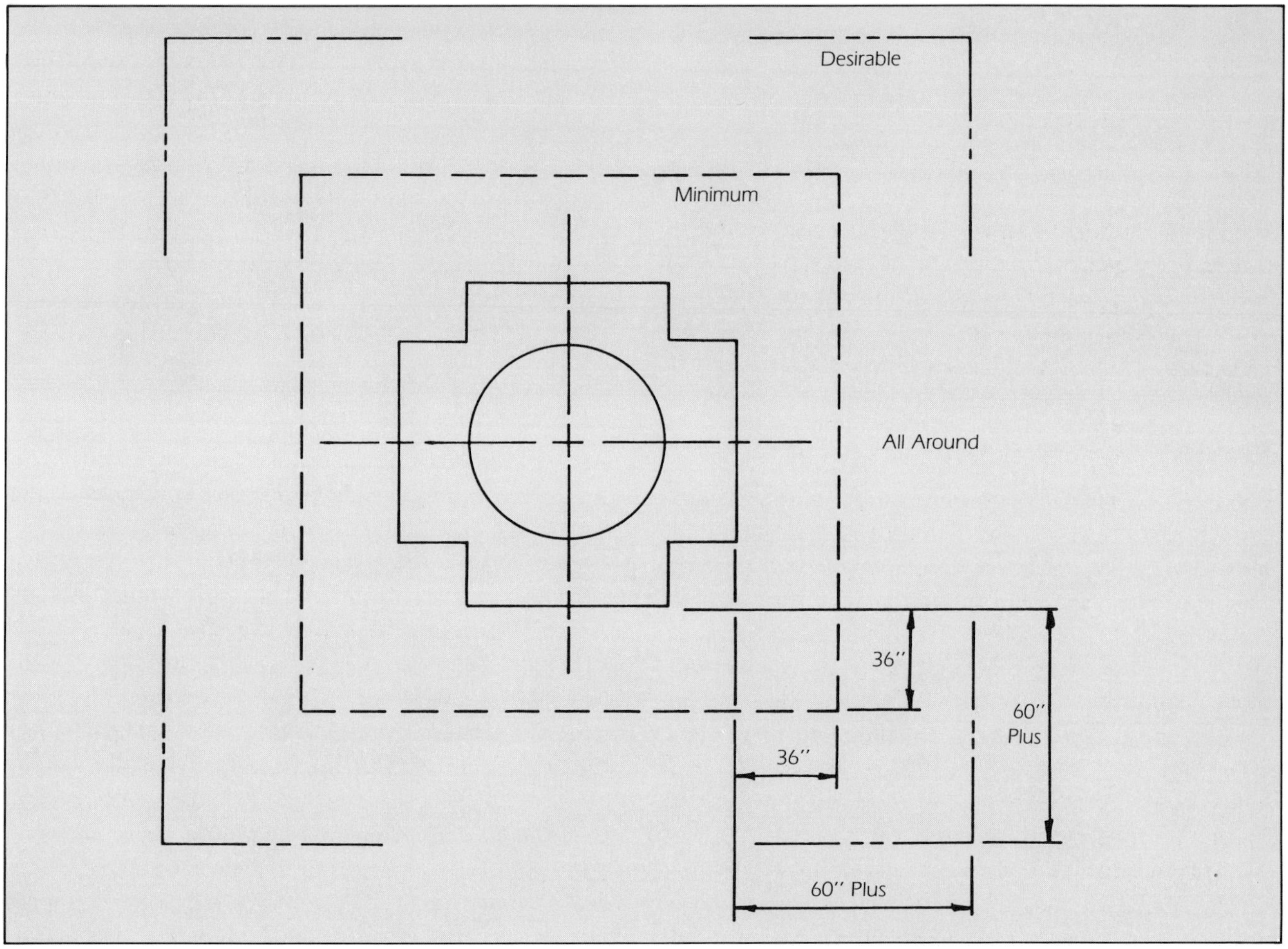

true of large pieces of equipment, such as cookers and dryers. In this case, a minimum distance of 36 inches would be inadequate, and 60 inches to 72 inches might be more reasonable. Conversely, small equipment, such as small pumps, might require less than the 36 inches minimum; although, in this case, something less than 36 inches is an exception to the rule. Ideally, there should be six to eight feet of free floor space around equipment.

When equipment is placed adjacent to walls, a minimum of 36 inches should be kept clear between the equipment and the wall. More distance is desirable and, in some cases, mandatory because of minimum aisle requirements. Life safety codes based on occupancy levels will establish the requirements for many main aisles.

Adequate space must be left above equipment. This space requirement is also related to the size of equipment, with larger equipment requiring more overhead space. A distance of 18 inches between a piece of equipment and a ceiling, an overhead platform, or another piece of equipment mounted in the overhead area, is the minimum needed to allow for cleaning.

Equipment should be elevated a minimum of six inches from the floor, or sealed to the floor. Again six inches is considered a minimum with more elevation suggested for larger pieces of equipment. Dryers, toasters, proofing boxes, and other similar large pieces of equipment might require an elevation of 24 inches or more to make the cleaning job easier. Once again, it is not unreasonable to expect that equipment be elevated 10 to 12 inches or more. *(Figure 2-9)*

FIG. 2-9 The area occupied by a piece of equipment determines how far the equipment should be elevated from the floor; 10 to 12 inches or more is reasonable.

FIG. 2-10 Access doors should be located so they can be reached easily.

Access Door Orientation. Equipment with access doors should be arranged so that personnel can reach these doors without crawling over or under equipment. *(Figure 2-10)* Floor space should also be provided to allow for unobstructed access and for the use of portable service platforms to reach doors located high above the floor. Whenever possible, doors should be located to allow for convenient floor access.

COMPOSITION

Equipment layout is concerned with the arrangement of all the equipment, as well as the single piece. Composition takes advantage of equipment to provide a balanced and well-proportioned arrangement. A well-composed arrangement of equipment can be obtained through repetition or the identical placement of similar equipment or equipment components. *(Figure 2-11)* For example, identical pieces of equipment, such as spouts, pipes, or motors, can be lined up together.

SERVICE LINES

Many fine layouts have been ruined by the careless placement of supporting service lines, such as:

- Steam and condensate piping
- Air and vacuum piping
- Conduit and electrical raceways
- Dust collection and pneumatic conveying lines
- Ventilation ducts

The placement of these service lines requires careful planning and detailed layout drawings. These lines should not block off access to the equipment or aisleways. *(Figure 2-12)* All minimum clearances must take service lines into account, such as 36 inches minimum clearances around the equipment or six inches below. The arrangement of the service lines themselves should not create a difficult cleaning problem. *(Figure 2-13)*

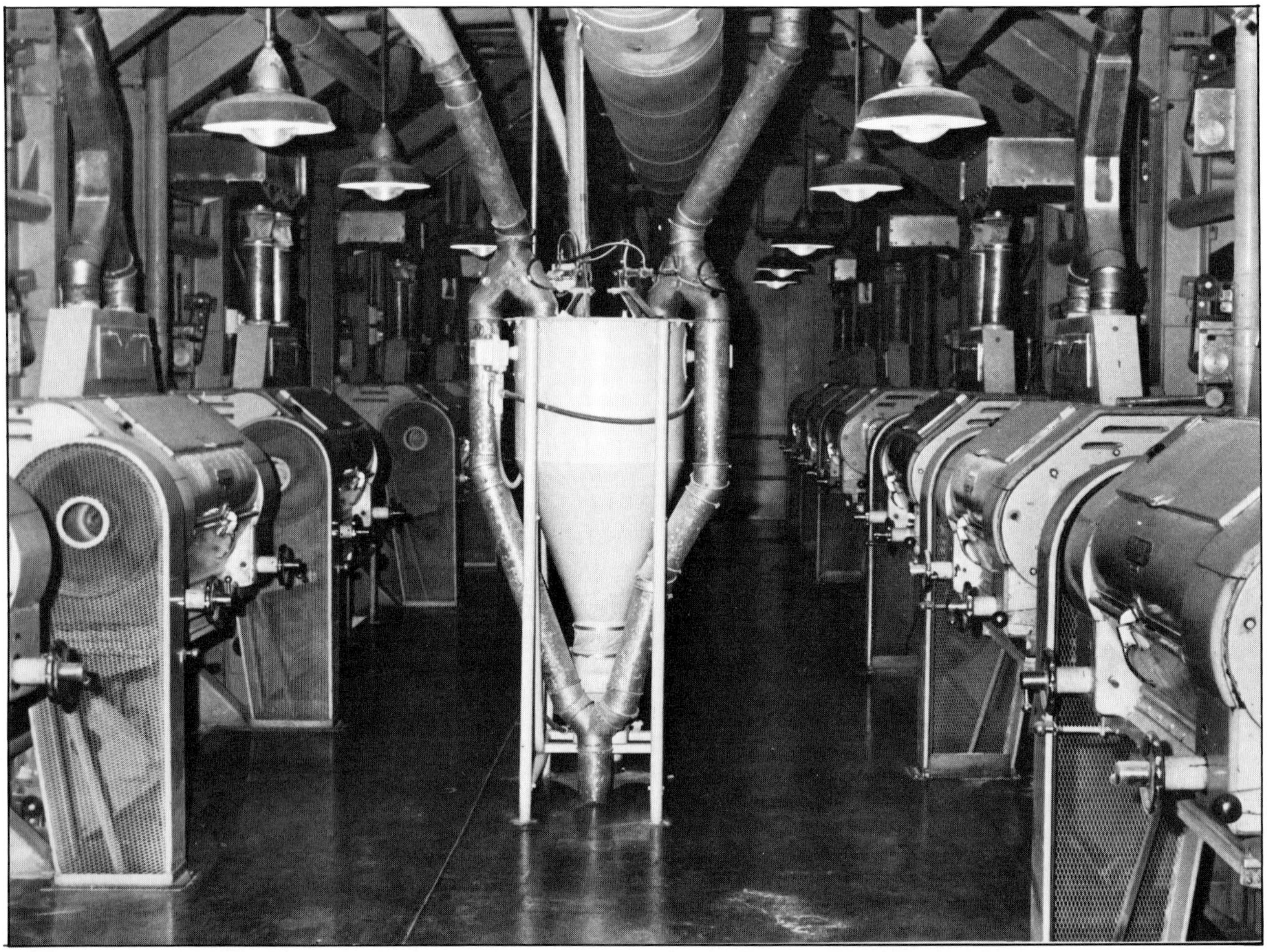

FIG. 2-11 An orderly layout that takes advantage of repetition looks good, instills pride in the workforce, and favorably impresses visitors.

FIG. 2-12 Advantages from planning adequate space for aisles and around equipment can be quickly lost through misplaced pipes that block aisles or reduce floor space, such as those shown.

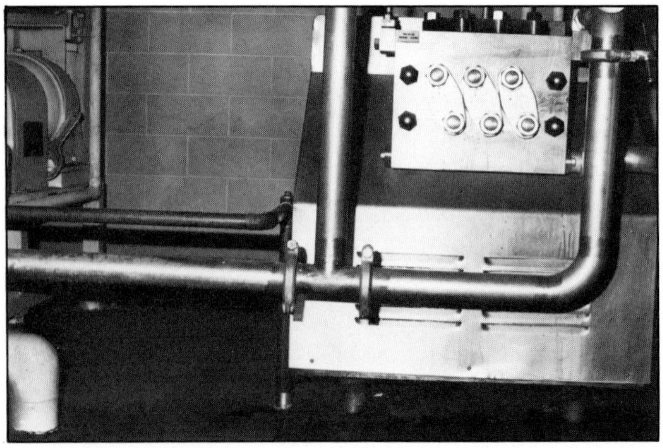

FIG. 2-13 Arrangements of groups of pipes or conduits must be done with sanitation in mind. The conduits in this arrangement are grouped too closely together.

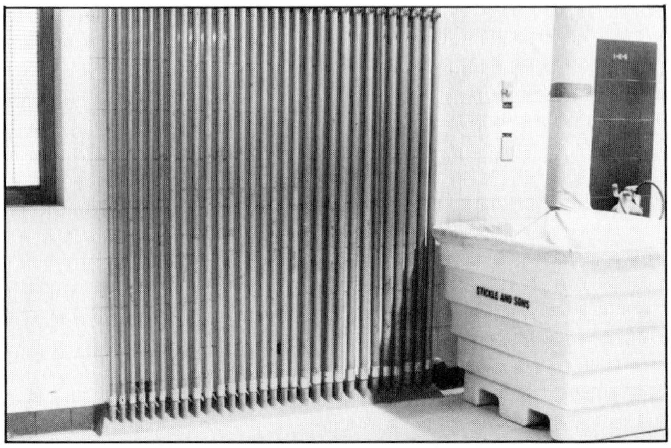

OVERHEAD AND CEILING AREAS

Layout planning applies to overhead or ceiling areas as well as the floor level. Unfortunately, this area is frequently overlooked and as a result, the field engineers or contractors are left to install pipes, conduits, ductwork, and other equipment without benefit of drawings. Without installation drawings or preplanning, the only criterion for the installation is an open space. This results in a crowded network of pipe and equipment which is extremely difficult to get at or clean. *(Figure 2-14)*

Overhead layouts should be planned and where applicable, pipe chases should be considered. Pipe chases enclose the pipe in a confined area, making it somewhat easier to maintain. Access doors should be provided in the pipe chase for inspection and cleaning.

SERVICE MACHINERY FLOORS

Single-story structures in particular lend themselves very well to second-story machinery floors or lofts. This design concept might be a way to have all of the advantages of the single-story plant, while at the same time eliminating the maze of overhead piping and equipment that can be a problem and a disadvantage of single-story design. The whole loft concept is basic and not necessarily revolutionary. It won't win any unusual building design awards, but it will solve what is, for many food plants, a very costly overhead cleaning problem. As for the loft area itself, it can be pressurized so that it will require little cleaning.

Machinery floors conveniently house heating and ventilating equipment and ductwork in addition to all the required service piping such as steam mains, condensate returns, water lines, and compressed air manifolds. Electrical conduits along with the lighting systems can be neatly installed in this area. First floor lights can be recessed into the ceiling with the light enclosure easily serviced from the upper floor. A central vacuum system as well as dust collection equipment can also be located in this area. The opportunities to take full advantage of this design concept are many, and of course, the first floor manufacturing equipment service line connections are easily made from vertical pipe or conduit drops directly from the overhead machinery floor. The actual amount of horizontal pipe runs found in the overhead area of the first floor is at its worst minimal. Building specifications can assure that only short horizontal runs will be permitted with simple statements that specifically limit the length of the horizontal run.

The architect's rendering clearly shows the concept of the single-story/machinery loft food plant. *(Figure 2-15)* With just a little imagination, this vision of a neat, well-laid-out and easy-to-clean food plant can be readily seen. With a sound overall engineering plan it can quickly become a reality. The following picture shows a machinery loft area

FIG. 2-14 Overhead ceiling areas are too often overlooked in layout planning.

FIG. 2-15 Single story plants with specially designed machinary loft areas eliminate the maze of piping and equipment found in the overhead areas that have caused cleaning, maintenance, and product safety problems in food manufacturing for several decades. Their advantages are noteworthy:

- **Improves plant appearance**
- **Reduces cleaning costs**
- **Allows for unobstructed lighting**
- **Minimizes overhead contamination concerns from leaking pipes, torn insulation, and peeling paint**
- **Improves access for maintenance**

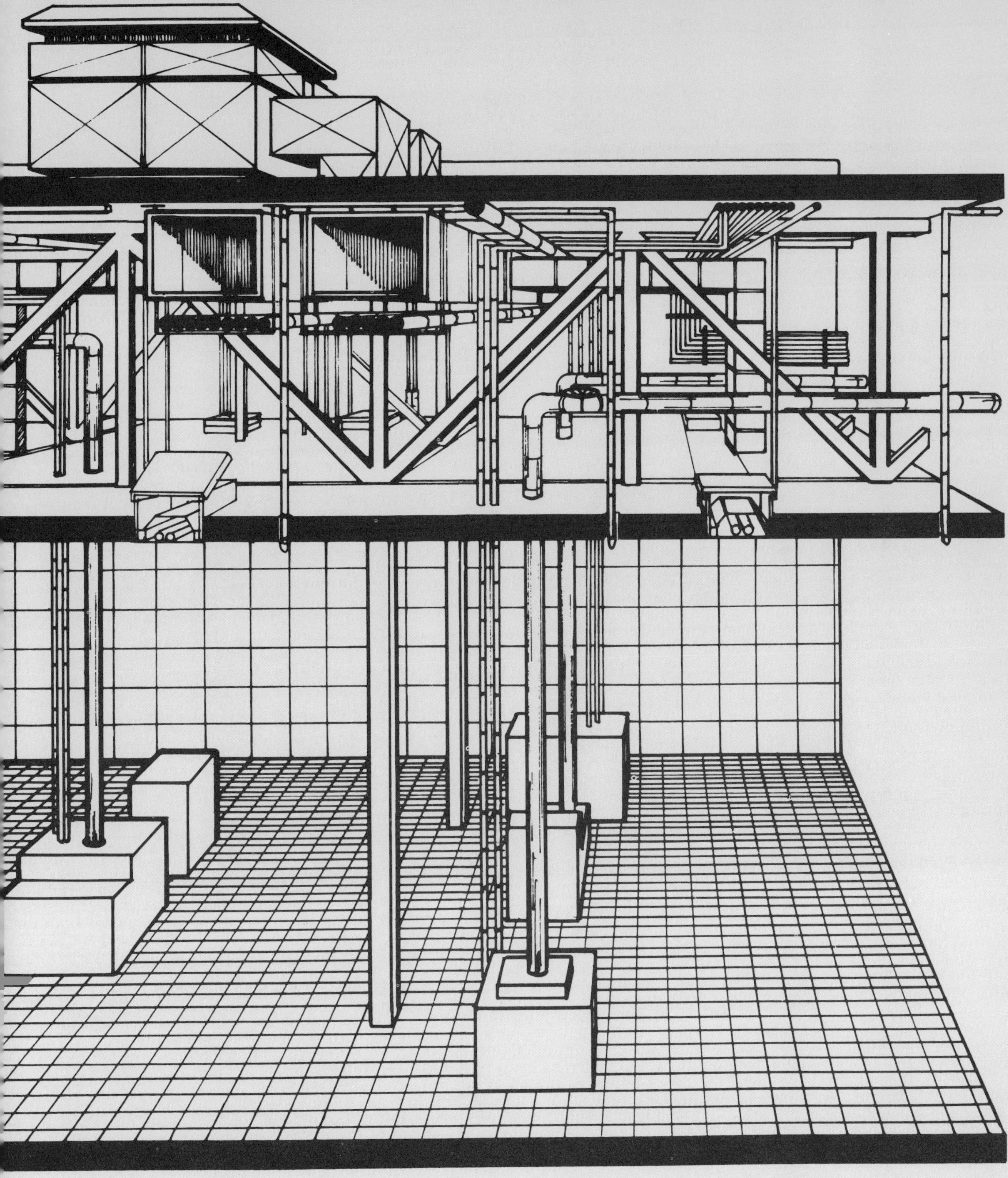

located directly over a single-story food manufacturing operation. *(Figure 2-16)* Since these areas are not intended to house heavy processing or packaging equipment, lightweight concrete floors and metal deck ceilings can reduce the building cost and in some situations make it possible to justify the design concept on sanitation and maintenance savings.

FALSE CEILINGS

The use of false or suspended ceilings to contain the overhead piping and equipment is generally not acceptable. The void areas between the upper ceiling and the suspended false ceiling is very difficult to maintain. In addition, this space often becomes heavily infested with insects. When this happens the solution may require removal of the ceiling.

Nonetheless, certain design features might make false ceilings more acceptable. If false ceilings were designed with catwalks to permit complete inspection and maintenance of the void area and were dust-tight, they might be more readily accepted. The design should also include installation of automatic pesticide misting applicators to prevent or control potential insect problems.

OVERHEADS/LAYOUT DRAWINGS

Good planning and detailed drawings of the piping and equipment to be located in the overheads is still one of the better methods of satisfying sanitation needs. Layout planning for overhead areas should include the following features:

FIG. 2-16 Typical loft area

- Pipes and conduits should be grouped in areas where floor access is possible and in such a manner that the group itself does not create a cleaning problem.
- Sanitary lines or lines subject to dripping should not be routed over open product zones or equipment access doors.
- Ductwork should be located so access doors may be reached from the floor.
- Pipe, conduit, or ductwork should not block access to suspended equipment or interfere with the lighting.
- Suspended equipment should be located in an area where floor access is possible.
- Access platforms and catwalks should be provided as required to permit the overhead cleaning of pipe, conduit, ductwork, and equipment that cannot be serviced from the floor.
- Pipe or conduit vertical drops to the floor should be located so they do not exceed the 36-inch floor clearance limit which is to surround the equipment. The 36-inch clearance should be calculated with all pipe and pipe drops in place.

FLOOR TRENCHES

Enclosed floor trenches or the simple placement of some service lines directly in the floor should also be considered as a way to keep the overhead areas as open as possible. When floor trenches are used they should be watertight, backfilled with fine sand, and a concrete knockout slab should be poured over the sand. This would allow for future access should it become necessary. Open trenches or those covered by wood or metal plates inherently create serious sanitation problems and, in general, are not to be considered.

EQUIPMENT SUPPORTS

Equipment supports should be kept simple and designed so they can be cleaned. They should not obstruct the equipment or the cleaning area surrounding the equipment.

The sanitary design of supports requires that we look at:

- Construction materials
- Structural shapes
- Arrangement of the structural shapes

CONSTRUCTION MATERIALS

Black iron, mild steel, stainless steel, and plated metals can be used satisfactorily to support equipment. The type of material chosen is determined by the manufacturing process and the selected cleaning methods. If the process is a wet process or is a process requiring wet cleaning, then stainless steel, aluminum, or plated steel support materials are preferred. Support applications involving fruit or vegetable acids should require the use of stainless steel. Painted mild steel supports are rarely satisfactory for a wet environment, since peeling paint and corrosion are a recurring product safety problem.

For dry process equipment, mild steel supports are very satisfactory. Mild steel supports should be painted with epoxy base paint.

STRUCTURAL SHAPES

One of the basic objectives of support design is to eliminate ledges and flat surfaces where dust, dirt, and product spills may accumulate. Structural steel members are available in the following shapes, most of which are satisfactory for equipment supports.

Angle Iron. The angle iron support is acceptable, but when used in the horizontal plane, the vertical leg should point down. *(Figure 2-17)* It can also be used in a horizontal plane with the heel pointing up. It is often used in this manner as a lateral brace. *(Figure 2-17)*

Channel. Channels should be used with the web in the vertical plane. If it is used with the web in the horizontal plane, the flanges should point down. *(Figure 2-17)*

I or H Beams. I or H beams should not be used with the web in a horizontal plane. *(Figure 2-17)*

Tee. The vertical leg of the tee should point down. *(Figure 2-17)*

Tubing. Tubing is one of the preferred support materials. The surface is smooth and easily cleaned. Round tubing has the added advantage of reducing flat surface areas where product spills can accumulate. When tubing is used, all open ends must be sealed. *(Figure 2-17)*

Piping. All open ends of piping must be sealed. Pipe which is used for supports should not have threads because threads are both difficult and hazardous to clean.

Formed Channel. Formed channel with a J-like hook at the base of each flange should not be used in a horizontal plane. *(Figure 2-17)* Dust, dirt, and product spills can accumulate in the J-like hook area.

Perforated Angle. Formed perforated angles should not be considered for support fabrication as they are too difficult to clean.

All-thread. All-thread rods should not be used as support hangers. They are exceptionally difficult and hazardous to clean. Hanger rods should consist of plain mild steel threaded on each end as required.

Flat Plate Steel. Plate steel should be limited to gussets, mounting pads, and lateral bracing. Flat plate steel used in a horizontal plane as a base plate for mounting motors, gear boxes, and other such equipment provides a large flat surface for dust and dirt to accumulate. Also, crevices,

FIG. 2-17 The choice and orientation of structural shapes are seemingly small details, but they have a big impact on the sanitation program.

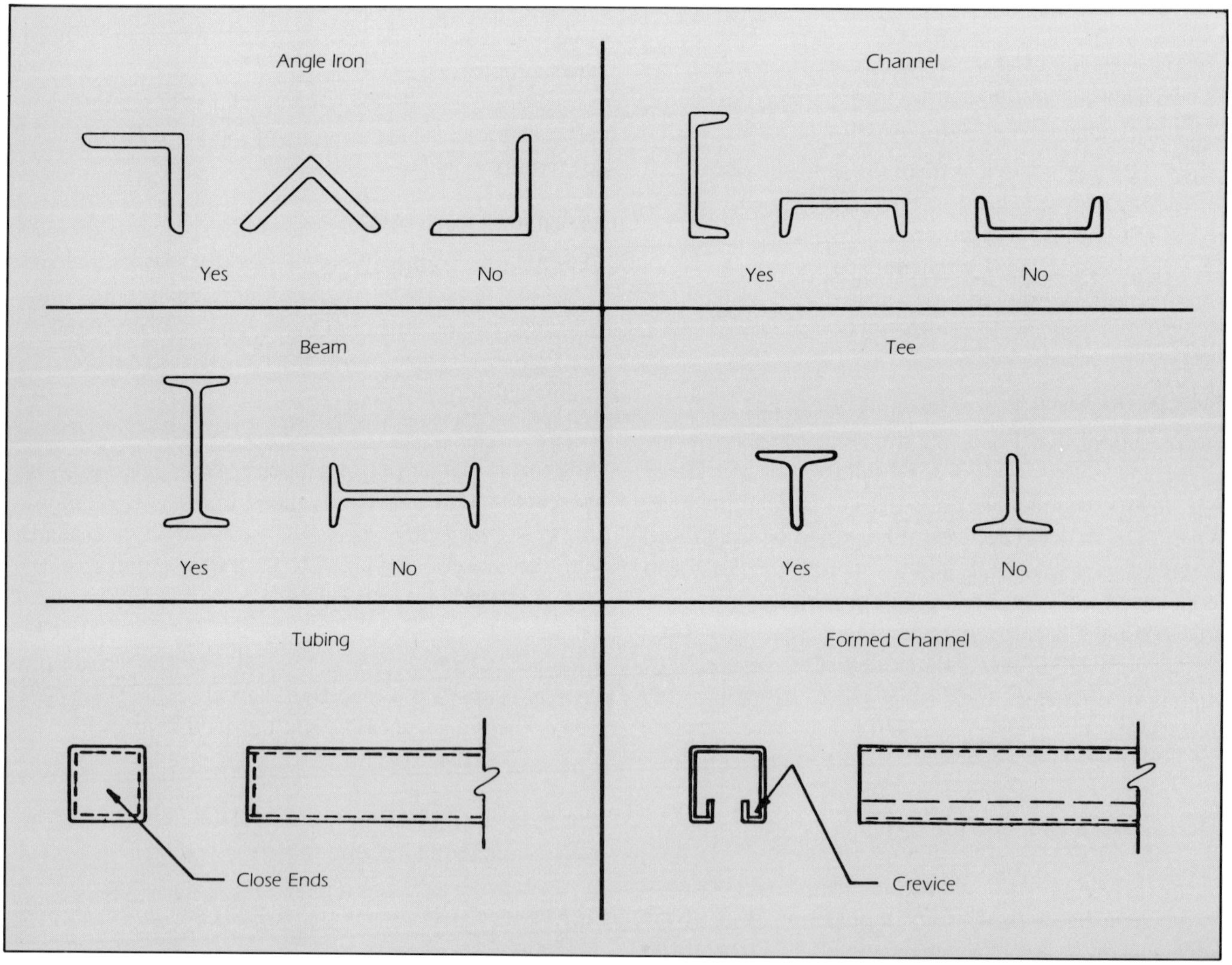

cracks, and void areas are formed between the plate and the mounted equipment. Flat plates do not create objectionable surfaces when placed in a vertical position to support other equipment.

ARRANGEMENT OF SHAPES

Not only is the selection of structural shapes important to sanitation, but so is the arrangement of the shapes. Many ledges, cracks, crevices, and dead ends can be eliminated with the proper use of structural members.

Angle/Channel Legs. Angle iron or channel used in a vertical manner to form a leg or a hanger creates a difficult-to-clean crevice at the floor junction. *(Figure 2-18)* Tubing is a better choice for this application.

Pedestal. A single pedestal floor support constructed from a tubular shape is preferable to a multiple leg arrangement. *(Figure 2-18)*

Horizontal Members. Horizontal members should be used for bracing only when necessary. When used as part of a floor-mounted support, the horizontal member should be at least 12 inches from the floor. *(Figure 2-18)*

Shipping Braces. Many floor-mounted equipment supports

FIG. 2-18 The object of using and arranging structural shapes is to eliminate ledges and minimize crevices, cracks, and void areas which collect dust and dirt.

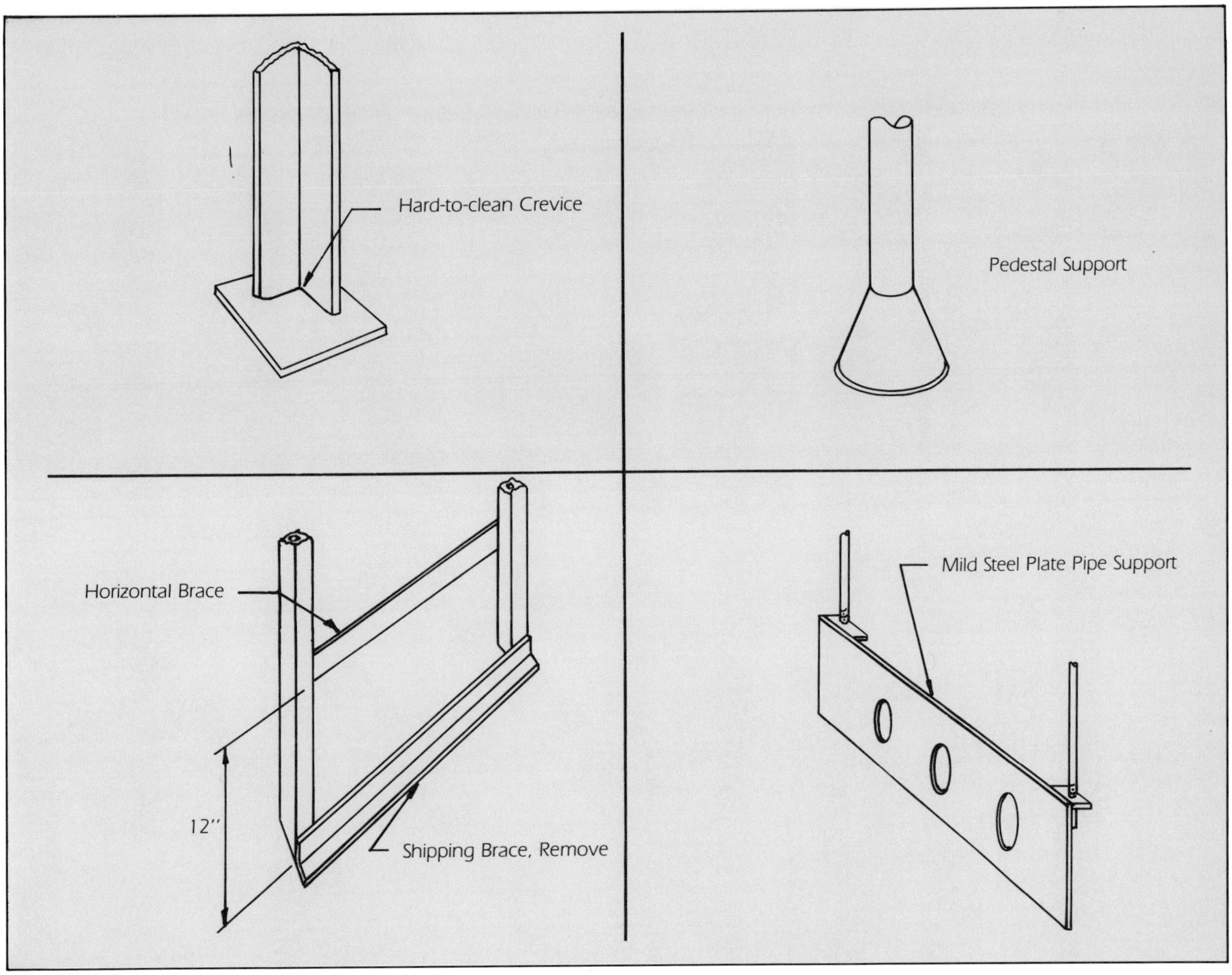

are shipped with angle iron shipping braces at their bases. In many cases, the support is installed with the shipping braces in place. These braces are unnecessary and create cleaning problems; they should be removed. *(Figure 2-18)*

Trapeze Supports. Trapeze supports commonly used to support pipes or ductwork should be kept simple. Mild steel plates or bars used in a vertical plane reduce the flat surface area. *(Figure 2-18)* Angle iron and round pipe can also be used, but they also increase the ledge surfaces. The ends of pipe and the drilled hanger holes must be sealed. The use of channel should be minimized. Channel used in the vertical plane doubles the surface area to be cleaned.

CATWALKS AND ACCESS PLATFORMS

To fulfill the access requirements of a well-planned layout, catwalks and access platforms will be required. *(Figure 2-19)*

Access equipment such as platforms and catwalks can present cleaning problems. They can also be the cause of product safety problems. Insects tend to harbor in such structures; paint peels easily. The sanitary design of catwalks and platforms involves selection of the proper construction materials and construction details.

CONSTRUCTION MATERIALS

The surrounding environment will influence the choice of materials used to construct catwalks and platforms. In a wet environment, stainless steel, aluminum, and plated steel are most suitable. Painted mild steel is satisfactory for the dry manufacturing environment.

Deck Plate. Platform or catwalk decks can be fabricated from smooth plate, checker plate, or open grating. Smooth

FIG. 2-19 Catwalks and access platforms are needed for equipment access and must be designed so that they can be cleaned.

plates are more desirable than checker plate for the dry environment. Dust, dirt, and product accumulations on smooth surfaces are easier to clean. Checker plate is preferred in the wet processing environment for safety reasons.

Open grating should never be used in overheads or in other suspended areas because cleaning is very difficult and time-consuming. A far greater concern is that dirt from shoe soles may be scraped off by the grate and fall from the overhead to contaminate product zones below. *(Figure 2-20)* Open grating does not belong in overheads or at any suspended level inside the plant. It can be used on the floor in wet areas as this application poses no contamination threat and it can be more readily wet cleaned. Fiberglass grating offers a noncorrosive feature which is desirable for an application of this type.

Open grating can be used outside as an equipment service catwalk or platform and from a safety standpoint, this should be considered. However, if equipment is installed below the deck, the possibility of contamination again appears, and a solid deck should be used. An outdoor installation of this type is another application for fiberglass grating.

Superstructure. The supporting superstructure materials should be compatible with the environment in which they are installed. Wet processing environments will require corrosion-resistant materials, while painted mild steel is satisfactory for dry processing areas. Framing details of the superstructure should eliminate as many free ledges as possible to minimize accumulations of dust and dirt. This can best be accomplished through the selection of structural shapes such as square, rectangular, or round tubing, or angle iron. If channel or I-beams are used, they should be boxed in with a light gauge sheet metal strip to form a tubular shape. In this way the bottom dust-catching ledge of the channel or beam is eliminated.

CONSTRUCTION DETAILS

Construction details must include kick plates four inches high. Kick plates are more sanitary when formed in a press break as an integral part of the access deck. The corner crevice between the kick plate and the deck surface is easier to clean because of the coved junction. *(Figure 2-21)* If kick plates are welded to the deck, they must be four inches high and continuously welded in place to eliminate the crack at this juncture. Kick plates of this type should never be installed slightly above the deck surface. This type of installation creates a gap between the bottom of the kick plate and the deck, and it is very easy for dust and dirt to be kicked through the gap and onto manufacturing equipment below. *(Figure 2-22)*

FIG. 2-20 Using open grating overhead in manufacturing areas can result in product contamination from dirt scraped off the soles of shoes by the grate itself.

Where the deck is to be wet cleaned or is to be located in an insect-sensitive area, the deck plate joints should be continuously welded. Tack welding or bolting the plates to the supporting steel frame can be a satisfactory alternative for an insect-sensitive area, if all the seams are carefully caulked.

All open ends of tubular superstructure members should be sealed. Most catwalks and platforms will require guard rails that meet safety requirements. All open ends of the rails must be sealed to meet sanitation needs. All construction welds must be chipped and brushed. Weld slag is also a contamination concern.

Catwalks and access platforms are extremely important to the sanitation program. Properly located and carefully constructed, they help make cleaning tasks easier and safer.

SANITATION SPECIFICATIONS

Good engineering practices fully utilize written specifications to establish precisely what is to be included in a specific project. Specifications are prepared to establish required structural details of the plant buildings, including all service equipment such as plumbing, lighting, and heating and ventilating equipment. Specifications also spell out equipment details. These include drive components, production capacities, and special features such as temperature-sensing devices or pressure switches. The same specifications are used to define the required sanitation design detail.

As with other design criteria, it is not always economical to draw every sanitation requirement in graphic form on the working drawings. A certain portion of the detail can best be defined in written specification form. Many pieces of equipment are standard or are simply ordered without drawings. Specifications should still be prepared, however. Equipment considered standard or "off the shelf" might still require minor changes prior to installation. Specifications as well as working drawings should be legally bound to the project through signed contracts or purchase agreements.

Written Specifications. Written specifications should include every requirement that is not covered on the drawing. In some cases, specific details need to be covered in complete detail in the specifications as well as on the working drawings. Written specifications should refer to corresponding drawings including vendor drawings as needed.

Sanitation specifications address most of the items

FIG. 2-21 Coved kick plate floor junctions are easier to clean.

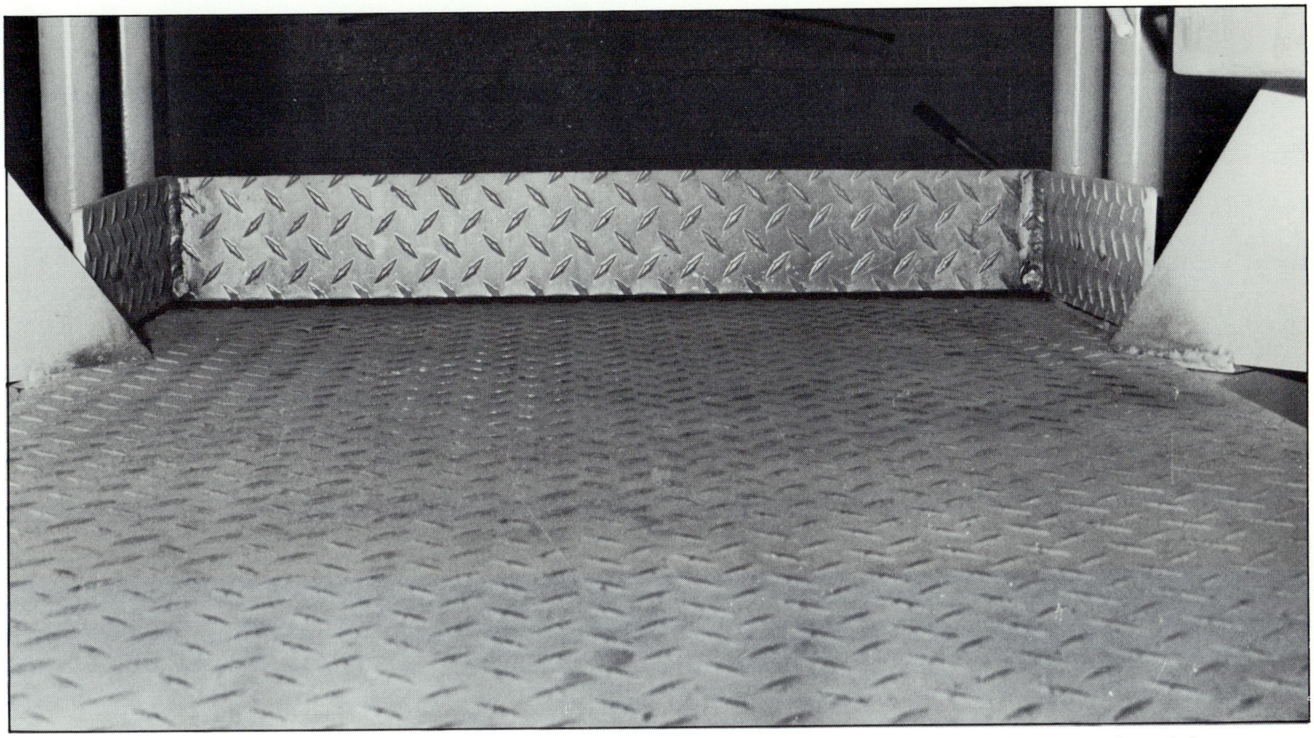

FIG. 2-22 Raised kick plates allow dust and dirt to be kicked through the gap, causing contamination to surfaces or products below.

we will be covering in this text. Certainly such things as materials of construction, fabrication detail and assembly instructions need to be spelled out in a very detailed way. Specifications for welding requirements must include references to any samples submitted. Emphasis must be placed on the need for food grade, nontoxic materials of construction.

Surface finishes, paints, lubricants and cleaning solvents also require specification. Many times, special protective coatings are placed on equipment surfaces to protect them while they are in transit. Specifications make certain that these materials are nontoxic and easy to remove.

In many instances, vendor drawings or brochures are not specific enough to spell out the material or assembly details of every component or accessory. A good specification attempts to provide protection against such unknown conditions by making general statements such as *All components must be constructed of food grade materials* or *The equipment must be constructed so that all product zone surfaces can be exposed.*

Often overlooked components or accessories frequently furnished with equipment include motors, motor rails, bearings, seals, gaskets, and fasteners. All too often pressure gauges, thermometers or recording instruments have glass faces. Time clocks and standard dial face clocks for telling time are also equipped with glass lenses.

Beware of vendor specifications that refer to government standards and imply that their equipment and all of its components have been approved by a certain regulatory agency. Beware also of statements on your own specifications that call for approval from a certain government regulatory agency. Most food regulatory agencies, such as the F. D. A., do not approve specific products or design details. The U. S. D. A., on the other hand, has recognized certain products and has established a system for acknowledgment.

A great deal can be accomplished in the sanitary design of a food plant and food plant equipment with a well-written specification. *(Figure 2-23)* A well-written specification only requires a true understanding of the sanitation design needs. Consultation with a professional food sanitarian can be helpful in identifying these design needs.

FIG. 2-23 Equipment specification

A B C FOOD CO.

EQUIPMENT SPECIFICATION SHEET

PAGE 1 OF 1

ITEM NAME: Enrober — JOB NO.: 102982

BY: T.J. DATE: 10/29/82 PLANT: Lancaster — SPEC. NO.: 182

CHK BY: B.A. DATE: 10/29/82 PROCESS AREA: Coating Area — MARK NO.: 1118

MFGR: Anderson and Anderson — DRAWING NO.: M218 & M219

VENDOR: Anderson and Anderson — REQ'D DEL. DATE: 1/10/83

POWER SOURCE INFORMATION: MTR. HP.: 1 MAKE: U.S. FRAME NO.: QTY REQ'D: 1

MOTOR RPM: 1800/184 VOLTS/PHASE/CYCLES: 440/3/60 TYPE:

---

* Sanitary Requirements

Scope:
All work is to be completed in accordance with specifications and drawings M218 and M219. No changes are to be made from these specifications without the written consent of A B C Food Co.

Frame:
The enrober frame is to be constructed of square tubing per our drawings M218 and M219. *All open ends of the tubing to be closed. *All bolt holes drilled in the tubing are to be spooled and/or drilled and taped. *All welds to be chipped and brushed. *Finished frame is to be zinc plated, .003 thickness.

Drum:
*The enrober drum shall be constructed of 304 stainless steel with a number 4 finish on the outside and a number 2B finish on the inside. *All welds shall be continuous and ground flush per weld sample number 1.

Drive:
Motor to be furnished as specified above. *Drive components, drive mount, and drive guard to be furnished per your special sanitary design detail #657 & 659. *All electric components to be Nema 4 suitable for wet environment application.

CHAPTER 3 BUILDINGS

CHAPTER CONTENTS

GENERAL CONSIDERATIONS IN BUILDING DESIGN
FOUNDATIONS AND FRAMING
FLOORS
WALLS
ROOFS
CEILINGS
BUILDING INSULATION
WINDOWS
ENTRANCE AND EXIT DOORS
LIGHTING
VENTILATION
PLUMBING
UTILITY PIPING
EMPLOYEE FACILITIES

Having a good building design and selecting the proper building materials are extremely important. These important design requirements have a big effect on product safety and on the cost of the sanitation program. The sanitation failures that commonly occurred in food plants built many years ago can serve to illustrate the importance of good design and suitable building materials. During those building periods, wood was used to frame buildings and to lay floors. Rough wood beams, columns, and ceilings were difficult to clean; wood also dried out and cracked with age, which opened up numerous insect harbors. Walls were constructed of rough bricks with many mortar joints and they included many windows. The brick surfaces collected dust and dirt and, of course, the windows were an ongoing product safety problem. Lighting was inadequate and the electrical enclosures lacked dust-tight and watertight features. The presence of so many sanitation problems in these early food manufacturing plants were the direct result of deficient design and building materials.

There is no doubt that the design of food manufacturing plants has come a long way. Overall improvements in building design concepts and building

materials have resulted in a better product safety environment and have slowed the rapidly rising cost of sanitation programs. Despite this progress, a great deal of improvement is still needed. Many of these needed improvements are related to the way we are selecting the new building materials and the way we choose to use them. For example, the incorrect choice and placement of structural steel framing members can add many hours to the cleaning schedule. The absence of such easily overlooked, yet necessary items as insect screens and air filters, or the failure to properly seal a wall can seriously complicate the insect control effort.

Many of these problems are the result of simple oversight. Still others result from attempts to cut corners to avoid additional building costs. These attempts rarely pay off. Most sanitation design details, such as the addition of a few extra inches between a structural steel member and a wall, or the selection of a good viscous concrete block sealer instead of a conventional paint primer add little or nothing to the overall costs. Even in those instances when sanitation design criteria require more expensive materials or a more involved design scope, the additional expense can be cost justified as a return on investment.

Finally, it should be noted that while engineering for food safety and sanitation has improved, it is by no means perfect. New, untested materials and increased productivity demands are just a few of the challenges that face engineers, sanitation managers, and the many others involved with the design, construction, remodeling, or upgrading of plants. The search for easier, better ways to produce safe, wholesome food products will never end.

GENERAL CONSIDERATIONS IN BUILDING DESIGN

SIZE AND TYPE OF STRUCTURE

The size of the structure is heavily influenced by the plant's output or capacity, while the type of structure is influenced by the type of products and the operations carried on within. The plant layout process described in Chapter Two takes these and many other factors into consideration and is, in essence, a systematic process which determines both the size and type of building needed. This process is useful for both newly planned facilities and those being remodeled.

ARCHITECTURAL DESIGN

Not every food plant has to be an award winner, but from a sanitation viewpoint, there is little wrong with that objective. *(Figure 3-1)* The architecture of food plants should complement the surrounding community. In this way, they are more readily accepted and leave a favorable impression on all who have an opportunity to see them. Attractive plants help in establishing good employee relations, as well.

CONSTRUCTION MATERIALS

The large number of building materials available makes choosing among them difficult. Add to this process the evaluation of sanitation needs, and it becomes increasingly difficult. The problem does not lie in a lack of suitable materials; rather it lies in selecting the proper materials for the job at hand. This problem is compounded by the cost of certain materials. Many of the preferred materials cost more than the many alternatives that on the surface appear to be satisfactory. The difference, of course, is in the durability and cleanability. In general, the higher quality materials may cost slightly more but they also last longer. Substituting materials of a cheaper quality can result in higher maintenance expenses and could result in some serious product safety problems. By considering the food product risks, the proposed cleaning methods, and the market life of the product, good decisions can be made when selecting building materials.

PLANT SERVICES AND EMPLOYEE FACILITIES

Adequate room must be planned for administrative, maintenance, and engineering facilities. Failure to do so often results in these support service areas being crowded into the areas designated for manufacturing. More important are employees' facilities such as locker rooms, lounges, cafeterias, and various personal services. Although these facilities add to the building cost, their return on investment should not be underestimated. More will be said about employee facilities later in this chapter.

LANDSCAPING

To avoid problems from water runoff, buildings should be located on the higher elevations of the property, or the terrain should be reshaped to elevate them. The grounds should also be evenly graded to eliminate low spots where water may accumulate. The buildings should be oriented

FIG. 3-1 Attractive food manufacturing plants are more readily accepted by the community.

so that prevailing winds do not blow directly into the manufacturing areas. Accessways and parking lots should be paved, as dust from gravel surfaces can be blown into the plant by high winds. *(Figure 3-2)* Railroad track beds should also be paved. Paving prevents weeds from growing and paved areas are easier to maintain. Crushed rock ballast should be avoided as it is difficult to keep clean and rodents easily form burrows by nestling the rock together.

Additional landscaping should include grass, shrubbery, and a few trees. Grass and shrubbery should be kept neatly trimmed for vermin control. Plants, shrubbery, and trees that are known to attract certain insect species should be avoided; an established botanist and an entomologist can be helpful in making the proper selection. Trees and shrubs should be planted so they do not grow up against the buildings.

To accommodate anticipated rain water, adequate storm sewers should be provided. The grounds and paved areas should be sloped toward the storm sewer inlets in accordance with accepted engineering standards for ground water runoff. Dormant water puddles naturally attract insects and they frequently become breeding areas. They can also become a source of foodborne microbiological contamination.

For rodent control, a strip of pebbled rock a quarter inch in diameter should encompass the entire building or buildings on the exterior side of the foundation wall. *(Figure 3-3)*

One-quarter-inch pebble rock is recommended because larger rocks are easily nestled or interlocked together by rodents to form burrows. Concrete or asphalt are also acceptable materials to form this strip, although these materials are somewhat more costly. The strip should be 30 inches wide and four inches deep. Roofing paper or three- to five-mil polyethylene sheets placed in the bottom of the excavation help control weed growth. This strip also provides an unobstructed flat surface for the placement of bait boxes next to the foundation walls where rodents like to travel.

SECURITY LIGHTING

All lights attract insects to some degree. Security lighting systems for grounds, parking lots, and accessways should be designed with this consideration in mind. Ultraviolet light sources attract more flying insects than sodium vapor lights. In addition, sodium lights are more energy-efficient. Both high and low pressure sodium lights are available. High pressure sodium is more widely used indoors because low pressure sodium produces a very annoying monochromatic light that badly distorts all colors. High pressure sodium distorts colors also, but to a lesser degree.

FIG. 3-2 Paved parking lots and accessways reduce plant dust problems.

FIG. 3-3 A 30-inch wide strip of small rock ballast or a paved surface at the building foundation contributes to a sound rodent control program.

Security lighting should consist primarily of high or low pressure sodium lights. Mercury vapor lamps produce an ultraviolet light and should be used to reduce the flying insect pressures on the plant. In this case, ultraviolet mercury vapor lights should be used away from the buildings as a counter attractor to keep insects away from the plant. High or low pressure sodium lights should be used in the area close to the plant. Many high-flying insects are attracted to lights they see below them. Consequently, outdoor lights should be shielded so they cannot be seen by insects from above. Of course, they should also be directed toward the area they are to illuminate.

Light intensity should be limited to the need. Too often there is a tendency to overlamp with higher wattage lamps. Lower wattages attract fewer insects.

FOUNDATIONS AND FRAMING

FOUNDATIONS

In some sections of the country, the subterranean termite causes a great deal of structural damage. One of the most common species, the eastern subterranean termite, is found from Ontario, Canada, south to Florida, and west to Arizona and Utah. In areas of the country where these pests are common, consideration should be given to termite soil treatments. The major objective of termite control is to prevent the insect from moving its nest from the ground to its source of food, the wood in a structure. A complete control program requires treatment of the soil, of the foundation, and of the wood in the structure with pesticides.

Many industrial complexes feature concrete footings and concrete foundation walls that normally do not require treatment unless cracked. In such cases, holes are drilled in the crack and treated with pesticides.

Foundation walls constructed of concrete block are more difficult to treat for termites, because the hollow block cores require pesticide treatment. Because of this extra effort, a concrete block foundation is less desirable than solid concrete. However, with either type of foundation, the soil on both sides of the foundation should be treated at the time the excavation is backfilled.

Treatments for termite control vary according to types of pesticide chemicals used and the environmental circumstances. Many of the chemicals used, including Chlordane, Dieldrin, and Aldrin, are highly toxic and have a residual value exceeding 25 years; the termite control program must consider the proximity to watersheds discharging into community reservoirs or navigable waterways. Because of the serious threat to water sources posed by these substances, they are best used with the guidance of a pest control specialist.

Basement. When the structure includes a basement, the basement walls must be waterproofed. Concrete walls should be waterproofed by applying a heavy coat of asphalt to the complete exterior surface of the wall which is below grade. Sloping the flat surface of the footing on the exterior side of the wall will prevent water from accumulating on the footing. Drain tiles should be placed around the entire foundation to cope with any anticipated ground water problems. *(Figure 3-4)* All excavated areas should be backfilled and compression tamped as required to minimize the problems created by the ground settling.

STEEL FRAMING

There are several framing methods used in food plant construction. The most common is conventional steel framing with H or I beams and columns. In many cases, particularly in warehouse construction, bar joists or trusses are substituted for steel H or I beams. This method of framing is very acceptable for the most part. However, when the walls go up, the beams will automatically be

FIG. 3-4 Damp and wet basement problems are minimized with proper waterproofing techniques and soil drain tiles.

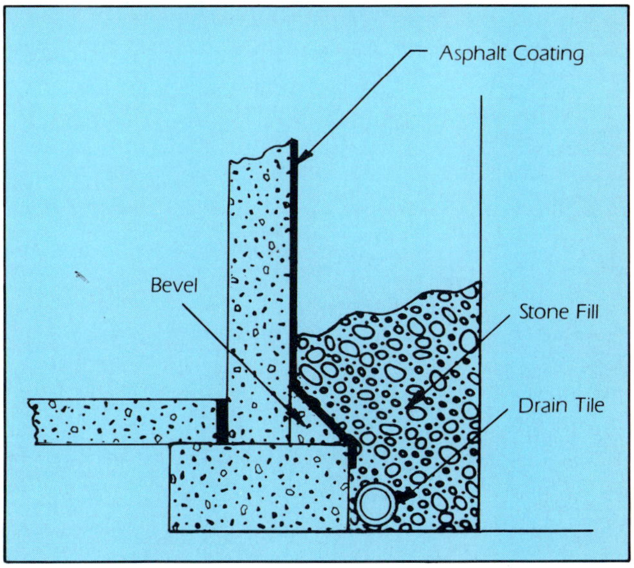

located too close to the wall, creating a linear cavity between the web of the beam and the wall. This cavity is impossible to clean because there is not enough space between the edge of the beam and the wall to permit access to the cavity area. *(Figure 3-5)*

Beams should be held out from the wall four to six inches.*(Figure 3-6)* Another approach involves boxing in the backside of the beam with a piece of sheet metal welded to the flanges to form a tubular shape on one side. The tube ends must be totally closed and all welds should be continuous. *(Figure 3-7)*

Stair stringers also tend to be placed too close to the wall. *(Figure 3-8)* When these stair stringers are fabricated with channel, as they frequently are, linear cavities once again appear. These cavities are created between the channel web and the wall surface. Channel stair stringer floor junctions are also difficult to clean because the flanges and web of the channel form a pocket at the floor line.

Channel stair stringers can be held out from the wall four to six inches and the pocket formed at the floor line can be cement-grouted. The grout should be sloped 60° toward the web of the channel. An alternative approach is to construct the stringers of tubular shapes or mild steel plates.

Void areas created by channel or beams can be filled with polyurethane foams and sealed with two thick coats of epoxy paint. The sealing is important as insects thrive in the foam. Lightweight concrete can also be used to fill these void areas, but be aware of cracks from shrinking. These cracks will need to be sealed with a caulking material.

The H- or I-shaped configurations of the steel column floor junctions create small, difficult-to-clean floor areas on either side of the web of the column. These should be grouted with nonshrinking grout. *(Figure 3-9)* The grouted area should be sloped 60° toward the web of the column.

FIG. 3-5 Steel beams mounted too tightly against walls produce linear cavities that are uncleanable.

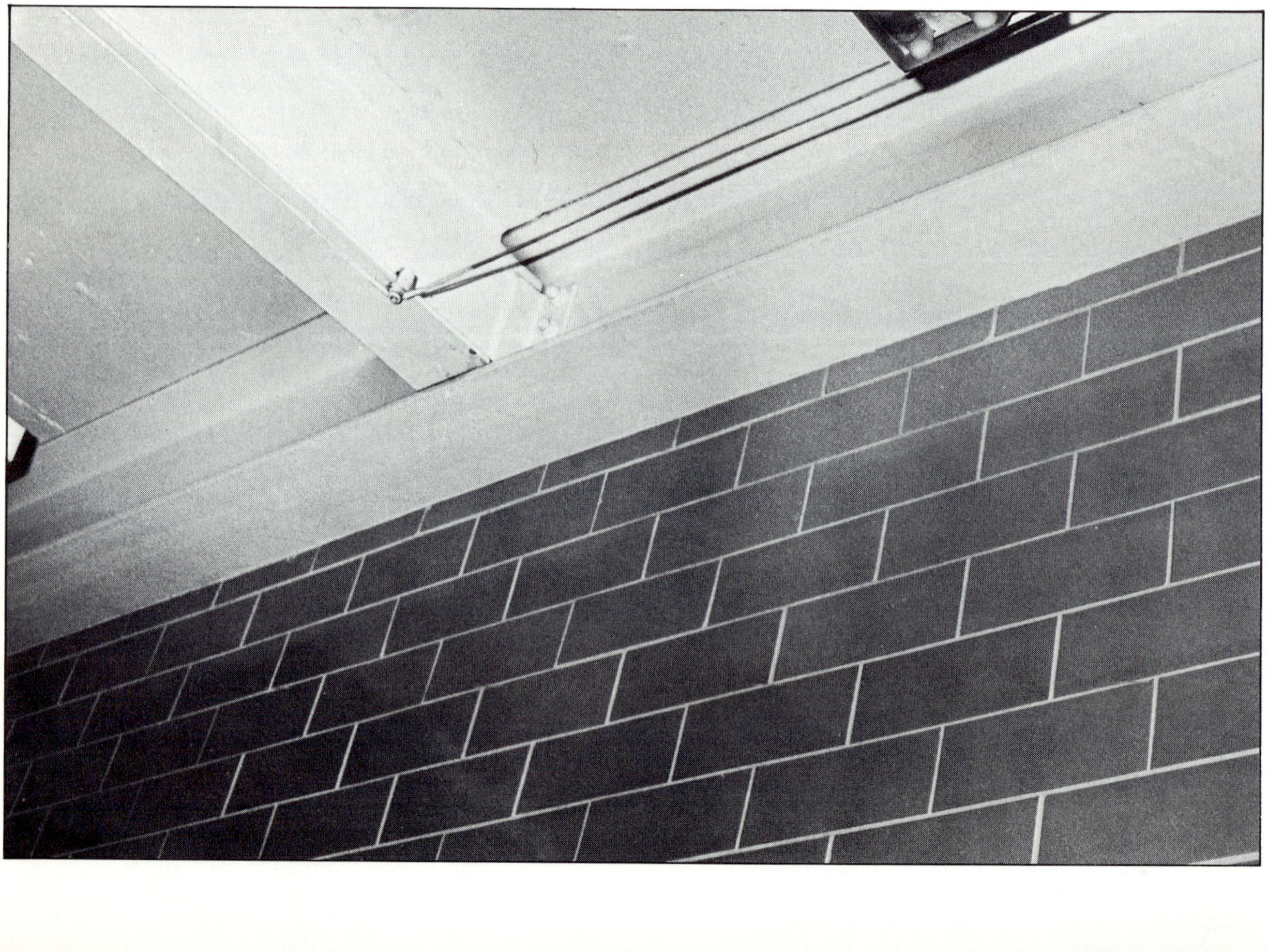

FIG. 3-6 The back side of a beam can be cleaned if held out from the wall four inches or more.

FIG. 3-7 Beams which are boxed in on the back side or constructed from tubular shapes eliminate the cavity problem altogether.

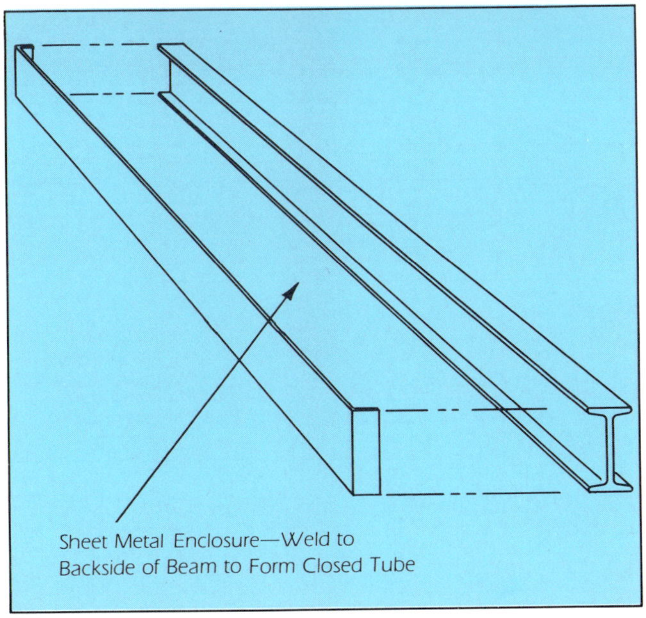

FIG. 3-8 Stair stringers constructed of channel placed tightly against the wall create another uncleanable linear cavity.

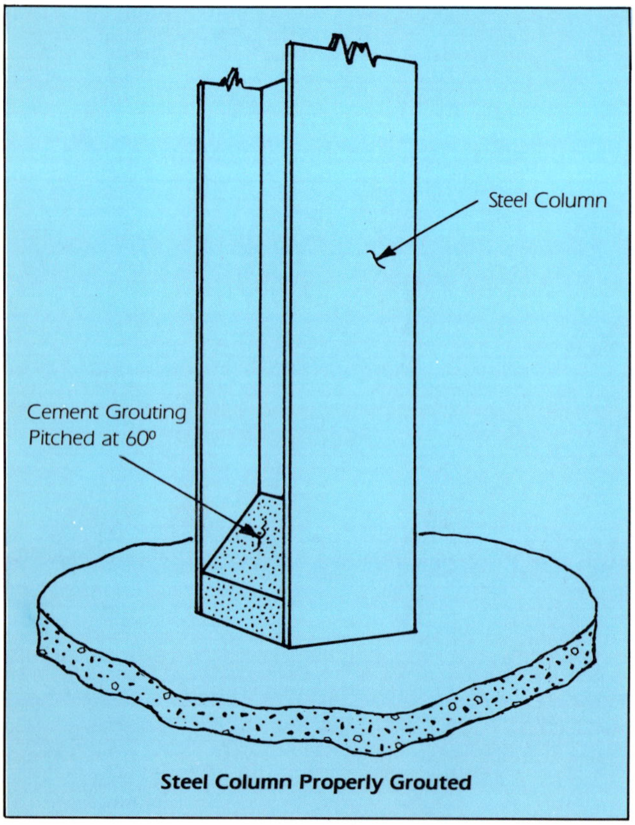

FIG. 3-9 Column floor junctions grouted between the web and flanges make floor cleaning much easier.

In warehouses, building columns are subject to damage from lift trucks, and they create cleaning problems at the column/floor junctions. To prevent these problems, the columns should first be surrounded by a 42-inch high protective pipe sleeve. The void area between the column and the pipe sleeve is then filled with a nonshrinking concrete mixture. At the top of the pipe sleeve the concrete should be sloped upward to the web and beam flanges at a 45° to 60° angle. *(Figure 3-10)*

Steel H- or I-shaped columns also tend to be placed too close to the wall. Boxing in the backside of these columns at the edges of the flanges with a light gauge sheet metal which is welded in place to form a tubular shape eliminates the void area created by the web, flanges, and wall. An alternative method of dealing with this problem is to construct a pilaster around the column. *(Figure 3-11)*

Install channel wall girders so that the flanges of the channels are pointing down. A large cuplike cavity running the full length of the channel is formed when the channel flanges point up. This cavity is difficult to clean.

All-thread rods should not be used as girder support rods. All-thread collects dust and is hazardous to clean. Mild steel rods threaded at each end, as required, is a better way to support the channel girders.

Many bar joists or roof trusses are built with angle iron intermediate members and their lower and upper members are built from structural tees. Because the angle iron intersects at various points along the lower and upper structural tees, inaccessible or difficult-to-clean areas are formed. Substituting round pipe for the angle iron intermediate members will make the design somewhat more sanitary.

Tubular Framing. Square and rectangular steel shapes for framing members offer a number of sanitation

FIG. 3-10 Warehouse columns need extra protection to prevent damage from lift trucks.

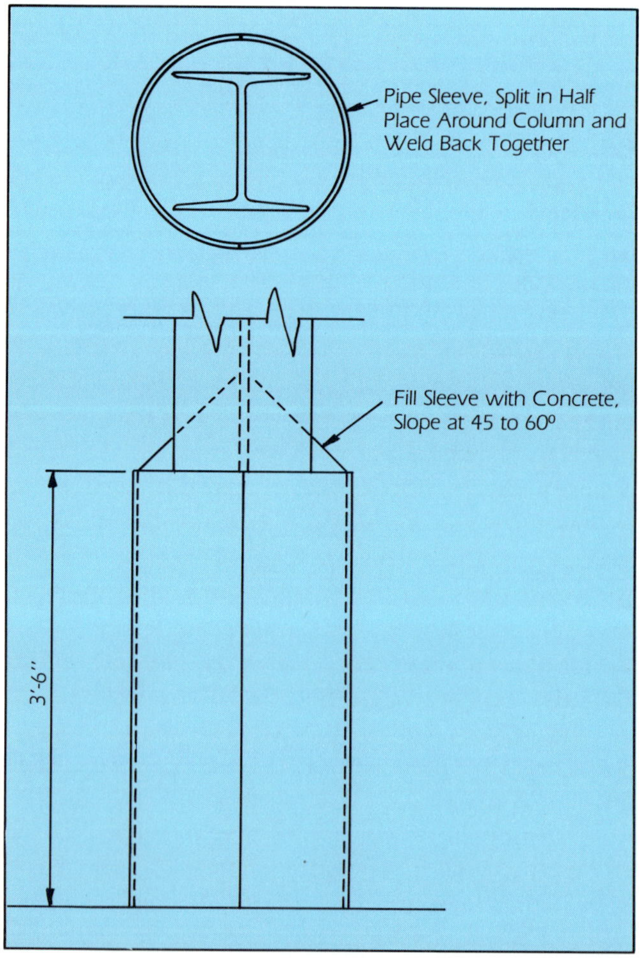

advantages. Using tubular shapes for beams eliminates many dust and dirt collecting surfaces. Cavities and void areas created when beams are placed too close to walls are also eliminated.

Square and rectangular columns eliminate the difficult-to-clean floor/column junction as well as the void areas created when the columns are placed too close to walls. Round tubing or pipe is also an acceptable sanitary design for structural columns.

When tubular shapes are used, it is important to close all open ends. When drilling holes in the tubing for various attachments, the holes should be threaded or other measures should be taken to seal them so insects cannot enter. All unused holes must be sealed.

Structural Steel Plating. In wet manufacturing environments, such as those found in meat processing or canning plants, peeling paint and corrosion of structural steel building framing members pose serious product safety problems, and they detract from the overall appearance of the plant. The heavy-duty cleaners and sanitizing solutions used to wet clean these plants cause further peeling and corrosion. Numerous painting systems and structural steel surface preparation techniques have been used to eliminate this situation. In most cases they have enjoyed little or no success and the threat of product contamination continues.

Wet manufacturing environments or manufacturing process areas that require the use of large quantities of hot water, detergents, and sanitizers for cleaning should not be painted. The structural steel building frame members should be plated instead. Zinc or galvanized plating surfaces are more than adequate for applications of this type. On the whole, plated surfaces stand up extremely well for long periods of time in the wet processing environment. Unfortunately, the connecting hardware such as nuts, bolts, and washers used to bolt the members together are frequently overlooked, though they also require plating. Such things as door frames, window casings, and miscellaneous equipment supports should also be plated or constructed from corrosion-resistant materials such as stainless steel.

Plating large structural steel building framing members can be both costly and troublesome. Large members are hard to handle and require a lot of plating material. In addition, not every plating facility has a dipping tank large enough to do the job. For this reason, plating specifications for building steel should be limited to those environments where it is needed. Structural steel plating applications are generally limited to newly planned additions or to major remodeling projects.

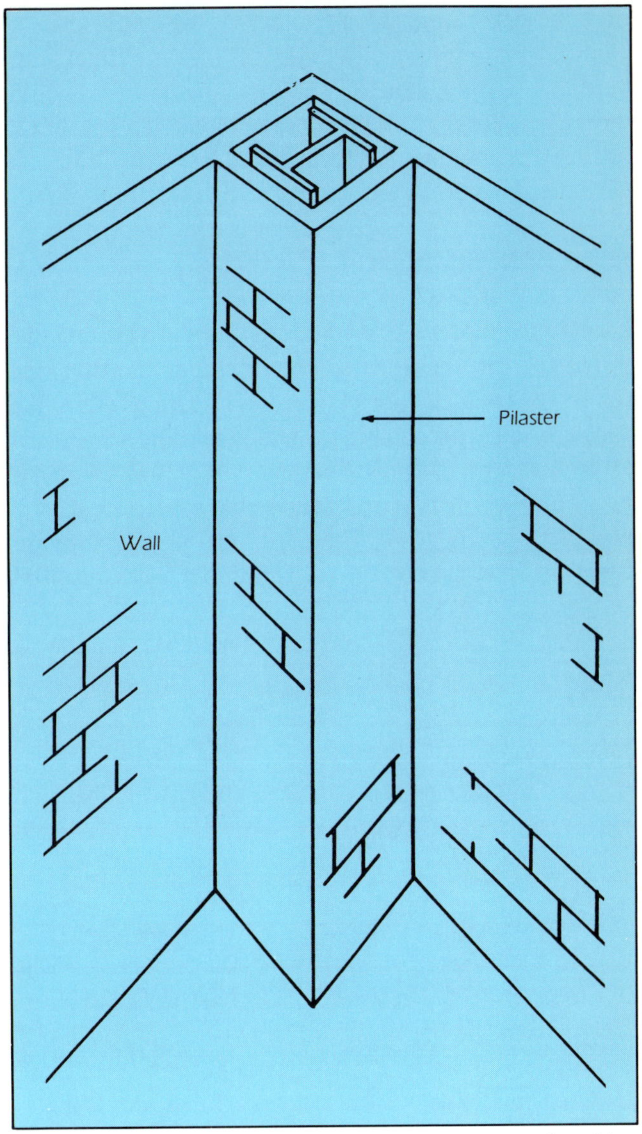

FIG. 3-11 Columns placed too close to the wall create uncleanable cavities; pilasters correct this problem.

It should also be noted that structural steel plating is not compatible with some foods. Plating is therefore limited to nonproduct zones. This point is covered further in Chapter Four.

Because of the variety of plating materials and finished surfaces available, it might be a good idea to examine a few samples before committing a large area to plating. Plating materials vary in surface thickness, hardness, and appearance. Selection should be based on the various activities, operations, and materials which will be found in the chosen area. As previously stated, zinc and galvanized materials are adequate for most applications.

CONCRETE FRAMING

Two other methods of framing are reinforced concrete beams and columns, and composite superstructures. The difference between these two methods is that the reinforced concrete method places reinforcing rods in the concrete, while the composite method reinforces with steel H or I beams. The two methods have the particular advantage of eliminating ledges that collect dust and increase cleaning costs. These methods of framing also fireproof steel beams and columns, which can result in lower insurance costs. *(Figure 3-12)*

You should be aware that when forms are stripped from concrete work, the finished surface is not totally smooth and free from pits and checks. In some cases, extensive honeycombing has occurred. These imperfections can become insect harbors and should be rubbed out with a rubbing stone and a mixture of plain cement and water mixed to the consistency of very thick paint filler. This should be done while the concrete is still *green* or uncured.

FLOORS

The selection of proper flooring material is directly related to the area of the plant involved and the operation to take place within the area. For example, concrete with an

FIG. 3-12 Concrete framing produces a virtually ledge-free framed structure, a big plus for sanitation.

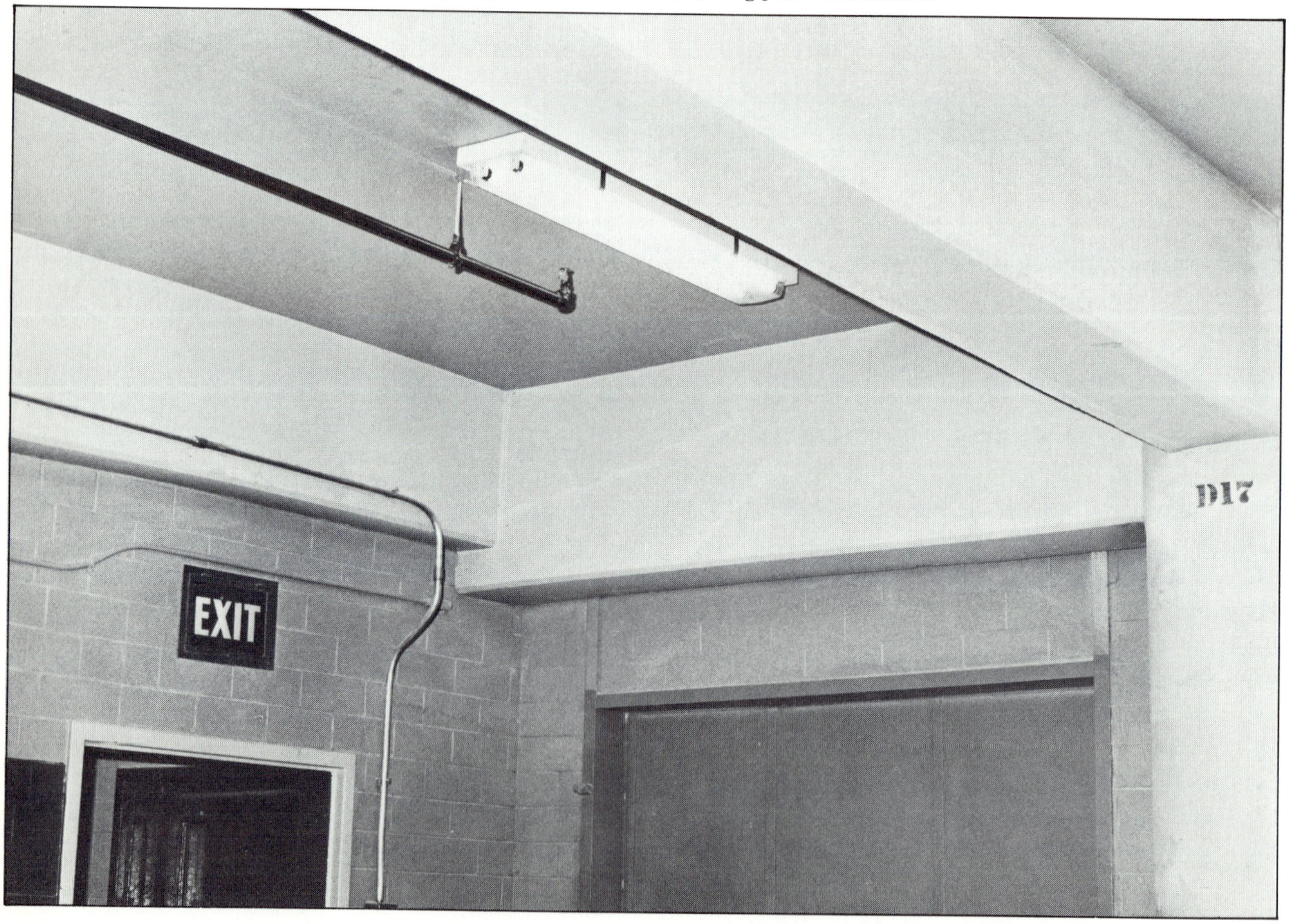

applied floor hardener might be selected for a warehouse floor. A wet manufacturing area may require concrete with a monolithic surface coating or an acidproof brick or quarry tile floor. The administrative area may call for still another type of flooring material. In this section we will discuss a number of flooring materials and the areas for which they might be considered.

All flooring surfaces must be suited for the conditions of exposure they are likely to encounter. They must also be easy to clean and be able to stand up to cleaning and sanitizing chemicals. If pesticides and heat sterilization are used to control insects, select flooring materials that can withstand chemical pesticides and heat sterilization temperatures.

The sanitation dollars spent on the floor surfaces are dollars well spent. The appearance of the floor is very crucial to the overall appearance and efficiency of the plant. If the floor is neat, clean, and well-maintained, it will make a lasting and constructive impression on all who see it, including visitors and regulatory inspectors. It will also have a lasting impact on plant performance.

CONCRETE FLOORS

Concrete is a very suitable flooring material for many areas of the food manufacturing facility including the warehouse, maintenance shop, storeroom, and some manufacturing floors. However, if we look at a finished concrete floor under a microscope, we find nothing more than a series of tiny pores. These pores support dust, and for that reason, no matter how hard we sweep the floor in these areas, they always continue to look dirty. For this reason, uncoated concrete floors are not acceptable for dusty areas.

One of the most serious challenges for concrete floors comes from food acids, particularly citric acids. These acids can quickly break down both new and old concrete floors, depending upon the exposure level. Once the acids break through the finished surface layer of the concrete, the stone aggregate is exposed and the floor becomes rough and difficult to clean. It also has an adverse effect on the appearance of the floor.

Concrete floors used in wet processing areas will eventually break down, particularly if they are exposed to continuous running water or harsh chemicals such as sanitizing solutions and some pesticides. Also, concrete floors can be easily stained by leaking lubricants and spilled paints. Trouble really begins when concrete floor surface coatings or floor paints are applied. *(Figure 3-13)* It seems that many surface coatings are not what they are advertised to be. However, the reality of the matter is that we may not have selected the proper floor coating or prepared the surface properly to receive the coating. Improper surface preparation is the single largest reason for floor coating failures! There are many good monolithic coatings available, but each of them is designed for a specific job, and all are contingent upon good surface preparation.

Coating Exposures. The success of monolithic floor coating begins with the selection of the proper coating. In other words, we must first identify what stresses the floor will be exposed to and select a coating which can stand up to those requirements. Possible stresses include forklift traffic, cleaning chemicals, sanitizing agents or pesticides. After we have identified the level of traffic and type of cleaning method the floor will be exposed to, we can select the appropriate coating material.

Surface Preparation. Once the coating is selected, the coating manufacturers will specify a number of surface preparation conditions, including concrete curing, sandblasting or scarifying, acid etching, and primer application.

- Concrete Curing. Trapped water results in the failure of many floor coatings. As the trapped water evaporates and moves toward the floor surface, it tends to lift or bubble the paint. Peeling paint then results. Proper concrete curing prior to coating applications can solve this problem. Curing can take as little as a few days or as long as 90 days or more. Normally 15 to 30 days are adequate.

- Sandblasting/Scarifying. Sandblasting or scarifying is required for most coatings. Old floors often contain spilled lubricants and paints, while new floors are often treated with hardening materials that contain waxlike or rubberlike substances. These materials must be removed from the concrete, as they will affect the bond between the concrete

FIG. 3-13 All food companies are confronted with floor problems at one time or another. Painted surface coatings have their place. Floor surface satisfaction requires the proper selection and installation of the flooring system.

substrate and the coating materials. Old concrete floors might be weakened at the surface; this old concrete should be removed as well.

Sandblasting in a food plant is likely to cause serious dust problems, which can contaminate food or food ingredients. The addition of water to this process, called hydroblasting, reduces this contamination threat, but also introduces the problem of controlling the water. To avoid the dust problems associated with sandblasting and the water problems caused by hydroblasting, scarifying is the preferred method of conditioning floors. Most scarifying machines include a vacuum system that does an excellent job of controlling concrete dust. During scarifying, the floor can also be wetted down to help contain the dust.

- Shotblasting. Steel shotblasting is rapidly replacing sandblasting because it is considerably less dusty; however, the steel shot used in this method poses a contamination threat. This threat must be controlled by the use of a dust and shot collecting system.

- Acid Etching. Though scarifying is the recommended method of floor coating preparation, some surface coating manufacturers only require that a concrete floor be acid etched and rinsed to remove impurities. Newer floors, floors that are less soiled, or wood float finished floors may be candidates for acid etching only.

- Primer Application. As is the case with any paint or coating application, primers are very important in floor preparation. Primers contain adhesive binders that create a good bond between the surface to be coated and the coating. Two finish coats cannot be used in place of a finish coat and a primer coat. •

Concrete Floor Coatings and Paints. There are numerous coatings available that will do an acceptable job if the type of exposure the coating will be subjected to is properly identified, and the surface to receive the coating is prepared according to the recommendations of the coating supplier. You should request a written performance guarantee from coating manufacturers.

Most monolithic coatings are polymer-resin-based products composed of various binding and floor material compounds. When properly selected, these materials stand up well against citric, phosphoric and tartaric acids, sugar, and a number of other ingredients that create problems for concrete floors. Sugar is one of the more troublesome ingredients, because it reacts with and quickly weakens the cement binders in a concrete mixture. Polymer-based coatings also stand up reasonably well to the constant erosion capability of running water. These coatings may be applied in troweled-on layers ranging in thickness from a few mils to a quarter inch or more.

Some polymer-resin-based products contain styrene. In recent years, the U. S. Food and Drug Administration has encountered several instances where improperly applied and cured flooring systems of this type have produced styrene vapors. These vapors permeated packaging materials and food products in cold storage. According to the U. S. Food and Drug Administration's interpretation of the Federal Food, Drug and Cosmetic Act (21 USC 342(a)(3)), foods contaminated by styrene from floor surfacing applications are adulterated.

The United States Department of Agriculture requires advance approval of any monolithic flooring surface material proposed for use in plants under their jurisdiction. To prevent contamination from styrene vapors, all food products, particularly those in cold storage, should be removed before the coating is applied. Make applications at the required temperature, allow adequate time for the material to cure and finally, make certain that the area is thoroughly aerated.

Monolithic coatings normally cost less than other flooring materials such as quarry tile and acidproof brick. While lower cost is a significant advantage, it should not be the sole factor determining your choice of floor materials. First and foremost, the flooring material you select must satisfy the exposure and durability requirements identified.

In addition to monolithic coatings, there are very durable clear urethanes that can be used alone or in combination with epoxy floor paints. Because these surfaces are hard and abrasion-resistant, they are excellent for floors exposed to heavy pedestrian traffic.

Floor Openings. It is often necessary to cut holes of various sizes through the floor for equipment, ducts, and open stairs. Whenever a hole is cut, particularly in wet manufacturing areas, provide a curb at least four inches high that is watertight. *(Figure 3-14)* Once the equipment is in place, the openings can be grouted closed or closed with a cover of the metal flashing type to increase watertightness.

FIG. 3-14 Floor openings should be curbed, particularly in wet manufacturing areas.

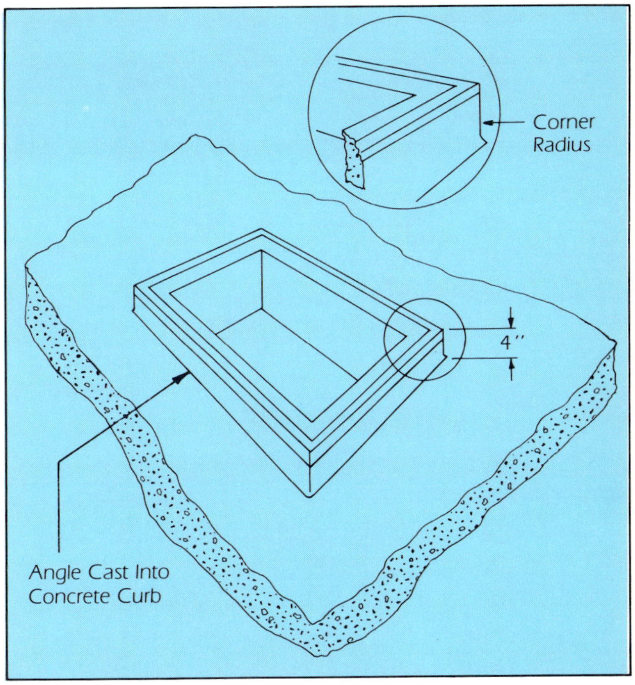

Control Joints. Use a pliable control joint compound with chemical-resistant properties to fill control joints cut in concrete. Food products can accumulate in unfilled joints.

FLOOR TILE/BRICK

Quarry tile and floor brick remain very popular for the wet processing areas of a food plant. *(Figure 3-15)* When properly laid, they provide an attractive and easy-to-maintain surface that is very functional. The key to successful installation lies with selecting the proper bed-setting cements and grouting mixtures.

There are several methods of laying tile or brick floors. Two of the more common methods are known as the *tile setter's method* and the *brick layer's method.* The tile setter's method involves placing the bed-setting cement mixture on the concrete substrate and then setting the tile. *(Figure 3-16)* Next, an epoxy or chemical-resistant grouting mixture is troweled into the joints. After the joints are struck with a striking tool, the excess grouting material is wiped up. This method is suitable for most light-duty wet processing applications where the floor will not be subjected to heavy vehicle traffic or to continuous running water and harsh chemical exposure.

The brick setter's method, used in conjunction with a waterproof membrane, is particularly suitable for areas subject to extremely wet conditions or to harsh chemical exposure. *(Figure 3-16)* The waterproof membrane is laid between the concrete substrate and the bed-setting

FIG. 3-15 For wet processing areas, tile or brick floors have many benefits.

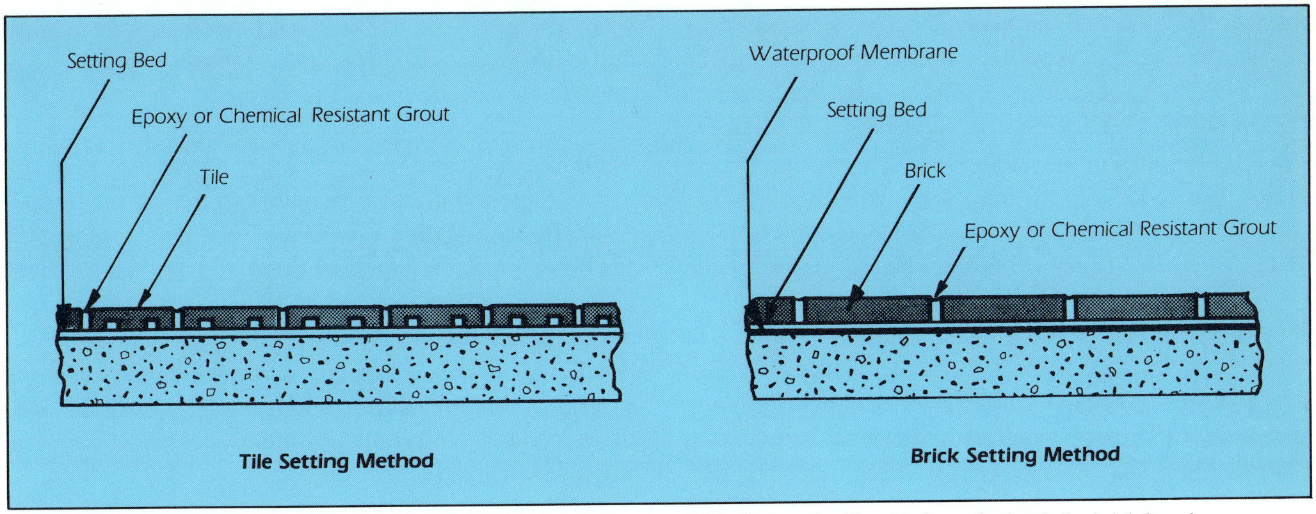

FIG. 3-16 Two installation methods are commonly used to install tile or brick floors: ***the tile setter's method*** **and** ***the brick layer's method.*** **Waterproof membranes can be used with both methods.**

material. Then the bricks are buttered on all four sides, laid in place, and the joint is struck. Waterproof membranes can be used with floor tiles as well.

Waterproof membranes are necessary because even the very best brick or tile joints allow some water or harsh chemicals to penetrate, regardless of the grouting material selected. If water or chemicals are allowed to reach the concrete substrate, they can cause the concrete to break down, which results in a cavity. When someone steps on a brick or tile directly over a cavity, the joint breaks. Mites, roaches, or other types of insects are attracted to broken joints and microbial problems can also occur. The waterproof membrane prevents water or chemicals from reaching the concrete substrate, thus preventing the concrete from breaking down.

The quarry tile or floor brick you select should have a nonskid, abrasive, or textured surface. Chemical- or corrosion-resistant red shale or fire clay floor brick that ranges in thickness from 1-1/8 to 2-1/2 inches should be considered for heavy-duty applications, such as those involving the operation of material handling equipment. Chemical- or corrosion-resistant quarry tile is more suitable for lighter-duty applications. Brick or quarry tile should be nonporous and have the low absorption characteristic required by some regulatory agencies.

When waterproof membranes are used, they should consist of an asphalt material. Hot asphalts are usually applied in several layers to form a thickness of one-quarter inch. Place a glass cloth reinforcing member between the layers of asphalt.

Grout and mortar selection is the key to the satisfactory performance of the total floor system. Select a grout for tile joints and a mortar for brick joints and for setting beds which are capable of withstanding the corrosive and thermal conditions you anticipate. Grouts and mortars are composed of fillers, binders, hardening agents, and special chemical- or corrosion-resistant ingredients. For the most part, these ingredients are thermal-setting plastics, such as epoxy and furan resin materials.

Make sure the tile or brick joints are held to a width of approximately a quarter inch. Joints must be completely filled with the specified joint material. Do not use spacer mix substitutes. For applying bed mortars, use a notched trowel. In some instances (usually involving a small area) bed mortar can be used for the joints as well. Be certain that the grooves on the bottom of the tiles are completely filled with the bed setting mortar to ensure tile floor performance. In cases where there are deep grooves on the bottom of the tiles, you may need to spread mortar across the back of each individual tile, in addition to applying it to the floor. Mortar applied to the floor should be troweled to make the notched trowel grooves run parallel to the grooves in the bottom of the tiles. This makes it easier for the bed-setting material to get into the tile grooves, as intended.

Normal building expansion and contraction require the use of control joints in tile and brick floors. The number and location of these joints is determined by design conditions. In general, expansion joints are required over building expansion joints in the base slab, around the floor drains, and at floor/wall junctions. The sealants used for control joints must resist chemicals, remain elastic, and bond well to the sides of the brick or tile.

Many grouting and mortar mixtures stain or stick to the top surface of tile or brick and become very difficult to clean. Prewaxing with a melted paraffin wax prior to installation will protect the top surface. Be careful so that no wax gets on the sides or bottoms of the tiles or bricks, as the wax causes a poor bond between these surfaces and the grout or mortar materials. After the floor is completed and the joints are set, the wax can be removed by steam cleaning.

The slope of a tile or brick floor toward the floor drain is determined by the base slab or by a special fill. The slope should be 1/8 inch to 1/4 inch per lineal foot; 3/16 inch per lineal foot is preferred. Floor/wall junctions or vertical curb/wall junctions should be coved, because coved corners are easier to clean.

Not all grouting mixtures adhere to metal drain flanges; because of this feature, drain flanges should be coated with bed-laying cement, or otherwise treated to create a good bond between the tile or brick and the metal wall of the drain. Make the final elevation of the floor drains 1/8 inch lower than the finished floor level. Locate drain or weep holes above the drain seepage pan. These weep holes allow water that seeps alongside the drain to find its way into the drain interior.

Throughout the food industry, quarry tile and brick floors are widely used in wet processing areas. Because these floors are exposed to different manufacturing conditions and to a variety of cleaning, sanitizing, and pest control chemicals, each installation should be looked at independently so that all service conditions can be identified. Only in this way can you choose proper flooring materials, including setting beds and grouting mixtures, that will meet the identified conditions. There are a number of helpful standards available that can also be applied to develop a concise floor specification.

METAL TILES

Consider metal tiles for loading and unloading areas and for staging points where vehicle traffic is heavy. Metal tile is constructed from tough rolled steel floor plate that has a raised lug pattern, for traction, permitting quick stops and fewer starting difficulties. Metal floor tiles are normally set in an epoxy cement bed.

WOOD

Many early food plants were built with wood framing and wood floors. Although wood is very comfortable for workers who are required to stand all day, it is difficult to maintain, as it readily stains and scratches. Of greater concern are the cracks and swellings that occur as these floors dry out or are exposed to relatively high moisture levels. For these reasons, wood is not a popular flooring material today. Unfortunately, some older plants still have wood floors.

Though some of these floors can be reconditioned by sanding and sealing with the new urethane sealers, many simply need replacement. When replacing these floors, a trick of the trade is to store the new hardwood flooring material in the same area where it will be used. In this way, the wood becomes conditioned to the atmosphere, and any shrinking or swelling takes place prior to installation. As a result, less cracking and swelling appear later. *(Figure 3-17)* Wood subfloors in these old structures should be replaced with plywood underlayment. If the old underlayment is in good condition, make certain that all cracks and joints are caulked with a good pliable caulking material.

RESILIENT FLOORING MATERIALS

Floor materials such as vinyl, vinyl asbestos, and asphalt tile or linoleum should be used only in administrative areas, cafeterias, locker rooms, or hallways. *(Figure 3-18)* When used in manufacturing zones, they become very difficult to maintain, as they stain and scratch readily. It is also extremely difficult to maintain a good adhesive bond between the substrate and the resilient material. As a result, some serious insect problems can develop under these flooring materials.

FIBERGLASS

Fiberglass floors feature polymer resins used in conjunction with fiberglass blankets which are placed between resin coats. Once again, problem identification and final surface preparation are the keys to a satisfactory job. Failure to meet these requirements will cause unsatisfactory results.

FIG. 3-17 Old wood floors can be reconditioned and made to look like new.

FIG. 3-18 Resilient floor applications are satisfactory for plant locker rooms, cafeterias, administrative areas, and hallways.

Some fiberglass floor resin mixtures give off toxic fumes when they burn, although the burning characteristics of most fiberglass mats may be no worse than wood. Ask the manufacturer about this problem. Resins that give off toxic fumes should not be used.

BITUMINOUS/ASPHALT

Except for warehouse railroad track wells or bulk unloading shed track beds, bituminous or asphalt floors are not satisfactory surfaces for manufacturing areas. *(Figure 3-19)* Steam and lubricants weaken this surface, making it easy to damage. Nonetheless, it is an excellent material for warehouse railroad track wells or bulk unloading shed track beds, as it remains pliable to some degree, and it will give or flex with the movement of train tracks. Sealing the porous surface of this material with a good asphalt sealer produces a smoother and easier-to-clean surface. Sealers also make asphalt more impervious to lubricants. Though durability and maintenance can be problems with asphalt, when used in the areas prescribed, wear and tear will be minimized, as there is little traffic in these prescribed areas, aside from the trains. In addition, cold asphalt repairs are easily made.

CRUSHED ROCK

Occasionally, crushed rock, sometimes called *rock ballast,* is used in track wells or unloading sheds. This material is unsatisfactory, as it freely catches various types of debris or railcar dunnage and removal is a very labor-intensive cleaning task. In addition, rodents find it easy to form burrows and nests in crushed rock.

FIG. 3-19 Although bituminous surfaces have limited plant application, they can be used for railroad trackwell surfacing.

FIG. 3-20 Food product inventories can be closely controlled if warehouse storage areas are identified through floor striping.

WAREHOUSE FLOORS

Concrete is the preferred material for warehouse floors. Seal these floors with a concrete sealer to make them stain resistant and easier to clean.

Warehouse floors should be striped to identify storage locations. This measure facilitates a first-in, first-out stock rotation program, an important phase of product safety. *(Figure 3-20)*

To identify storage limits, paint an 18-inch-wide white strip around the entire perimeter of the warehouse floor at the wall junction. *(Figure 3-21)* Do not permit storage within the confines of the white area; this is essential for an effective rodent control program.

Sometimes, a concrete curb 18 inches wide and 6 inches high is used as an alternative means of establishing the storage limit at the wall. However, this practice merely compounds rodent control problems by establishing two rodent runways: one on top of the curb at the wall junction and one on the floor at the six-inch curb wall junction. It is important to remember that because rodents do not see well, they must feel their way along walls with their whiskers. The presence of these two runways necessitates the placement of traps on top of the curb at the building wall and on the floor at the six-inch-high curb wall.

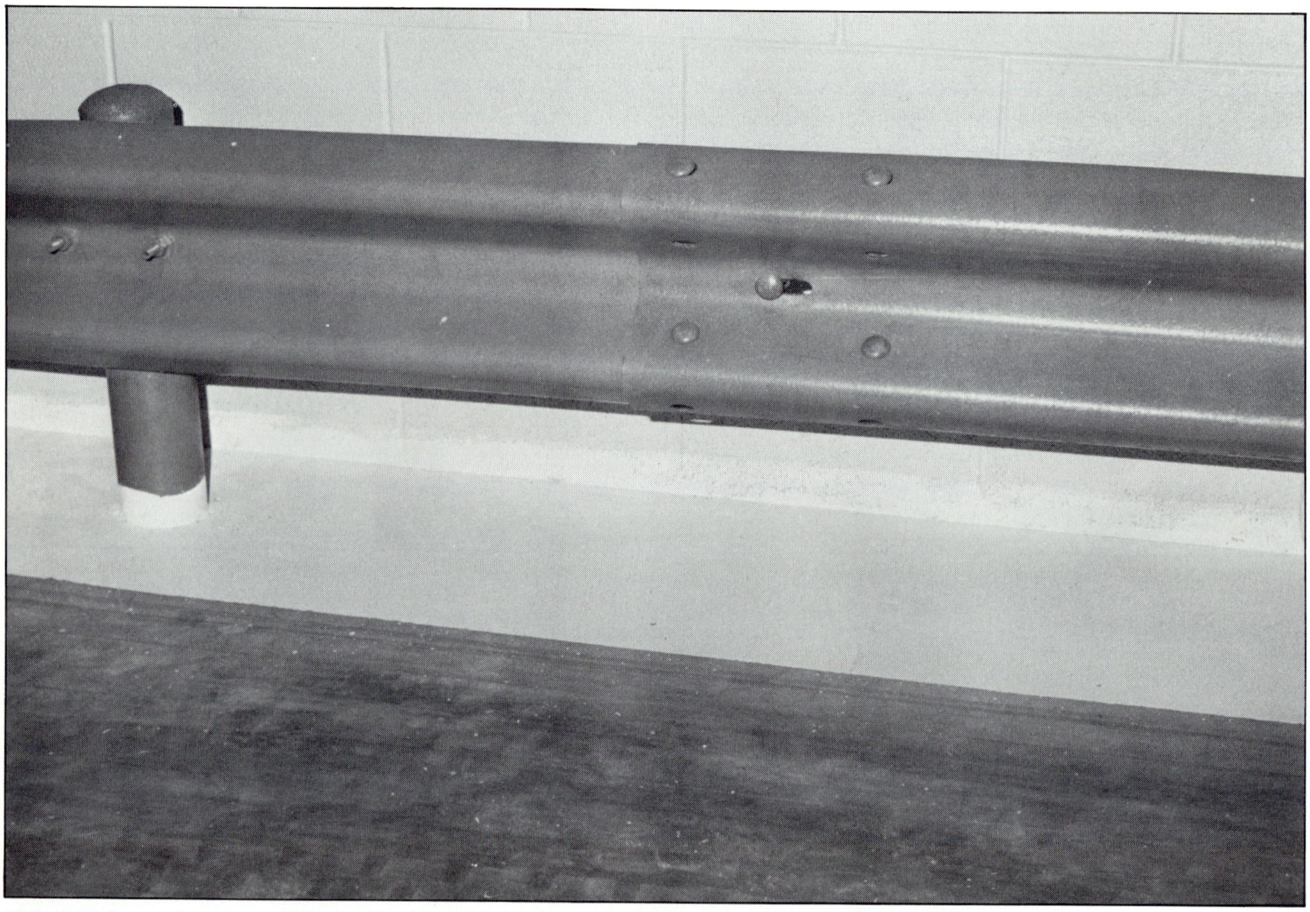

FIG. 3-21 In warehouses, an 18-inch wide white perimeter strip identifies storage limits at interior wall junctions and provides a place for rodent traps. A steel guard to protect the wall is recommended in lift truck aisleways.

WALLS

Selecting wall surfaces is difficult because of the variety of available materials. In addition, the finished surface requirements of these walls will differ according to their location in the plant, and the type of abuse they suffer from cleaning chemicals. Wet process manufacturing areas require wall surfaces that are impervious to moisture and will stand up against the harsh actions of cleaning chemicals. Dry, dusty manufacturing processes require a hard, smooth finish that is free of pits, checks, and cracks. Warehouse walls require a surface that will not harbor insects and other vermin. All wall surfaces must be built with materials that are easy to clean.

There are several wall materials that can be used alone or in combination with a number of surface coatings to obtain a wall surface that is smooth, nonporous, nonabsorbent, and impervious to moisture. A vast number of new wall-coating systems and facing-tile materials are available to upgrade the walls of many older and well-established food plants. Prepare the surface carefully to receive these new materials. In some cases, removing all of the old surface material, repointing the joints of blocks or brick walls, and patching the cracks and holes will be necessary.

When you choose new wall materials, consider their ability to be recoated or resurfaced in the future. Cleaning and recleaning takes its toll on wall materials, and at least some repainting will be required. Anticipating the maintenance requirements of the original wall materials can avoid some later resurfacing expenses.

CONCRETE

There are three methods for constructing concrete walls:

forming and pouring into place, pouring on the floor (tilt-up), and precasting. Precast wall panels are normally cast at a precast plant in unspecified lengths and then saw-cut to the required length. Then the panels are hauled to the job site and set into place. If properly installed, this type of wall is very satisfactory for most parts of a food plant. *(Figure 3-22)*

In many food plant areas, the finished concrete surface may require paint. Remember, if you want satisfactory results, the same conditions that apply to floors apply to walls as well. You must identify the problem or exposure the coating is likely to have and properly prepare the surface to receive the coating according to the manufacturer's recommendations. For example, if you plan to apply a coating to a wall that is formed and poured-in-place concrete, take into account the common practice of applying lubricants to the metal concrete forms. These lubricants must be removed from the metal concrete forms because they can permeate the concrete and affect the bond of the coating primer. Acid etching can be used to remove form lubricant but sandblasting is probably more effective.

Precast Joints. Although precast concrete panels are widely used and have economical advantages, their joints require careful caulking. A good caulking compound, made of acrylic to retain its elasticity, should be used. Other pliable caulking materials may also be suitable.

Precast panels must also be secured and seated at the foundation with appropriate grouting or caulking materials.

CONCRETE BLOCK

Concrete block is satisfactory for many food plant wall surfaces. Blocks are available in different densities or weights; light and heavy are the most common, though in some sections of the country a medium-weight block is also available. Lighter blocks are undesirable from a sanitation viewpoint as their pores support insect and bacteria growth. Effective block fillers can eliminate the

FIG. 3-22 Concrete wall surfaces are satisfactory for many areas of the plant.

disadvantages of more porous blocks.

Avoid the use of volcanic ash and cinder blocks. These very porous blocks have pin holes extending all the way into the hollow block cores that can be penetrated by insects. Volcanic ash blocks also make fumigating troublesome, as obtaining a good gas-tight seal is difficult. In addition, ash materials absorb some fumigants and later give them up in the form of offensive odors.

Mortar Joints. When laying concrete block, use standard cement and sand mixtures in the mortar joints. The finish coating materials will provide some protection from food acids, sugar and other ingredients, and cleaning materials. The joints should be struck more lightly than usual to minimize the deeper crevice where dust or food particles can lodge. Do not eliminate striking altogether, as a less than satisfactory appearance can result.

Concrete Block Finishes. You will want to choose a concrete block finish appropriate for its location. Administrative, process, dusty, and bacteria-sensitive areas require different finishes.

- Administrative areas. Concrete block used in an administrative area or storeroom could be finished with two coats of epoxy paint. *(Figure 3-23, upper-left corner)*

- Process areas. For a relatively dust-free and bacteria-insensitive food process, one latex

FIG. 3-23 Concrete blocks are satisfactory wall materials. The choice of surface finish from block fillers and paints depends on the manufacturing environment.

1. Two Coats Semi-Gloss Epoxy Coating,

2. One Coat Latex Block Filler Plus Two Coats Semi-Gloss Epoxy Coating.

3. Two Coats Latex Block Filler Plus Two Coats Epoxy Coating.

4. Two Coats Latex Block Filler with Cement Additive Plus Two Coats Semi-Gloss Epoxy Coating.

block filler coat and two coats of epoxy paint are satisfactory. *(Figure 3-23, upper-right corner)*

- Dusty areas. A coating featuring two coats of latex block filler and two finish coats of epoxy paint is satisfactory for a dusty area. *(Figure 3-23, lower-left corner)* Notice that the pores of the block are nearly all filled.

- Bacteria-sensitive areas. These areas require a smooth, pit-free, moisture-impervious surface. As many pores as possible should be completely filled. The finish for a bacteria-sensitive area features two latex block filler coats. A 20% mixture of plain cement is added to the filler material to thicken it. Two finish coats of epoxy paint are then applied. *(Figure 3-23, lower-right corner)* The epoxy paint used for all four of these applications should be semi-gloss or glossy enamel.

Tile-like Coatings. There are a number of coating products on the market that produce a hard, smooth, tile-like finish for use on various wall materials. The coatings are applied as a two- or three-coat system. These products are made with polymer (polyacrylate and polyesters) and urethane materials. Apply these coatings quite thickly, using a brush, squeegee, or trowel. The final surface has the appearance of a smooth, hard-surface ceramic or glazed tile. These coatings are particularly well-suited for concrete block walls and can also be applied over other wall surfaces, such as brick or poured concrete.

OTHER WALL MATERIALS

Brick. Brick interior walls have lost a great deal of their popularity. This can be attributed to cost and to inherent cleaning difficulties caused by the many mortar joints and the rough surface of the brick.

Glazed Tile. For wet processing or bacteria-sensitive areas, vitreous china or glazed tile walls are your best choice. These walls stand up well, have a very attractive appearance, and are very easy to clean. Choose a finished surface that is glazed, not dull and rough. The block joints should be grouted with an acid-proof or epoxy grout. The grouting material should be applied after the wall is laid up with the normal sand/cement mortar, and the joints are struck slightly deeper. After the joint material has set, trowel on the epoxy mixture, restrike the joint and wipe off the excess grout from the block. *(Figure 3-24)*

Laminated Block. A less expensive wall material that provides a glazed tile surface is concrete block with a vitreous china lamination on the outer surface. The vitreous material can be laminated to one or both sides of the block at the time the block is made. This finished surface performs well in wet processing zones. *(Figure 3-25)*

Ceramic Facing Tile. Ceramic facing tile provides an exceptionally fine wall for many wet processing areas. Make sure you choose ceramic instead of plastic tile, which is much less durable. Use epoxy or acid-proof grouting mixtures with this tile to minimize the vulnerability to moisture. When moisture seeps behind tiles, it can loosen them.

Fiberglass/Plastic Wall Systems. When properly applied to existing wall surfaces, fiberglass and plastic wall systems can result in a new finished surface that is smooth, durable, impervious to moisture, and easy to clean. The success of these walls depends on the degree of exposure to high temperatures and harsh chemicals, as some of these surfaces do not stand up under these conditions. Also, the surface of the existing wall must be properly prepared to receive the new material or separation will occur.

Free-standing plastic panel wall systems usually consist of 1/32- to 1/16-inch thick plastic sheets bonded to a 5/8-inch fir plywood sheet. Two and one-half inches of foam insulation are bonded on the backside, then another piece of 5/8-inch fir plywood is bonded in place. The bonding of still another 1/32-inch or 1/16-inch plastic sheet forms a sandwich panel. The panel size most commonly used is 4 feet wide by 8 feet long. Joints are covered with a plastic snap-in molding strip.

Plastic panel wall systems are often used for cold rooms or freezer room construction. In these applications, cold does permeate the wall to some extent. A lower temperature results, which makes the wall less desirable to most insects.

Plastic panel systems are sometimes used in wet manufacturing areas. If you do so, be sure that all horizontal and vertical joints are completely watertight. Water that

FIG. 3-24 Glazed tile or vitreous china wall materials are easy to clean and stand up well in wet manufacturing areas.

FIG. 3-25 Vitreous china materials can be laminated to concrete blocks.

permeates the wall through the joints will saturate the fir plywood and cause rotting. The foam insulation also traps and holds moisture. This damp, soggy condition creates an attractive environment for many types of insects. When this condition develops, it becomes extremely difficult to control and, in some cases, it necessitates removal of the wall panels.

In wet manufacturing areas, it is also important to apply the proper water-tight joint sealants. Make sure you closely follow the installation instructions for the system you choose. Remember that many bacteria-, mold-, or yeast-sensitive processes require daily use of water, detergents, and sanitizers and that the integrity of most joint sealants are severly tested by the exposure to harsh chemicals and large quantities of water resulting from this kind of cleaning.

When considering plastic wall panels or any other wall system, you must carefully balance the sometimes conflicting factors of product performance, low cost, and sanitation enhancement. Economic life, the all-important sanitation requirements, the need, and the application must all be considered.

Metal Panels. Metal walls can present a number of sanitation problems. Metal has a high expansion and contraction rate which makes wall panel joints difficult to keep sealed. Metal also has a high heat transfer rate, resulting in condensation problems. Metal wall systems, particularly single- and double-face sandwich prefab types, should be selected with these problems in mind. Most manufacturers offer panels in varying widths. Choose wider panels for fewer vertical joints and smooth metal panels for easier cleaning. *(Figure 3-26)* Do not drill in sandwich panels to fasten equipment and pipe supports. Equipment and pipes should be supported from the steel framing or the ceiling instead. Use a good pliable caulking material in the panel joints to obtain a good seal. Insect problems can thus be avoided. Sandwich panel fabrication joints are not dust-tight and insects can easily crawl into the hollow area of sandwich panels.

Rodent infestation can be a serious problem with prefab or field-assembled sandwich panel walls. Once they gain access, rodents are able to move freely in the inside core area of sandwich panel walls. A vertical runway for rodents is created from access points at ground level all the way to building access points at the top of the wall, when

FIG. 3-26 Metal panel walls with smooth interior surfaces enhance sanitation.

insulation does not completely fill the metal corrugated inner core. With their great climbing ability, rodents can easily dig their claws into the insulation materials and climb vertically to the eaves without any difficulty. From there, they are free to enter the plant and to move along beams into any area of the building they choose. This type of a problem usually begins when the panels begin to rust at the bottom or ground level junction. It can also occur if the ground junction is not properly sealed, as a mouse only needs a quarter inch opening to gain access.

The solution to this problem is to use metal end caps for both the bottom and the top of the panels. *(Figure 3-27)* These end caps must conform to the corrugation pattern of the panel and must fit tightly so they can be spot-welded or riveted in place and then sealed to keep insects out.

Do not confuse these end caps with closure strips used at the bottom fastening junction between the base fastening cleat or angle and the inside wall surface. End caps are used to enclose the open ends of the panels. The closure strip is used to create a seal between the panel and the fastening

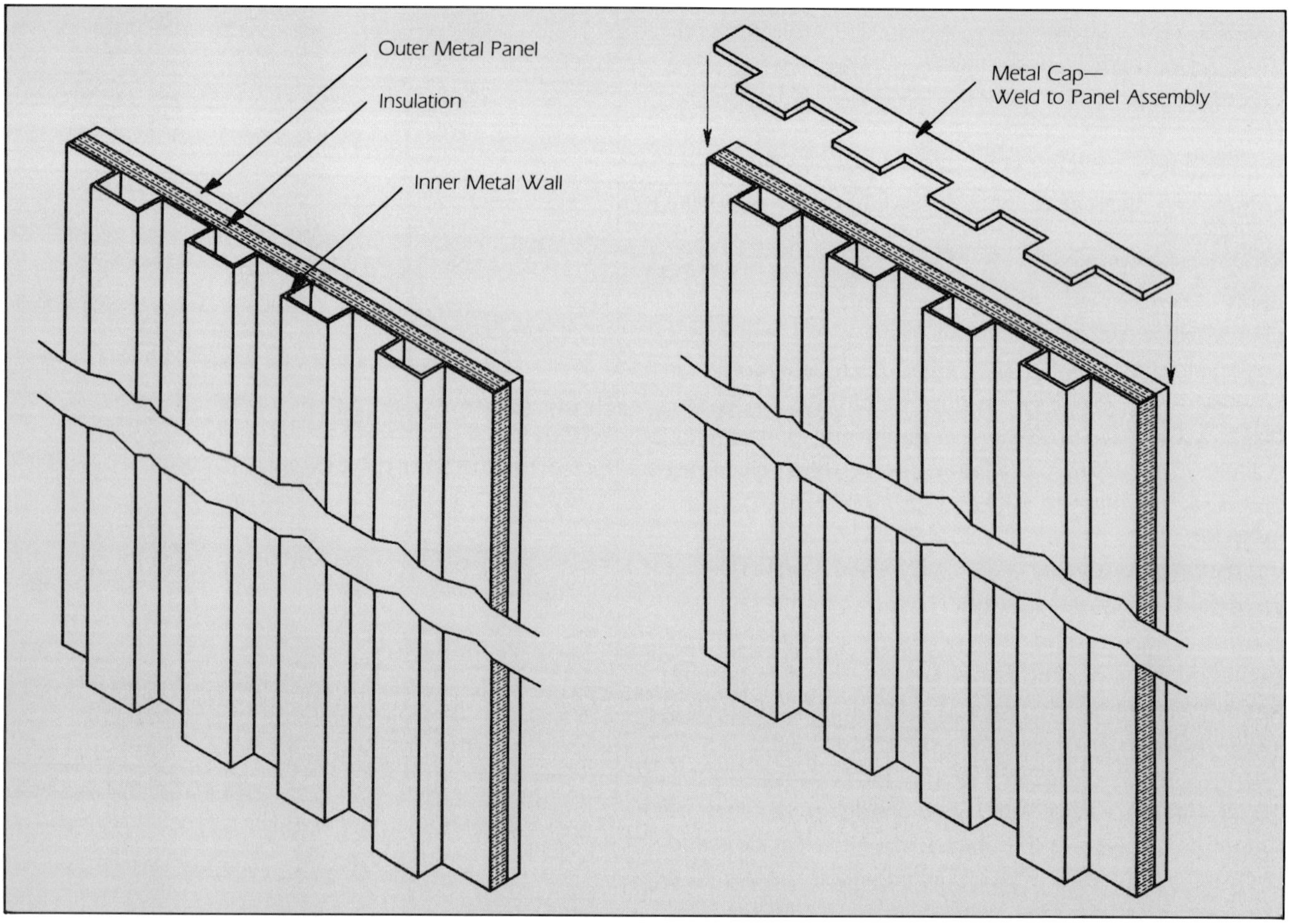

FIG. 3-27 Corrugated metal wall panels must be capped at the top and bottom to prevent rodents from entering the inner hollow core created between the outer metal panel and the insulation. Rodents can easily scale insulation surfaces and enter buildings at the ceiling of the top floor.

cleat. In the same way, the conventional closure requirements at the eave juncture are not effective in and of themselves. The panel end caps on the top or uppermost panels are required as well.

Do not underestimate the importance of the panel end caps. Rodent problems in metal panel walls of a sandwich type are not universal; however, once rodents gain access, the problem becomes widespread in a short period of time. Rodent-proofing can require a complete and costly overhaul of the wall panels at the floor and eave junction. The problem is quickly compounded by the rapid reproduction capabilities of rodents, who can use the wall cavity to nest as well as for plant access.

The economic life of metal panels should also be considered, as metal is subject to corrosion. This is an important concern when pesticides are to be used at floor wall junctions, or if the panels are used in or adjacent to a wet process. These conditions will shorten the life of the metal panels.

Plaster/Plasterboard. Plaster and plasterboard walls are not impervious to moisture and are easily damaged. Limit the use of plaster walls to administrative areas.

Wood and Miscellaneous Wall Materials. Wall materials such as plywood panelings, pressed woods, and insulating boards should be limited to administrative areas. Most of these materials are neither moisture-proof, nor very durable.

WALL CONSTRUCTION DETAILS

Top Block Course. The top course of all block walls must be laid with a solid block to close off all open block cores in the layers below. If top courses are not closed off, a vast network of insect and rodent harbors are created throughout the entire wall. *(Figure 3-28)*

Coved Floor Junctions. Use coved base tile blocks or facing tiles at the floor junction to improve cleanliness. *(Figure 3-29)* Sanitation inspectors frequently list poor corner cleaning on inspection reports. Avoid square corners, which are difficult to clean.

Wall Openings. Openings cut in panel walls for ductwork, pipe, and air intakes should be framed and sealed.

Closure Strips. Rubber or synthetic rubber closure strips are generally required with corrugated panels at the floor junction. These rubber materials must be impervious to pesticides or other chemicals that may be used, as some pesticides can dissolve certain rubber compounds.

Eaves. All eave and wall junctions must be weather-tight to keep out the elements as well as pests, birds, and rodents. Metal wall eave junctions, in particular, can be difficult to seal.

ROOFS

Roofs over food manufacturing areas should be easy to clean. Because many food manufacturing processes use a great deal of air, numerous exhaust stacks are vented through the roof. Sometimes process equipment extends through or is mounted on the roof, resulting in many instances in product spills on the roof, due to choke-up or particle entrapment in the exhausted air stream. Left unattended, these spills quickly attract birds, rodents, and insects. How rapidly and thoroughly you are able to clean up the roof depends on the type of roof you select.

FIG. 3-28 Solid block layers should be used as the top course to close off pest animal and insect access to the core area of the blocks.

FIG. 3-29 Coved floor wall junctions are much easier to clean.

FIG. 3-30 Pitch and gravel roofs are a better choice for warehouses than for manufacturing areas.

PITCH AND GRAVEL ROOFS

Pitch and gravel, in particular, should not be located over process areas where product spills can occur. It is impossible to pick up a product spill on a pitch and gravel roof without taking gravel along with it. *(Figure 3-30)* Sometimes lime is used to clean up a spill, but the roof is frequently damaged in the process. Pitch and gravel roofs are more suitable for warehouses or other areas where spills are not likely.

SMOOTH ASPHALT ROOFS

Smooth asphalt roofs using various durable felt roofing papers and insulating materials are very suitable for installation over manufacturing areas where spills might be anticipated. These roofs are cleanable, though extra care must be taken to prevent damage with cleaning tools. *(Figure 3-31)*

CONSTRUCTION DETAILS

Roof Downspouts. Roof water accumulations attract vermin. To avoid this problem, all flat roofs should be designed with downspouts to handle rainwater. Equip the downspout drains with bullet-nose grates that project upward. Take extra care to prevent product spills from washing into the drains. This will result in contamination of the storm sewers that flow into natural water tributaries. In regions with a cold climate, downspouts should be located inside the building to prevent freezing.

Roof Openings and Caps. All openings through the roof should be curbed and flashed. The curb should extend 12 inches or more above the finished roof. Insulation and roofing materials should extend up the outside of the curbed side walls of the opening. *(Figure 3-32)* Do not place insulation on the inside of the curb wall, as it is difficult to clean and frequently becomes infested with insects.

Insulate all roof opening metal caps, and make sure they are constructed as panels of the sandwich type. Place insulation material between the inner and outer panel wall. *(Figure 3-32)*

FIG. 3-31 Product spills or deposits from manufacturing vents or exhaust stacks are more easily cleaned off smooth asphalt roof surfaces.

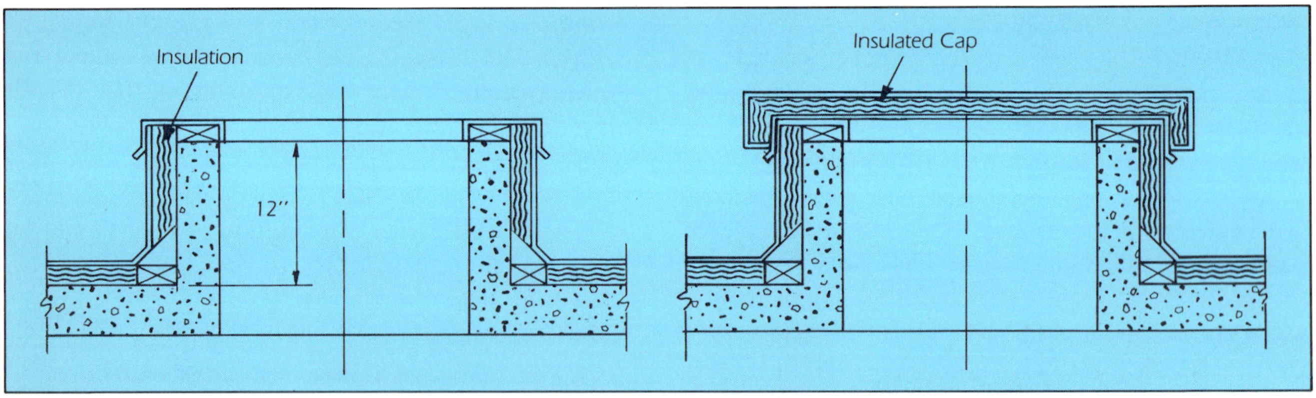

FIG. 3-32 In many geographic areas, roof openings require insulation to prevent condensation problems; insulation should be placed on the outside surface to minimize sanitation problems.

FIG. 3-33 Corrugated metal ceilings commonly produce many hard-to-clean cavities at the ceiling framing junction.

FIG. 3-34 Concrete ceilings are acceptable, particularly when rubbed smooth free of pits and checks.

CEILINGS

Selecting the proper ceiling materials and choosing the correct ceiling design play an important role in plant sanitation. Because they are hard to reach, ceilings accumulate many kinds of airborne matter including smoke, steam, condensation, dust, wet particulate matter, mold, bacteria, and insect pests. The proper selection and application of ceiling materials and the use of the correct construction detail makes a big difference in how easily these problems can be solved.

METAL CEILINGS

Using corrugated metal panel ceilings in production zones can create problems. *(Figure 3-33)* The cavities formed by the panel corrugation at the beam are hard to clean. In addition, these panels have linear or longitudinal joints on 12- to 18-inch centers which occasionally harbor insects. Unfortunately, these joints are very hard to seal and keep sealed. Metal ceiling panels have a high heat transfer rate that can cause considerable condensation problems. In view of these difficulties, avoid the use of metal ceilings in food manufacturing areas. If you must install this kind of ceiling, do not choose corrugated panels; use the smooth, flat ceiling panels instead.

CONCRETE CEILINGS

Poured concrete slabs or ceiling panels are your best choice for ceilings as they allow for the suspension of equipment. *(Figure 3-34)* Their surface should be smooth, pit-, and check-free. If you wish to paint the ceiling, condition it properly first. In environments where peeling paint is a concern, stain the ceiling instead. Stain penetrates concrete rather than adhering to the surface, and it allows the concrete to breathe or release accumulated moisture.

If you select a precast, poured concrete panel ceiling, make sure you seal the longitudinal joints with a good, pliable caulking material. However, the use of caulking materials in the overhead poses the problem of direct contamination to equipment mounted directly under the ceiling. Be careful to prevent this from happening, or instead, use a formed-and-poured-in-place concrete ceiling for the manufacturing areas of the plant. Warehouse, maintenance, and similar areas are better choices for precast concrete panel ceilings requiring caulked seams.

GYPSUM CEILINGS

A gypsum board ceiling can be used satisfactorily in some warehouse applications, particularly when combined with a ledge-free framing plan. Make your framing plan consist of tubular shapes or standard structural beams for main members that are boxed-in, by welding sheet metal to the edge of the beam flanges and submembers of channel. Channel can also be boxed-in by sheet metal welded to the flanges. Use bulb tees on four foot centers for the submembers. Then lay a bead of pliable caulking material on the flanges of the tees and put gypsum sheets in place. Next, place the selected insulating material over this structure and then place a conventional roofing system on top. However, do not use this type of ceiling in manufacturing areas. It lacks strength for supporting equipment and is not impervious to moisture.

PLASTER CEILINGS

Plaster ceilings are neither impervious to moisture, nor durable. These ceilings should be limited to special applications such as those in administrative areas, locker rooms, and administrative supply storage areas.

BUILDING INSULATION

Many types of insulating materials are satisfactory for food plant buildings. However, asbestos or asbestos-containing insulation materials should not be used because of the hazardous nature of asbestos fibers. These fibers are injurious to health; they cause asbestosis, and they have been associated with various kinds of cancer.

From a sanitation viewpoint, the kind of insulation material selected is not as important as how and where the material will be used. Most insulation materials provide a natural habitat for insects, particularly with the presence of a nearby food supply. Therefore, the sanitation emphasis should be on using insulating materials in such a manner that they can be sealed off from insects. This means that wall and ceiling insulation should only be used where it can be sandwiched between inner and outer panels. Ceilings are insulated by placing the insulation material on the roof base and placing the roofing materials directly over them. Emphasis should be placed on sealing off the interior and exterior surfaces.

Using soft insulating blankets or sprayed-on insulating materials as a finished wall or ceiling surface is unsatisfactory. Insulated surfaces are both difficult and costly to clean and maintain, and they represent a highly attractive natural habitat for all types of insects and other vermin such as rodents.

Using insulating boards as a finished surface is also unsatisfactory. Insulating boards are usually hard to clean, difficult to keep sealed, and rarely impervious to moisture. Also, the unsealed joints can harbor many insects.

As previously mentioned, most insulating materials, including fiberglass, can be used in a food plant. However, remember that the presence of any type or any quantity of insulation in the food product contaminates the product. All insulation materials should be chosen and used with the intention of minimizing the threat of contamination.

WINDOWS

WINDOWS

Because of the many problems created by windows, most modern food plants have eliminated them altogether. Windows are a costly maintenance item both from a mechanical and a cleaning viewpoint. Keeping them closed is also a constant problem. Regardless of the sophistication of the heating, ventilating, and air-conditioning systems, there are always those who are uncomfortable with surrounding temperature conditions and feel the need to open a window or a door. Open windows and doors necessitate the use of screens which also present maintenance problems.

Glass Block. If windows are required for natural light purposes, use glass block. It is easy to maintain, and cannot be opened.

SCREENS

The superiority of mechanical ventilation systems cannot be overstressed. However, the need for ventilation through open windows and doors will probably exist for some time. Windows and doors that can be left open should be screened. Standard mesh window screen is satisfactory. Screens should be designed so they can be removed for washing and repair.

In some instances, it might be necessary to reinforce insect screens which are located within three feet of the ground with a heavy-gauge wire 1/4-inch mesh screen to keep rodents out. Rodents can gnaw through light-gauge insect screens.

ENTRANCE AND EXIT DOORS

All personnel entry and exit doors should be of hollow metal construction. Metal doors selected should offer good, tight, spot-welded seams. *(Figure 3-35)* Loose-fitting seams can result in an infestation of the hollow cavity with insects. Although solid core is available, wood doors should not be selected since they generally require more maintenance and are vulnerable to the gnawing teeth of rodents.

Do not locate night lights over or near outside entry or exit doors because they attract night-flying insects. *(Figure 3-35)* Instead, place them 30 or 40 feet away to detract insects from the points of entry.

FIG. 3-35 Entrance and exit doors should be constructed of metal. Night lights should not be located directly over these doors as they attract night flying insects.

TRUCK DOORS

Truck doors should be one of two types: the metal roll-up or the hollow metal panel lift-up. Make sure they provide a good, tight seal at the door frame or door track. *(Figure 3-36)* As with any food plant door, your object is to be certain that it is completely rodent-proof. Since a mouse needs only a 1/4- or 3/8-inch opening for access, this is sometimes a problem. The roll-up type of door frequently has a housing or sheet metal enclosure just above the door frame that serves to contain the door when it is up. Because the enclosure traps dust and dirt, modify it slightly by cutting away the lower portion of sheet metal. You may want to order a door without the enclosure.

Dock Levelers. Dock levelers permanently installed in a special leveler pit are often used in conjunction with truck doors. This leveler pit is difficult to rodent-proof. Should a rodent gain access, it can easily crawl up the concrete pit wall, find the opening between the leveler and the leveler pit wall, and enter the plant. You can rodent-proof these pits by casting a steel box directly into the concrete. *(Figure 3-37)* This rodent-proofs the pit because rodents have difficulty scaling slick vertical metal surfaces. You can rodent-proof existing dock leveler pits by adding and eight-inch wide by eighteen gauge sheet metal frame on the front face of the pit opening and on the pit walls. In both cases, use the sheet metal in a vertical position. Then, cement the metal to the concrete with a suitable epoxy cement.

TRACKWELL DOORS

Trackwell doors must be constructed of metal as well. They are the most difficult doors to rodent-proof. These doors are usually corrugated, and they roll up and down in a vertical track. The door's threshold, the irregular shape of the rail track, and the vertical track that guides it in the up and down movement present problems for rodent-proofing.

FIG. 3-36 Truck doors should also be metal and should properly fit in the frame.

FIG. 3-37 Dock leveler pits must be rodent-proofed.

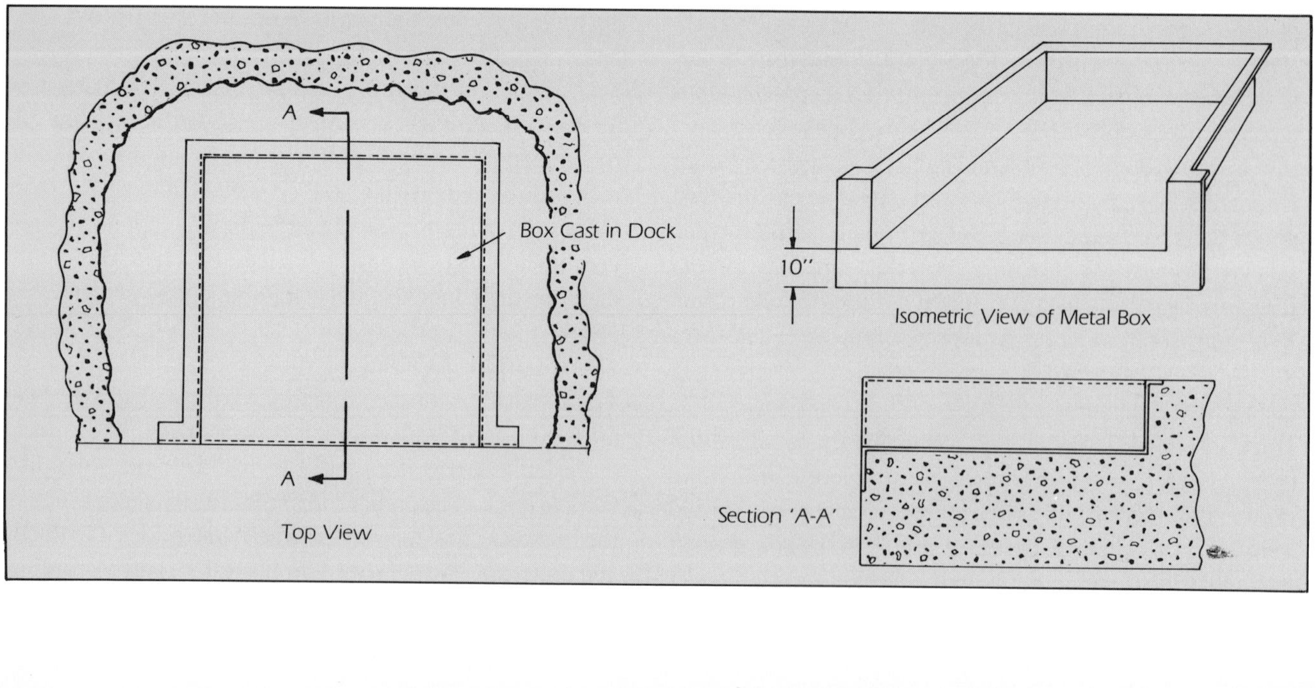

New Construction. Most thresholds on trackwell doors are constructed of concrete or asphalt. Over time, the salt dropping from the undercarriage of rail cars can spall the concrete and the dripping rail car bearing lubricants can cause the asphalt to break down. As a result, openings large enough for rodents can be created, allowing them to enter the plant. The threshold is also subject to distortion from ground frost in colder climate zones. A steel channel threshold can virtually eliminate all of these problems. *(Figure 3-38)* Note that this installation features a split track gap or opening. (This is the opening from a standard rail splice arrangement.) The installation is spotted so that the gap is directly in line with the fall line of the door. The channel threshold is at the same elevation as the lower portion of the railroad track bulb section. When the door is closed, there is a perfect seal all along the threshold. *(Figure 3-39)* The elevation of the threshold does not interfere with the rail car wheels since the flange on the wheel projects down only to the lower portion of the bulb section of the track.

FIG. 3-38 Railroad trackwell doors are often the most difficult doors to rodent-proof. The solution lies with the threshold and the provision of a gap between each track for the door.

FIG. 3-39 When closed, this door provides a rodent-proof seal, but only after tubing was added to reduce the clearance in the vertical door track (shown in the upper right-hand corner).

Electric heating cables are installed to keep any water or snow formations from freezing on the threshold. Should this happen with the door in the up position, there might be an unlevel surface on the threshold. Should freezing occur in the down position, which is more likely, the door gasket will be frozen to the threshold. A vulcanized or cemented rubber pad on the threshold might eliminate the need for heating cables in some cases as rubber does not usually freeze to rubber.

The threshold described above is costly, but so is its possible alternative: a citation for rodent contamination. The threshold is tied to the foundation or grade beam to accommodate the heavy loads of railroad equipment. This method of rodent-proofing trackwell doors is recommended for a new food plant, since it is superior to almost any alternative.

The last obstacle in making the trackwell door completely rodent-proof is solved by reducing the clearance between the vertical door track and the door itself. Sometimes this clearance is large enough to let a rodent work its way around. You can reduce this clearance by cutting out a section of the door and welding in its place a one-foot piece of square tubing that more closely approximates the inside dimension of the door track. *(Figure 3-39, upper right-hand corner)* A concept drawing showing the details of this threshold scheme has been provided. *(Figure 3-40)*

Existing Trackwell Doors. It would be exceptionally expensive to use the channel threshold system, previously described, on an existing trackwell door, since major preliminary work would be required before final installation. Another approach, featuring a rubber railroad track crossover pad system furnished by one of the major rubber manufacturers, has proved to be satisfactory in this rodent-proofing application. It features conventional rubber crossover pads, tie shims, replaceable rubber filler strips, and lag screws. *(Figure 3-41)* The system was originally designed as a railroad track crossover for lift trucks and other plant vehicles operating on the plant grounds. The replaceable rubber filler strip butts against the inside surface of the track, and leaves a small opening between the web of the track and the pad. This opening should be plugged with steel wool on each end. The filler strips are secured in place by the inner crossover pad which is lag screwed through the tie shims to the ties. The rubber filler strip is so resilient that it can easily be compressed by the rail car wheel flange. After numerous compressions, wearing and permanent setting of this strip in the compressed position occurs, and it must be replaced. To replace it, remove the lag screws and take up the inside pad, put in new strips, and reinstall the inside pad.

The pads on the outside of the rail are also lagged to the ties. Tie shims are used to bring the pads to a level flush with the top of the tracks. As a result, all the components are lagged to the ties, and any movement due to the train loads will allow everything to move in unison, a real plus.

Other rubber filler strip systems are available that are secured in place next to the inside of the track with removable steel retainer plates that are then, in turn, bolted to the threshold. When combined with steel channel cast in the concrete grade beam to form a level threshold, a rodent-proof seal is achieved. The removable retainer plate permits the filler strip to be replaced when necessary.

The track door clearance should be reduced using the tubular shaped additions previously described. Frost should not be a problem if the threshold on either side of the track is part of the grade beam.

Stair Gates. Storage areas in warehouses should be approximately 42 to 48 inches above the grade. Track wells and personnel access doors are located at grade level, requiring a stair to the storage level. The jumping and climbing ability of rodents enables them to use these stairs to gain access to storage areas. To prevent this, steel barriers with a gate should be placed at the top of the stairs. *(Figure 3-42)* The barrier should encompass the entire stair and stairwell.

STRIP DOORS

Strip doors have a number of applications throughout the plant. From a product safety view, they have useful application as thermal barriers for freezers and cold rooms where standard doors must be left open for prolonged periods of time. When they are used to back up outside doors that must be left open for long periods, they become effective in keeping out birds, flying insects, and unwanted dirty air. *(Figure 3-43)*

AIR CURTAINS

There are a number of truck doors that must remain open

FIG. 3-40 Concept drawings such as these are used to develop installation drawings.

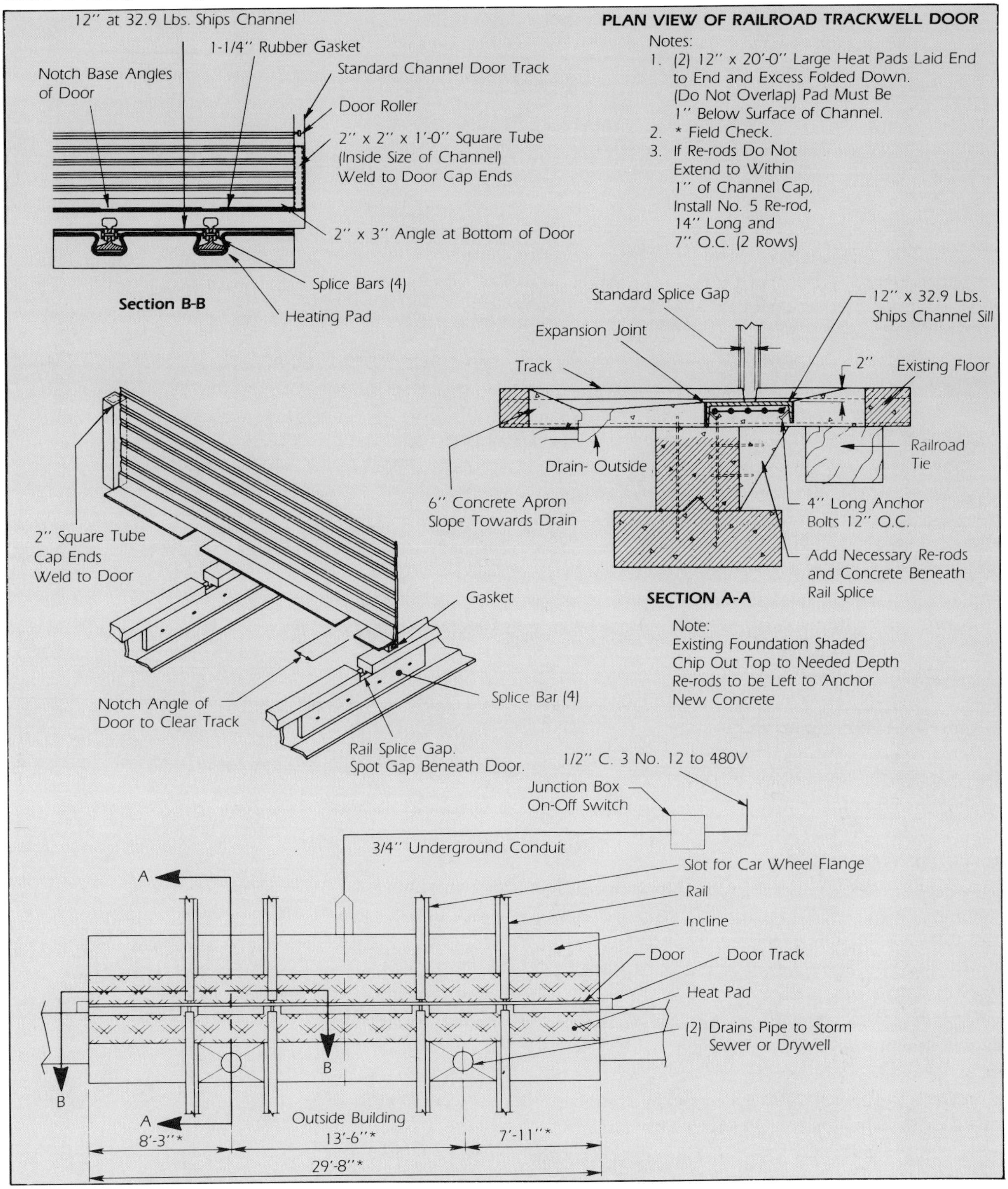

FIG. 3-41 A less costly way to rodent-proof existing railroad trackwell doors

FIG. 3-42 Rodents climb stairs easily. Metal gates at the top of the stairs are one obstacle they cannot overcome.

FIG. 3-43 Strip doors can keep birds and insects out of the plant and help control unwanted dust and dirt.

for long periods of time. Open doors create problems, most of which can be corrected with an air curtain. Those problems include loss of heat or conditioned air, dust, and flying insects. For air curtains to be effective, they must provide sufficient velocity to cope with the identified need.

Velocity requirements differ with the job the air curtain is required to do. Air curtains are influenced by other factors, including prevailing wind velocity, internal building pressure, and air temperatures. All of these factors affect the reliability of air curtains. Warm air attracts flies. Negative air pressure in a building can influence the design velocities of the curtain and prevailing wind velocities may be too strong. Air curtains might have an application and might be worth a try. However, they can be successful only if all these important factors are taken into consideration.

SHIPPING AND RECEIVING DOCKS

All shipping and receiving docks should be constructed so that they are rodent-proof. Rodents have tremendous climbing and jumping capabilities and can easily crawl up many vertical concrete walls. Truck dock walls can be easily scaled by most rodents. New docks should be constructed with a 10- to 12-inch overhang. A channel member should

FIG. 3-44 Rodents can climb many vertical concrete surfaces such as dock walls. Dock walls, however, are easily rodent-proofed.

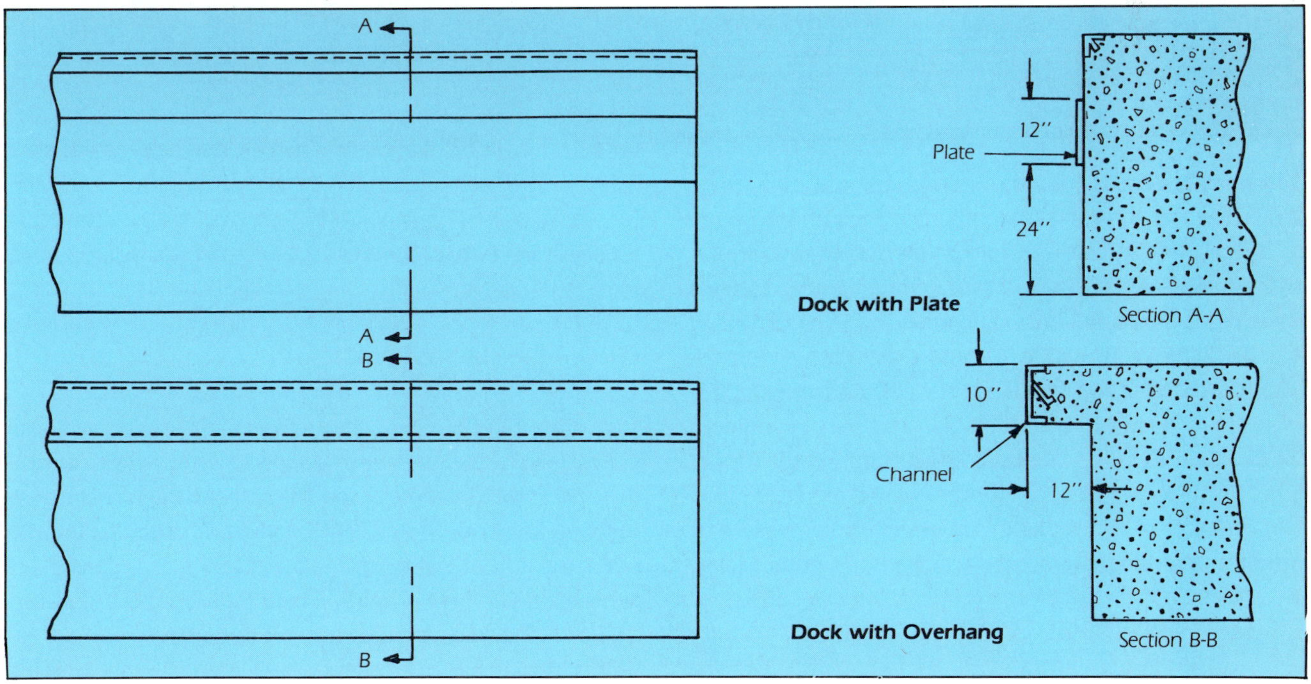

be cast in the concrete to form the vertical face of the overhang. *(Figure 3-44)* On existing docks, an 8- to 12-inch wide light-gauge sheet metal strip should be glued to the dock wall. *(Figure 3-44)* Rodent-proofing should be provided for both interior and exterior docks. You may question the need to rodent-proof an interior dock when the trackwell door has been rodent-proofed. Remember, however, that these doors are left open for long periods of time for rail switches or when truck trailers are being spotted.

Dock Shelters. Open or unprotected dock doors are a problem. Food products being loaded or unloaded can be damaged by the weather, and dust, dirt, insects, and birds can easily get into the plant. Properly selected and installed dock shelters should be standard equipment for exterior dock doors.

Canopies. The problems associated with most loading dock canopies involve bird and rodent control. The structural steel frame that supports the canopy provides many ledges for roosting and nesting. To eliminate this problem, dock canopies should be totally enclosed with facing panels at the ceiling, on the ends, and in the front. This enclosure must be very tight, as rodents, with their tremendous climbing ability, can easily gain access to this area and enter into the enclosure through even a quarter-inch opening.

LIGHTING

The benefits received from good lighting are numerous. Sanitation failures can be readily identified and cleaning tasks made easier. The ever-important overall appearance of the facility is greatly improved. Good lighting systems go beyond bright and pleasing illumination levels. Properly selected lights can be used to minimize insect problems. Insects respond to light in different ways. Fortunately, this response is somewhat predictable, and can be used to control them. Unlike humans, insects can see ultraviolet light and are attracted to it. Fluorescent and mercury vapor lights produce ultraviolet light; sodium vapor lights produce virtually none and therefore attract insects to a lesser extent.

In addition to the type of light, we are concerned with construction of the light fixture. Many fixtures are not dust- or insect-tight and they subsequently become natural incubators, as insects are attracted to warmth as well as light. A number of fluorescent fixtures fall into this category.

FACTORS TO CONSIDER IN LIGHTING DESIGN

Glass contamination is a serious concern and should be considered when designing lighting. Light fixtures must be provided with protective plastic shields to reduce possible contamination should a tube or light globe break. Glass contamination is undoubtedly one of the most serious, if not the most serious, physical contamination problem confronting the entire food industry.

The quality and quantity of illumination are the key factors to consider in lighting design. The annoying aspects of light, including glare, reflected glare, shadows, and contrasts, are of as much concern as the absence of light.

Glare and Reflected Glare. Glare and reflected glare are most troublesome since they are normally offensive to the eye. Increased safety problems often result from blinding glare which tires the eyes. The selection and location of fixtures is important to eliminating direct glare; so is the selection of ceiling paint. Reflective glare is caused by light reflecting from walls, ceilings, equipment, and wet floors. It might not be possible to eliminate all glare, but light selection and location can be helpful in eliminating a great deal of it.

Shadows and Contrasts. Shadows and contrasts are undesirable, particularly deep shadows and heavy contrasts. These conditions make vision more difficult and are very tiring to the eye. Lighting selections that cast or distribute the light source over a wider area and diffuse the light tend to reduce shadows. Sufficient background lighting tends to reduce contrasts which are normally caused by specific task or operation lighting.

Lighting Levels. The illumination level should be determined by the operation and the activity associated with the operation. In some cases, the activity might call for a very intense inspection of the product in process, and, in other situations, a simple observation of a piece of equipment at work. There will nearly always be a need for sufficient lighting to readily observe and identify sanitation failures and to perform the various cleaning tasks. Cleaning requirements and frequency vary from department to department or job to job, but every cleaning job must be done under a good lighting system. The following recommendations are suggested as maintained levels for some common operations or areas.

Area	**Foot Candles**
Raw material receiving	20 to 30
Ingredient warehouse	20 to 30
Bulk ingredient storage	30 to 40
Process departments	55 to 65
Product inspection station	110 to 130
Packaging department	70 to 80
Finished product warehousing	20 to 30
Maintenance area	70 to 80
Administrative area	60 to 90
Cafeteria	40 to 50
Locker and rest rooms	30 to 50

Lighting Loss. Lighting will be at its peak level when an area has just been equipped with new lighting and has been freshly painted or cleaned. However, lighting levels begin to drop off almost immediately and within six months the loss could be as much as 50% of the original output. These losses are attributable to normal depreciation (20%), dirty walls and ceilings (10%), and dirty fixture reflectors (20%). Because of these losses, it is recommended that lighting systems be oversized by 30% to 40%. Also, it is worthwhile to develop and implement a well-planned maintenance program. Proper cleaning done on a regular basis will greatly improve lighting. It is important that fixtures and reflectors be washed with a mild cleaning solution and not merely wiped off. The washed surfaces must be rinsed free of detergent residue. Relamping should be done on schedule with records maintained to determine life cycles and subsequent relamping schedules.

TYPES OF LIGHTING

Fluorescent Lights. Fluorescent lights continue to be a popular selection, although other light sources might be considerably more efficient. This type of light produces ultra-violet light which attracts night-flying insects. The fixtures themselves often become natural insect incubators; therefore dust-resistant fixtures should be selected. *(Figure 3-45)* The fixtures should be equipped with a polycarbonated plastic shield to protect the glass tubes, and, in the case of tube breakage, contain the glass and protect the food products in the immediate area.

FIG. 3-45 Select light fixtures that provide adequate light, are dust- tight and that have glass tube or globe protective devices.

Unfortunately, many older food plants still have a number of open style unshielded fluorescent fixtures in which the unprotected glass tube presents a serious contamination threat. Plastic tube sleeves and end caps are available from several manufacturing sources. *(Figure 3-46)* When the glass tube is encased in the sleeve and the end caps are in place, the tube is adequately protected. *(Figure 3-47)* Should accidental breakage occur, the glass would be contained. In some cases, these plastic sleeves tend to become yellow because of the intense output of heat from the tube. In cases where high heat output is anticipated, heat sinks should be included in the sleeve.

Fluorescent tubes which have been precoated with a shatterproof plastic material are also available as an alternate to the aforementioned reusable plastic sleeves.

High Pressure Sodium Lights. High pressure sodium is a very efficient light source that is suitable for many parts of a food manufacturing facility. Warehouses and outdoor uses, such as parking lots and accessways, would be good choices for this type of lighting.

Outdoor lights should be shielded so they cannot be seen by insects from above, and they should be directed toward the area they are to illuminate. Many high-flying insects are attracted to lights they see below.

High pressure sodium may also be considered for manufacturing areas if color quality is not absolutely essential. High pressure sodium does tend to distort colors, and where this is important to the quality of the product, an alternate type of lighting should be considered. In addition to energy savings advantages, sodium lights attract insects to a lesser extent.

Low Pressure Sodium Lights. Low pressure sodium lights have very limited application in the food plant. Color distortion is significantly worse than with high pressure sodium, and the globe contains free sodium which has hazardous explosive and flammable characteristics, under certain circumstances.

FIG. 3-46 Plastic sleeves provide protection for older fluorescent fixtures that do not lend themselves to conventional protective devices.

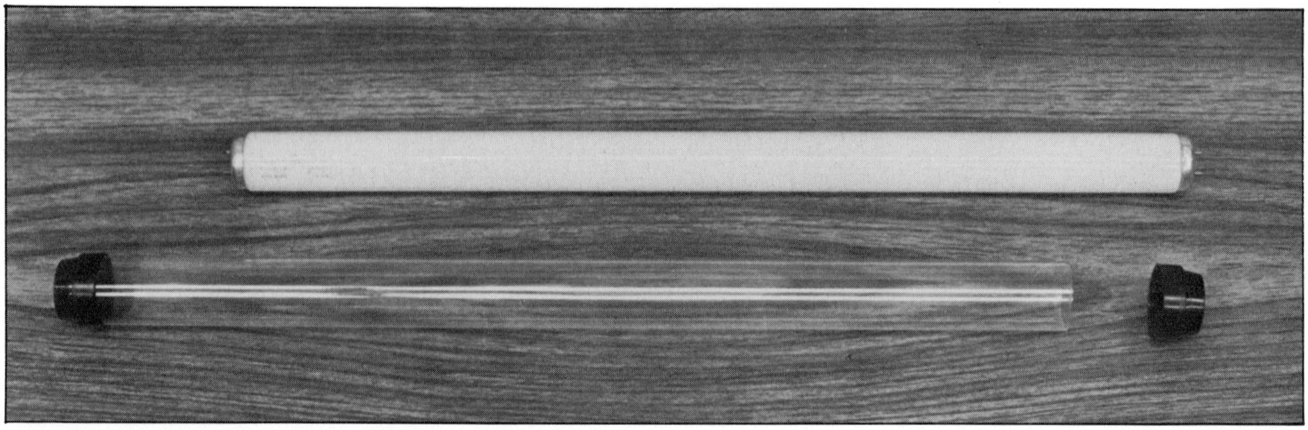

FIG. 3-47 The glass tube is encased by a plastic sleeve with two plastic end caps.

Both high pressure and low pressure sodium lights cast a slightly yellow haze. In some cases, this might be found to be objectionable, and the use should be limited. The yellow haze is more pronounced with low pressure sodium.

Metal Halide Lights. Metal halide lights produce a light that is very much like that produced by the sun. It is an efficient lighting system and can be used in many plant areas. It does not distort color; in fact, because the light quality closely approximates that of the sun, colors appear natural. The ultra-violet output of metal halide is also similar to that of the sun. For this reason, the attracting capability of fluorescent lights used in insect light traps is negated, making this kind of trap less effective.

Mercury Vapor Lights. Mercury vapor lights can be used in plants, although most applications are in warehouses where they can be installed at an elevation that will take advantage of their output capability. The mercury vapor light can also be useful as an insect counterattractor. These lights can attract insects away from the door when mounted on columns approximately 30 feet from the plant entrance doors or when mounted on the perimeter of the plant property as a security light.

Ultra-violet rays produced by metal halide and mercury vapor lights are believed to be harmful to the skin and eyes, causing burning, irritation, and, in some cases, skin rashes. This problem occurs only when the outer globe of the two-globe lamp is broken. Should this happen, the light is to be turned off immediately. To avoid this problem altogether, only fail-safe globes should be specified for these light fixtures. The fail-safe feature automatically causes the light to shut off when the outer globe breaks.

Incandescent Lights. Incandescent lights have a very limited application in manufacturing areas. Where they are used, they must be well protected with plastic enclosures. Special lamps with shatterproof plastic coatings are available and should be the choice for trouble lights used throughout the plant.

VENTILATION

Controlling the ventilation in the food manufacturing environment reduces the potential for many product safety problems. Poor ventilation results in the formation of overhead condensation, mold, and odor problems that threaten to contaminate food products and product surfaces below. Poorly designed ventilation systems can cause sanitation problems in other ways. For example, failure to filter or select adequate filtering media for incoming air streams can cause direct contamination of food products with numerous airborne contaminants. The design detail of the ventilating system can contribute to and cause sanitation failures by harboring dust, dirt, and various forms of insect pests. Environmental control is essential to producing safe, wholesome food products.

Ventilation can best be described as an exchange of unwanted air for a controlled and conditioned supply of healthy fresh air. Accomplishing this exchange in a manner consistent with good manufacturing practices requires the use of mechanical air handling equipment. Exhausters will be needed to remove undesirable air and make-up air systems will be needed to provide filtered, heated, or in some cases, cool air to the plant. Ventilation cannot be accomplished by simply opening doors and windows. To the extent possible, doors leading from the outside should not open directly into the manufacturing areas. As noted earlier, windows are inappropriate.

AIR CHANGES

In addition to the physiological requirements of air quality and quantity, food manufacturing areas of the plant will need sufficient air changes to maintain a healthy production environment. Normally this is far in excess of human needs and is generally dictated by the inherent characteristics of the manufacturing process itself. Steam or water evaporation from the process must be removed by direct ventilation or, if the situation calls for it, specially designed exhaust systems complete with collection hoods. Failure to remove this moisture will result in condensate and mold problems. Because the process does dictate air requirements, it is practically impossible to be specific. It could easily vary from just a few air exchanges per hour to thirty or sixty per hour. What is important is that sufficient changes are provided to cope with any anticipated problems. An industrial ventilation text can be helpful in selecting the proper air exchanges for certain circumstances. Food manufacturing areas should be maintained under a slightly positive air pressure condition. A strong case can be made for this design. To begin with, airborne contaminants cannot readily enter these areas. More importantly, when equipment is completely closed with a metal enclosure or housing and surrounded by a slightly positive air flow, it is difficult for dust from the product to escape from the equipment, particularly if there is a slight suction applied to the equipment itself. Dust collection systems perform best when used in an area under a slight pressure; in addition, combustion equipment, such as gas dryers and heaters, are safer and more efficient to run.

Maintaining a slight pressure in the manufacturing areas is not always easy. Many processes use large quantities of air for drying, baking, and conveying. This air should not be exhausted in total to the outside; it should be filtered where necessary and reused by blending the proper quantity of make-up air. A slightly positive pressure condition should exist when the designed air flow through the supply air equipment is about 10% greater than the designed exhaust capacity of the process and ventilating equipment.

AIR FILTERS

All incoming air provided for the manufacturing areas should be filtered to remove airborne particulates 50 microns in size and larger, unless special filtering requirements suggest the use of filters with the capability of removing finer particles. *(Figure 3-48)*

Renewable Media Filters. Automatic moving filter curtains are preferred. These units should be equipped with a timer, pressure switch control, or combination of both to advance the media on command. Pressure switch controls measure the pressure drop across the media which increases when the filter is dirty. Vertically applied filter curtain applications are suggested. Horizontal curtains are more susceptible to drooping, allowing particulate matter and insects to enter through the opening between the frame and drooping media. This is particularly true when the horizontal dimension of the intake opening is quite long, approximately ten feet or more.

Fixed Unit Filters. Replaceable or cleanable fixed unit impingement filters are also satisfactory for many air filtering applications. These units should be placed on a regularly scheduled replacement or cleaning program.

Formed Media Disposable Cartridge Filters. Formed media used in disposable cartridge filters may be bonded glass fiber, cellulose fiber, wool felt, synthetics, or other similar materials. The efficiency of these filters is normally better than that of the viscous impingement type. These types of filters should be considered for special filtering requirements.

Insect Screens. In some extreme cases, insect screens should be provided ahead of the filter media. *(Figure 3-49)* An extreme case would involve a dense insect problem, seasonal or otherwise, specific to a geographic location. Insect screens should be installed with good access and removable features so they can be easily cleaned. Eighteen-

FIG. 3-48 All air brought into the manufacturing area from outside the building should be filtered.

FIG. 3-49 Incoming air intakes require insect screens in areas where seasonal insects are a problem.

mesh screen constructed from corrosion-resistant material is generally satisfactory for an application of this type. Because screens are quick to blind from dirt and other foreign material, it is advisable to provide a total screen area that is 1-1/2 to 2 times larger than the air intake opening. This may seem to be a lot, but remember that the effective open area of the screen originally is only around 40% to 50% of the total area.

Intake Velocities. Air intake velocities should be as recommended by the American Society of Heating, Refrigerating and Air Conditioning Engineers, Inc. Excessive intake velocity can easily impair the efficiency of the filter media. Dirt and insects can be pulled right through the filters.

Odor Control. Parts of many food processes are odor bearing. Some odors are considered appealing, while others are very objectionable. However, what is appealing to one is not necessarily appealing to another, particularly when the odor is strong and, for all practical purposes, continuous with the process. Such a situation calls for frequent air exchanges and the use of air scrubbers on the air exhaust stack so odors do not become a problem for nearby neighbors.

Odors can be an internal as well as an external problem. Internally, odors can permeate other sections of the plant, contaminating products or packaging materials that have high sorption characteristics. Externally it represents an objectionable situation for people living in the area and can be a constant problem for them.

Ventilation Control. In many cases, odors can simply be controlled through good ventilation. With this process odor concentrations are reduced by dilution; excess air is vented outside, where it is further diluted until it is no longer an objectionable problem.

Air Scrubbing. Air scrubbers can be effective in removing or reducing odors from air streams. Scrubbers remove odors by absorption of the odor vapors in the scrubbing liquid and by condensation of the vapors out of the saturated air into a cold liquid.

Adsorption Control. Adsorption control of odors is effective in removing odors from sample air streams. Adsorbents commonly used are activated charcoal, silica gel, zeolite, and activated alumina. There are also a number of other adsorption materials available that are quite effective.

AIR HANDLING EQUIPMENT

Air handling equipment should be selected with good

FIG. 3-50 Access to the air handling system should be provided.

access in mind. *(Figure 3-50)* In medium- to large-size make-up air units, access doors should be included at all air plenums including mixing chambers, heating/cooling chambers, filter chambers, and in the case of large fans, the fan wheel. In smaller unit ventilators and unit heaters, the emphasis should be placed on access for complete inspection of key components and adjacent areas enclosed by panels. The filters should be easily removable. Drive guards and drive components should be provided with the sanitary design details recommended in the chapter on equipment design.

When fan drive motors are mounted on the vertical walls of the fan drive shaft housing, they are easier to clean than when they are mounted horizontally on a flat surface. In a flat position there is more surface for dust and dirt to accumulate in difficult-to-clean crevices and void areas. *(Figure 3-51)* If the drive motor is mounted horizontally or vertically, extreme care should be taken to eliminate the void areas normally created by motor rails.

Round Air Duct Design. Round duct design used on higher pressure (10 inch W. G. and up) applications should be fabricated with continuously welded longitudinal seams. Transverse joints should be flanged or welded. Lock-type spiral joints are undesirable. Although tightly formed, these joints still produce small crevices that collect dust and dirt and can harbor insects. Round duct fittings should also be fabricated with welded joints.

Rectangular Air Ducts. Rectangular or flat duct design should be fabricated with caulked longitudinal flatlock or Pittsburgh seams. *(Figure 3-52)* Flanged seams may also be considered. Transverse joints should be of the flanged or companion angle iron type. All flange ends should be rounded for safety reasons. The transverse joint should be caulked or have gaskets. The longitudinal joint should be caulked with a good grade of pliable silicone caulking material. This can be done before or after forming or assembly. However, fabricators are reluctant to do this before forming, as incorrectly applied caulking material can clog up the forming machine. Gasket material should be closed-cell neoprene.

Longitudinal seams should be continuously welded in rectangular ducts used in manufacturing areas of food plants manufacturing grain-based products with a high insect-sensitivity, such as flour, cereal, pasta, and some dry prepared mixes.

Intermediate Air Duct Reinforcment. Rectangular or flat duct reinforcement should be accomplished with externally mounted formed or structural angles. Formed hat sections, zee's, or channels are not considered good sanitary design. Beading as a method of reinforcement is also considered unsanitary, since the cavities formed by beading trap dust and dirt. Crossbreaking is an acceptable and a preferred sanitary method of reinforcing large, flat surfaces of ducts or plenums. *(Figure 3-53)*

Access Doors on Air Ducts. Access doors should be provided on all 18-inch and larger ducts. Access doors should be a minimum of 16 by 24 inches for good access. Doors should be on 20-foot centers and on each side of turning vanes, damper installations, or in each straight run separated by an elbow or tee. Die-formed, double-walled, insulated access doors with quick-opening handles are recommended. *(Figure 3-54)*

FIG. 3-51 Flat mounting plates for horizontal motor mounting commonly create crevices and small void areas that are difficult to clean.

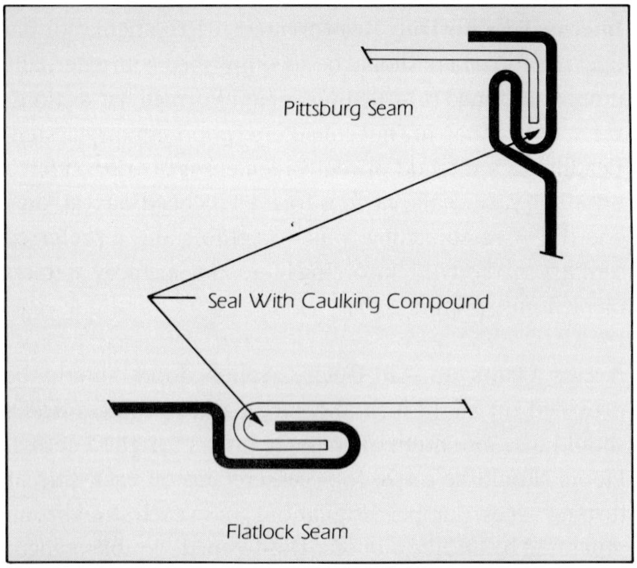

FIG. 3-52 Sealing ventilation ductwork seams with caulking materials or other sealants eliminates insect harbors.

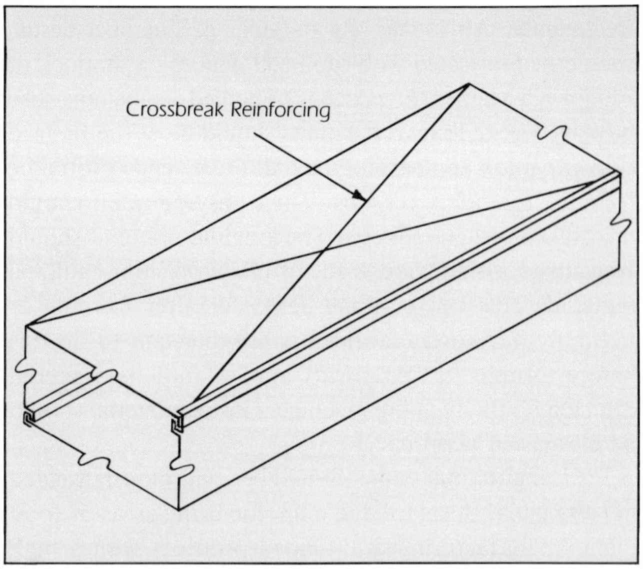

FIG. 3-53 Crossbreaking is satisfactory for reinforcing large ventilation ducts.

FIG. 3-54 Access doors should be provided in ventilation ductwork for inspection and cleaning.

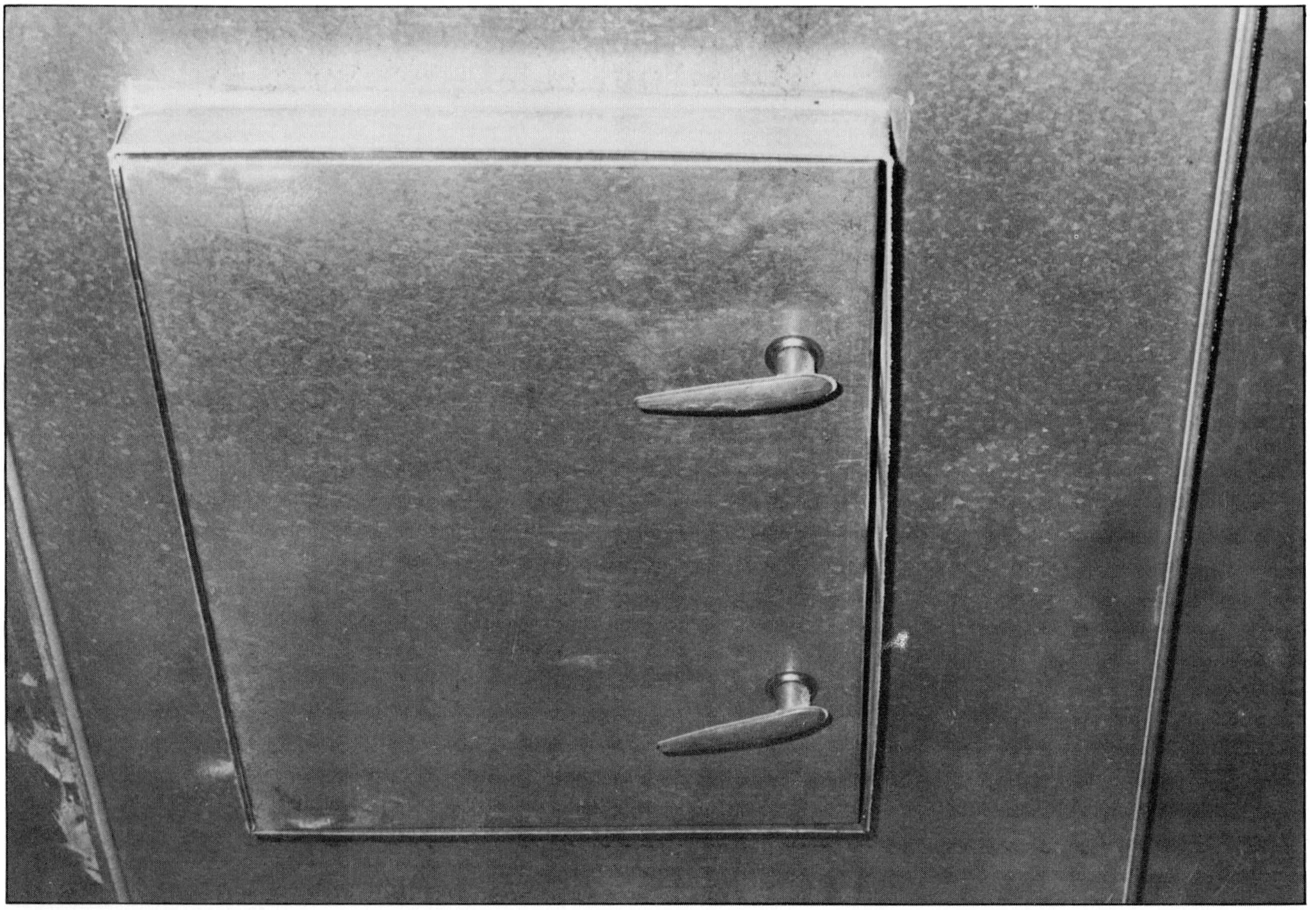

A removable section of duct should be provided for duct sizes smaller than 18 inches. The removable section should be 24 to 30 inches long and properly identified as a removable section. The removable section should be held in place with clamps or other fasteners that would eliminate the need for tools when removing it.

Intake or Exhaust Stacks. Intake or exhaust stacks should be designed to keep birds and insects, as well as rain and snow, out of the plant. Intake stacks should be provided with hoods, goosenecks, or similar devices, as well as half-inch mesh bird screens constructed of corrosion-resistant materials. If the intake air is not to be filtered, as is the case in many warehouse applications, the intake opening should be provided with a corrosion-resistant insect screen. For cleaning purposes, this screen should be designed for easy removal.

Exhaust stacks, too, should be designed with rain caps, vertical or offset discharges or similar devices to keep out the rain. For stacks exhausting objectionable fumes, rain caps might be undesirable since they deflect the fumes downward where they are not as easily dispersed. Stacks with vertical or offset discharges might be better suited for exhaust fumes of this type. Corrosion-resistant bird screens are important in good stack design. Self-closing dampers should be required, as needed, for both intake and exhaust stacks.

There are a number of desirable intake and exhaust stack design features which are designed to cope with the weather elements and various vermin. *(Figure 3-55)*

Gooseneck intakes should not be installed with the intake opening too close to the roof surface, but should be held off the roof 18 to 30 inches, depending on the velocity of the intake. When the opening is placed close to the roof, there is a chance that dirt from the roof will be picked up by the incoming air stream.

Intake and exhaust stacks should not be located in close proximity to one another since there is a danger that the exhausted air will be picked up and redistributed by the intake. Ventilation intakes, in particular, should not be located near process air exhaust stacks, such as those coming from baking ovens, dryers, and dust collectors. In general, exhaust air stacks should discharge at a point higher off the roof (six feet or more) than the inlet location of the intake stacks. The exhausted air will move from the immediate area faster at the higher elevation, and it is not as likely to be influenced by the intake stacks.

Duct Insulation. The selection of good, dense, closed-cell insulation with a well-sealed canvas fabric or vinyl-impregnated jacket will meet most sanitation needs. Insects can easily live in almost all types of insulating materials and must be sealed out by the insulation jacket. Dense insulation materials are preferred because they are more difficult to penetrate. More importantly, they stand up better against normal cleaning abuse. Insulating materials should not be used on the inside of ducts.

Insulated ducts that pass through wet processing areas should be jacketed with moistureproof materials that do not stain easily and also provide a smooth, easy-to-clean outer surface. Further information on insulation applications can be found in the following section on utility piping.

Acoustical Linings of Air Ducts. Acoustical interior duct linings are considered unsanitary. Accumulations of dirt and dust in these materials are very difficult to remove, and more often than not, become breeding grounds for insects. If it becomes necessary to use an acoustical lining in extremely noisy conditions, it should first be installed in a sealed plastic bag. This makes it less of a sanitation problem, and has little effect on the attenuation characteristics of the lining material.

Air Duct Hangers. The standard strap or trapeze hanger is an acceptable air duct hanger. Trapeze hangers fabricated with all-thread hanger rods are not acceptable. All-thread is hazardous to clean because of the razor-like sharpness of the threads.

EQUIPMENT AND DUCT INSTALLATION

The sanitary installation of floor-mounted air handling equipment and duct work requires that it be elevated from the floor 6 to 18 inches or sealed to the floor. The size of the unit will determine how far it should be raised. Small units or ducts occupying a small area need only be elevated the minimum six inches. Larger occupied areas will require 18 inches of elevation. If sealing the equipment to the floor is considered, be sure to take into account any anticipated vibration or expansion and contraction problems that might make retaining the seal difficult. By eliminating the solid bottom panels, the crevice between

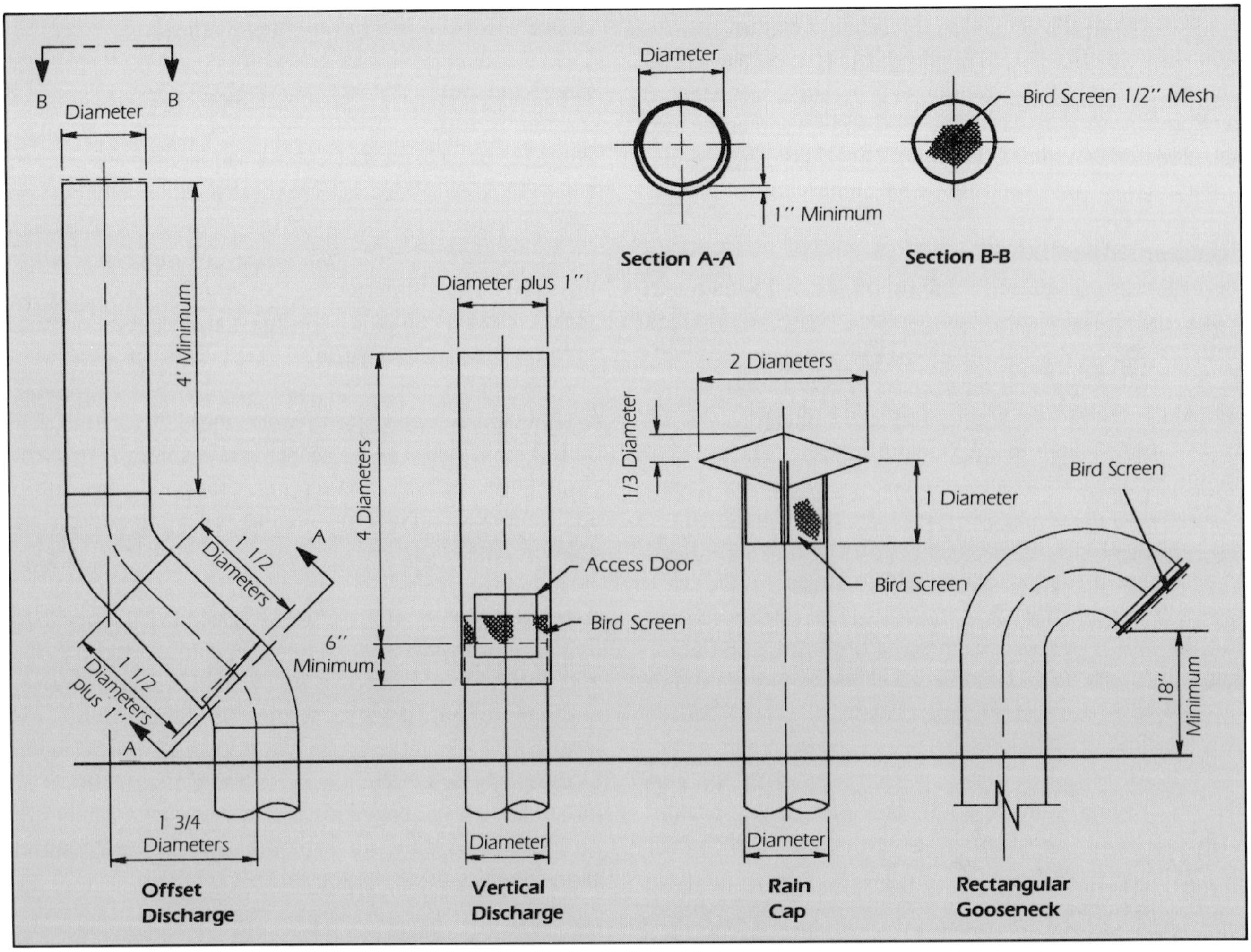

FIG. 3-55 This group of exhaust stack designs meets most sanitary design requirements.

the floor and the bottom panel can be avoided.

Supports for equipment installed in the overhead should be fabricated with as few ledges and crevices as possible. Angles and tubular shapes, instead of channel, will reduce ledge surfaces.

PLUMBING

The plumbing system is of paramount importance in the production of safe food products. Improperly installed water and sewage systems can lead to direct contamination of products or of manufacturing equipment. The plumbing system design must eliminate the possibility of cross-contamination of the potable water supply with raw sewage or liquid plant wastes or with the products of the manufacturing process itself.

Overhead sewage lines must be located so they do not represent a contamination threat to food products or product zone surfaces below, should a leak occur. Adequate supplies of potable water at desired pressures must be provided for the various plant cleaning systems. The plumbing system must be installed to meet the immediate and expansion needs of the food manufacturing process, the sanitation program, and the personal needs of plant employees.

The importance of strict compliance with local, state, and national plumbing codes cannot be overemphasized. There are a growing number of reported and unreported incidents involving food products or community water supplies having been contaminated by malfunctioning or poorly designed plumbing systems. This subject is too important to be treated lightly and demands the professional skills of a competent and experienced engineer who is considered a plumbing system design specialist.

WATER SUPPLY

Potable Water: General Considerations. An adequate supply of fresh, clean, potable water is essential to the manufacturing process and the sanitation program. *Potable* means that the water is safe for human consumption without further treatment. The responsibility for safe water lies with plant management. The plant water supply should be tested periodically to assure that it is in compliance with public health standards. If the source of supply is from an approved public system, annual certification of samples taken from within the plant is sufficient. Water from private wells should be tested every six months. The purpose of these water tests is to assure that no pollution of the supply has occurred and that the water is still safe. Test certificates should be kept on file at the plant.

Public water works personnel conduct frequent tests, and the supply to the plant can usually be accepted as safe. The major concern is with the plant system. When samples are gathered, they should be taken from various points within the plant, not at the meter or point of entry. A sample taken at the origin point does not indicate the quality of the plant water at the point of use. As a minimum requirement, the plant water supply should meet the *Drinking Water Standards* as established by the United States Public Health Service, Department of Health and Human Services.

The water supply should contain no offensive physical impurities that would affect the taste, smell, or appearance of the water.

The water supply should not contain any microorganisms that would represent a health hazard. Most diseases transmitted through water are from bacteria from human intestines. Besides being offensive, fecal pollution is very dangerous. Standard water tests include a test for coliform bacteria which is always found in fecal matter.

The water must be free from chemical impurities in harmful concentrations. This includes chemicals used to treat water, in addition to toxic chemicals. The water should not contain excessive corrosive properties. The water must also be free from any radioactive materials.

If the potable water supply comes from a private well, the well should be located on plant property and be protected from all pollution sources. The well should be built to prevent ground surface water from entering the well through the soil without sufficient filtration. Water should percolate through at least 10 to 15 feet of ground. Wells should be located on high ground and at a safe distance from all contamination sources, including septic tanks. The distance required between septic tanks and other contamination sources is normally specified by local health authorities.

Connecting a private water supply to a public water system will require the approval of local regulatory agencies such as the health department. However, there are a good number of regulations that prohibit any interconnection of public and private water supplies, regardless of the safety of the private supply.

If chlorinators are used to assure safe water, they should be automatic and equipped with alarms that indicate when the units are not functioning properly.

Nonpotable Water. Some manufacturing plants use a supply of nonpotable water for various purposes in and around the plant, including fire protection systems. A nonpotable water supply is a potential source of danger. If nonpotable water must be used, it should not be used in any area of the plant where food products are prepared, handled, or stored. The use of nonpotable water should be limited to applications such as cooling towers or similar equipment where there is no danger of product contamination. It must not be used in any way that it can contaminate the potable water supply.

Nonpotable water should not be used to perform any sanitation operations, such as washing floors, walls, or equipment. In all cases, nonpotable water lines should be clearly identified and should not be cross-connected with the potable supply.

Nonpotable water should not be used in any part of the fire protection system, including sprinklers that service any area that prepares, handles, or stores food products. When fire protection systems include the use of both potable and nonpotable water, they must be piped through completely independent systems. Isolating by valve arrangements is in no way permissible. The use of nonpotable water in fire systems in and around food plants normally requires the approval of local health authorities.

Potable Water Reuse. There are several situations where potable water may be reused for the same purpose. Water can be safely reused to perform cooling functions, such as cooling or chilling flaking rolls, cooling canned products from retort operations, and cooling sections of dryers. *(Figure 3-56)* All pipelines, tanks, cooling towers, and other such equipment used in the reused water system must be installed so that they can be readily inspected and cleaned. The water supply line to this system must be installed with a backflow preventor, or in situations where the plumbing code requires it, an air gap. The piping assembly should also include the necessary valves to completely drain the system down for cleaning and fresh water replacement.

Backflow Prevention. Backflow preventors or vacuum breakers should be used in the water system, where required, to prevent contamination of the potable water supply by plumbing fixtures, reused water systems, or other systems that could cause back-siphoning through a negative pressure condition. Negative pressure or partial vacuum can occur in any water supply system from time to time. The danger of back-siphonage can be minimized through the elimination of submerged water lines and the use of vacuum breakers at applicable points in the system. If submerged water lines cannot be avoided, vacuum breakers must be used. All backflow prevention valves or

FIG. 3-56 The reuse of potable water should be limited primarily to non-food and non-cleaning uses.

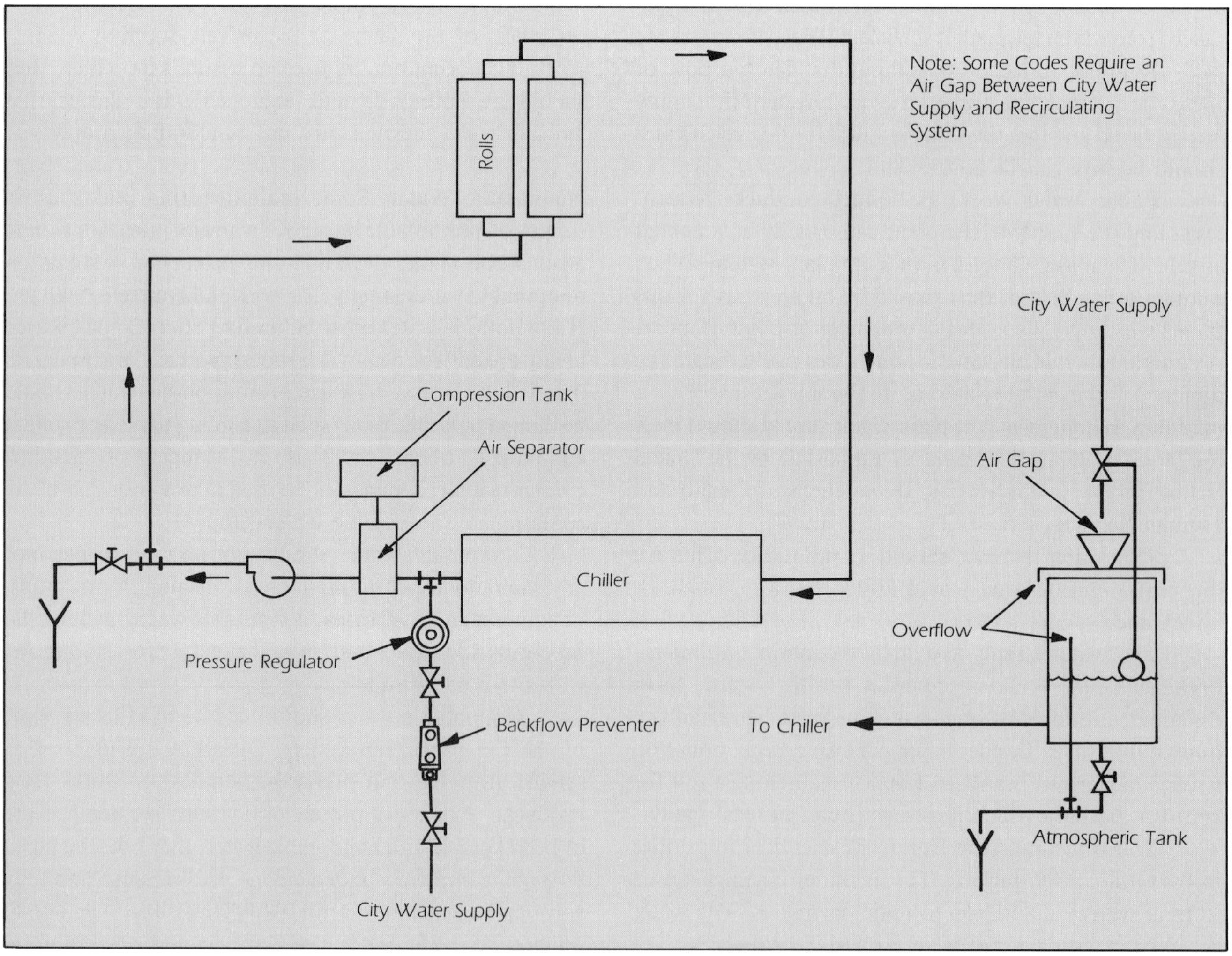

vacuum breakers selected should have the approval of local regulatory agencies. A program to assure that valves and breakers are functioning should also be established.

Backflow can also be the result of back-pressure. This condition occurs when a pressure higher than the supply pressure develops in the system and causes a reversal of flow into the supply. An example of this could be thermal expansion from a boiler. Appropriate backflow prevention equipment can virtually eliminate this problem. Appropriate equipment in this case is equipment designed for a pressure condition. Anti-siphon or atmospheric vacuum breakers should not be used against a back-pressure backflow condition. The mechanics of backflow back-siphonage and backflow caused by back-pressure are illustrated. *(Figure 3-57)*

The danger of potable water supply contamination from various cross connections should not be overlooked. Potable water supply contamination can come from a backflow of the food manufacturing process itself. In this case, food materials contaminate the clean water. You should be aware that there are some jurisdictional codes that do not allow for pressure connections between a food process and the water supply. An air gap and funnel connection to an atmospheric tank which is then pumped to the user equipment might be required. Where direct connections are permitted, they must be equipped with approved backflow prevention equipment.

Water Storage. Where it is necessary to store quantities of water in the plant, it should be done in such a way that the water cannot be contaminated. Only closed storage tanks constructed from corrosion-resistant materials should be used. The storage system should be designed with no dead ends so that water may circulate or flow freely. Water should not be allowed to stagnate. Water supply connections must be made in a manner that will prevent backflow.

Pipelines. Water supply lines should be installed to eliminate dead ends. Water lines no longer in use should be removed. Water that is not allowed to move freely stagnates, and a buildup of microorganisms can occur, leading to contamination of the water supply. Contamination from rust, scale, and grease can occur when repairs are made on old piping and pumps. New water piping installations should be made of copper or other corrosion-resistant materials.

It is very easy to take the safety of a water supply for granted. Many problems can be avoided with minor attention to one of the most important ingredients used in the typical food manufacturing plant, water.

DRAINAGE SYSTEMS

For obvious reasons, a food manufacturing facility must have two separate and distinct drainage systems. One, the sanitary system, should be connected to toilet bowls, urinals, and lavatories. The second system should be for general plant drainage, as from floor drains and process equipment. Special drainage systems should be considered for operations that produce an inordinate amount of liquid waste, with or without suspended solids. It is important that these systems are of adequate size and design so they will remove wastes quickly and efficiently. Accumulations of waste material in manufacturing areas create offensive odors, attract vermin, and, in themselves, represent a potential contamination hazard.

Sanitary Systems. Lines from toilets, urinals, and rest room lavatories should not be connected with any other plant process drainage lines, including floor drains. Sanitary lines should not discharge into grease interceptors or other liquid waste separating equipment. Sanitary lines must not be allowed to enter the main sewer lines at a point where there might be a possibility of the sanitary sewage backing up through the plant drainage system and flooding the floors in production areas. To the extent possible, sanitary system lines should be taken around manufacturing areas and/or arranged so that the routing from the area represents the shortest possible distance. Sanitary sewage back-ups of human waste in the production area would create a serious health hazard. Cleanouts should be located outside the manufacturing area.

Sanitary sewage lines must be installed so they do not pass over, or near any food manufacturing equipment, nor should they pass over food handling or storage areas. Vents from these lines should not be close to ventilation or process air intakes.

A complete set of updated sanitary piping drawings should be left on file with the plant maintenance and plant engineering department. Occasionally there is need for changes or additions to the system. These changes should be made in accordance with the design criteria indicated herein.

FIG. 3-57 Mechanics of backflow from back-pressure and siphonage

BACK-SIPHONAGE FROM BACKFLOW

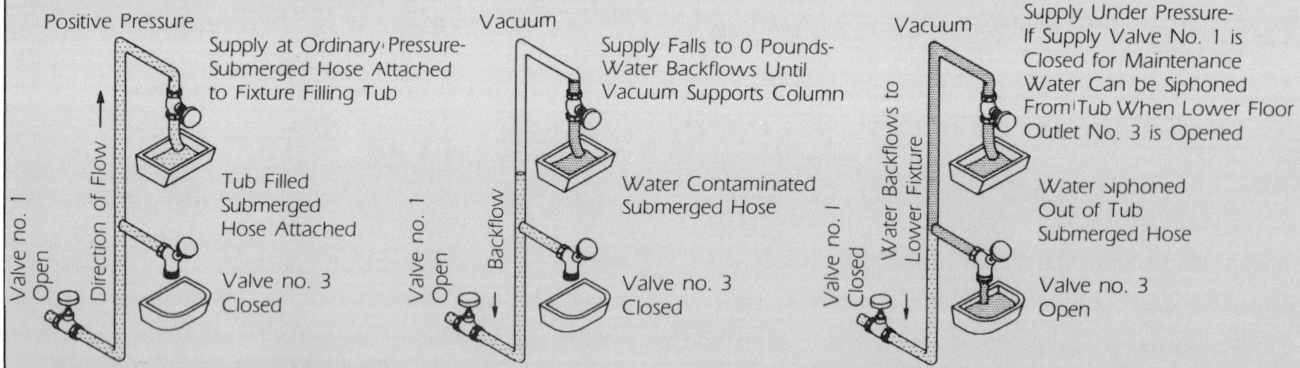

BACKFLOW FROM BACK-PRESSURE

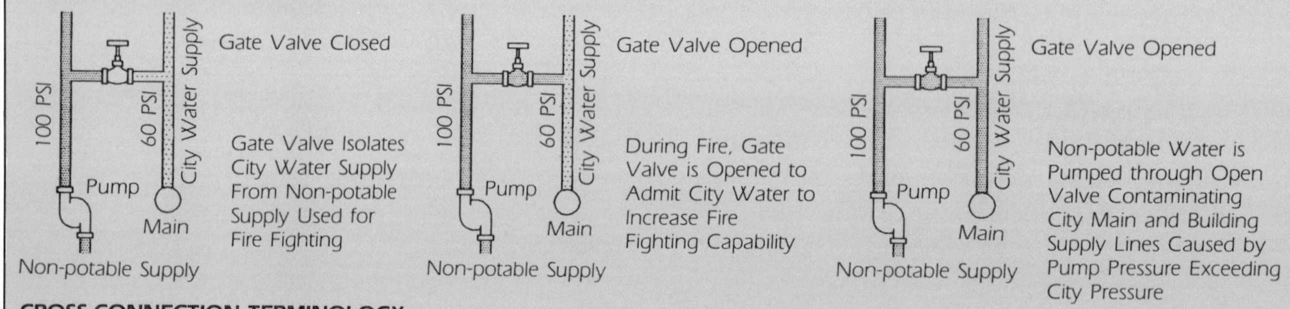

CROSS-CONNECTION TERMINOLOGY

BACKFLOW-
Unwanted reverse Flow of Liquids in a Piping System. Caused by Back-siphonage, Back-pressure or a Combination of Both.

BACK-SIPHONAGE-
Due to Vacuum or Partial Vacuum in Water Supply System. Caused by:

A. Ordinary Gravity: When Water Supply is Lost and a Fixture that is Elevated is Opened, Allowing Air into the System, Water Will, By Gravity, Reverse the Flow.

B. Undersized Piping: High Velocity Water Traveling Through Undersized Piping Can Cause an Aspirator Effect and Draw Water Out of Branch Pipes Causing a Partial Vacuum and a Reverse Flow.

C. Vacuum: Caused by Pumping Water From Supply System Which Results in a Pressure Drop or Negative Pressure in the System. A Break in the Main or Excessive Usage at a Lower Level in the System Can Also Be a Cause.

BACK-PRESSURE-
Condition Whereby Pressure Higher than Supply Pressure is Created on the Premises and Causes Reversal of Flow into Supply. Example: Pumps Thermal Expansion From Boilers.

CROSS-CONNECTION-
Any Actual or Potential Connection Between the Potable Water Supply and a Source of Contamination or Pollution. There are Two Types of Cross-connections: the Inlet Connection and the Direct or Pressure Connection.

INLET CONNECTION-
The Inlet Connection is a Connection Used for Filling a Receptacle That is Open to the Atmosphere. They Can Be Below Rim or Submerged, Such as a Bottom Inlet to a Plating Rinse Tank, Or They Can be the Over Rim Type in Which a Water Line Enters a Receptacle Over the Top Rim and Terminates at Point Below Top Rim. In These Types of Cross-connections, Backflow Occurs Due to Back-siphonage Only.

DIRECT OR PRESSURE CONNECTION-
Direct or Pressure Type Connection is One in Which Water Supply is Connected to Another Line or Pressurized Vessel. Good Examples are a Feed Water Line to a Boiler or a Primer Line to a Pump. In Cases of Direct or Pressure Connections, Backflow Occurs Due to Back-pressure When System Pressure Exceeds Supply Pressure.

PROPER SYSTEM PROTECTION-
To Protect the Portable System, You First Determine the Hazard Involved. Since All Cross-Connections Do Not Create the Same Degree of Hazard, We Should Make Our Choice on a Common Sense Basis.

Degree of Hazard is Dependent on Whether the Substance in the Non-potable System is Toxic or Non-toxic. A Toxic Substance is any Liquid, Solid, or Gas, Which, When Introduced into the Water Supply, Creates, or May Create, a Danger to Health and Well-Being of the Consumer. Example: Treated Boiler Water. A Non-toxic Substance is Any Substance That May Create a Moderate Hazard, is a Nuisance or is Aesthetically Objectionable. Example: Foodstuffs, Such as Sugar, Soda Pop, Etc. Secondly, You Must Select the Proper Device According to Type of Connection and Degree of Hazard.

Plant Drainage. Adequate plant drainage must be provided throughout the entire manufacturing area. The drainage system must be designed to meet all national and local plumbing codes. As a general rule, one four-inch drain inlet should be provided for every 400 square feet of wet processing area. Drains should also be provided adjacent to equipment requiring extensive wet cleaning. Floors should be sloped to the drain inlets at the rate of 1/8 to 3/16 inch per linear foot. Sloping should be uniform to avoid pocketing. *(Figure 3-58)* Floor drains are not usually required in dry storage areas and freezer rooms.

- Floor Drains. Floor drains should be equipped with metal traps having a minimum water seal of three inches. Full S traps, bell traps, and crown-vented traps are prohibited and should not be used. Floor drains should be a minimum of four inches in size and equipped with a removable secondary strainer to prevent the entry of cockroaches, rodents, and other vermin occasionally found in these systems. Drains having large removable buckets should be considered in areas where small quantities of sediment and solid waste are anticipated. Floor drains should have a relatively smooth surface and rounded corners.

- Trench Drains. There are still segments of the food manufacturing industry that have a need for the open style, trench drain. The sanitary design of trench drains can make them easier to clean. *(Figure 3-59)* Trench drains should be constructed with smooth, vertical walls and

FIG. 3-58 Adequately installed floor drains can assure proper drainage.

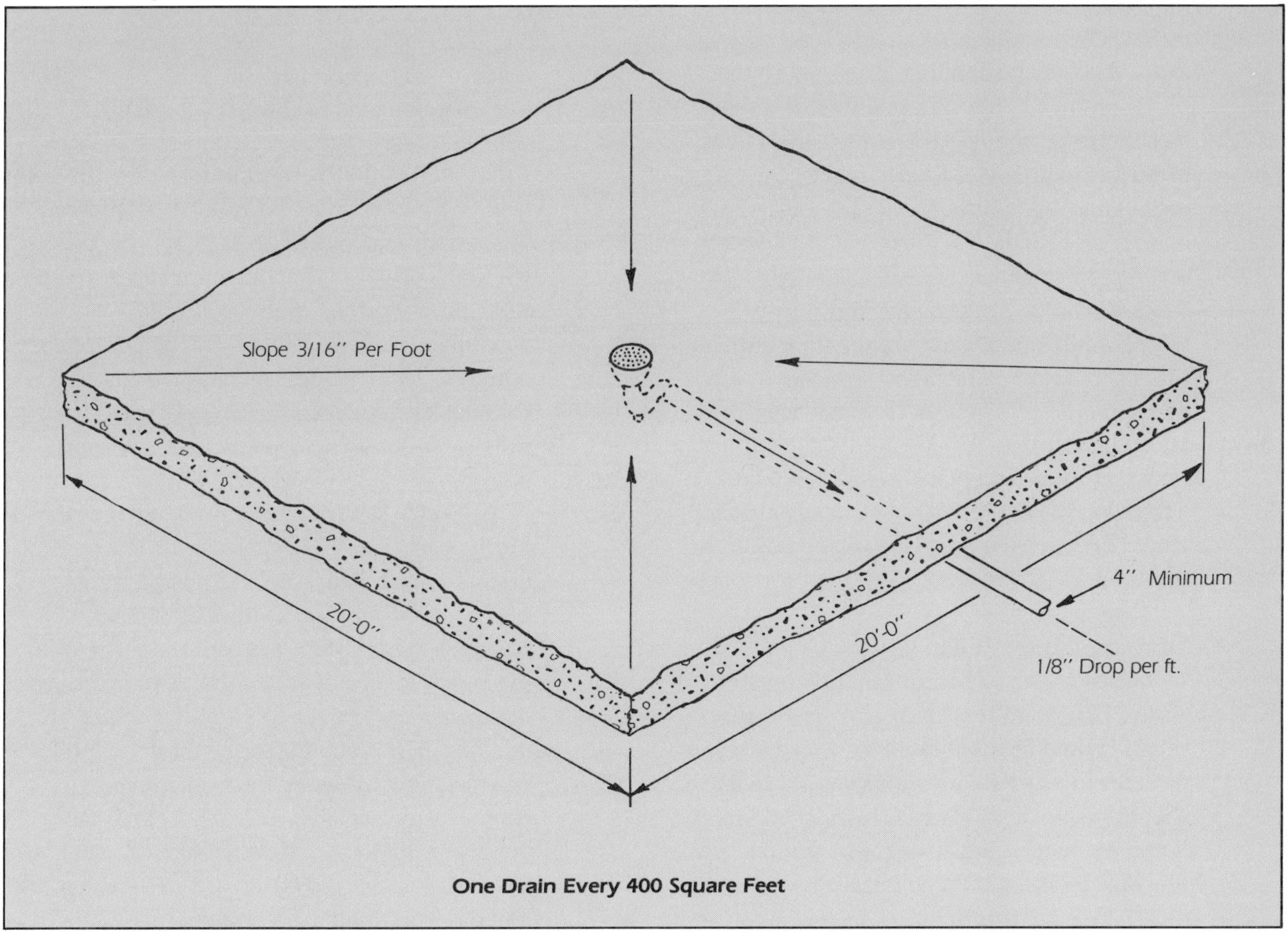

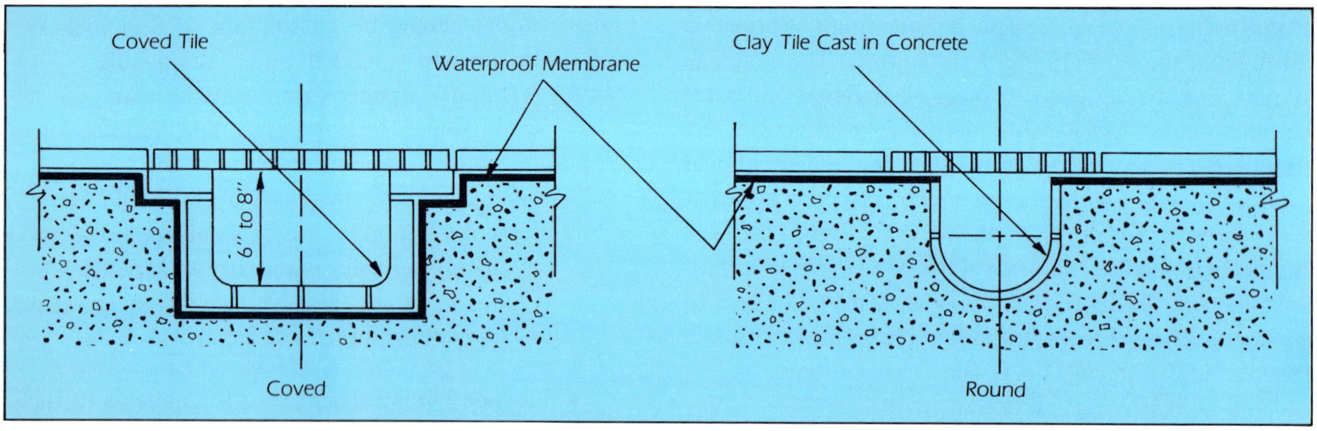

FIG. 3-59 Trench drains can be a major sanitation problem. Good design greatly reduces these problems and facilitates cleaning.

rounded or coved bottoms. The bottom of the trench should be pitched to the drainage inlet at a rate of 1/8 inch per foot. The depth of the trench should be limited to six to eight inches. Deep drains are more difficult to clean. The width should be restricted to that of conventional grating, usually six to twelve inches. Grates should be provided in sections of 36 inches or less in length so they can be easily removed. Grates should be constructed of steel or fiberglass of sufficient strength to accommodate anticipated vehicle weights. Inlets should be trapped with metal traps having a minimum water seal of three inches. Inlets should have secondary strainers to catch solids and prevent vermin from entering.

Trench floor drains are difficult to maintain at an acceptable sanitation level. Their use should be restricted to operations where they are required due to the nature of the food being processed.

- Drainage Lines. All drainage lines should be of sufficient size to permit rapid removal of waste. Individual branch lines from floor drains should be a minimum of four inches in diameter. Larger six-inch diameter lines should be provided where greater quantities of liquid wastes are anticipated. Horizontal drainage piping should be installed with a fall of 1/8 inch per foot for lines having a diameter larger than three inches, and one quarter inch per foot for lines three inches and smaller. The slope controls the velocity of the liquid flowing in the pipe. A velocity of two feet per second is usually the accepted practice for horizontal drain lines.

Drainage lines must be constructed with materials that meet national and local codes and should be installed so they do not leak. Cleanouts must be located throughout the system and placed so they can be used without posing a contamination threat to the food product; they must be readily accessible. Drainage lines must be vented, as required, to prevent sewer gases from backing up into the plant. Vents should not exhaust in an area adjacent to a fresh air intake.

Water service piping must not be run in the same trench as drainage or sewer lines. Plant drainage should not be cross-connected in any way with the sanitary sewage systems. Both systems should tie into the sewer main independently. Overhead drainage lines should be carefully routed to minimize the contamination threat to food products in process or their equipment contact zone surfaces. *(Figure 3-60)* Where necessary, extra precautions, such as catch pans, should be provided to catch leaking liquids. This should be a temporary

FIG. 3-60 Overhead liquid waste drainage lines must be carefully routed. The main drainage line parallels the wall, well away from the manufacturing equipment.

situation, and repairs of a permanent nature should be made as soon as possible.

INTERCEPTORS

When required, interceptors should be considered to handle excessive grease globules or other suspended solids which might be harmful to the local sewage system. *(Figure 3-61)* The size, type, and location should be approved by the local administrative authority; in any case, the location of the interceptor should not be in an area where it will represent a threat to the food product.

Interceptors should be installed where they can be readily maintained and cleaned. They should be installed in a manner to control any anticipated odors.

For grease interceptors to handle their rated capacities, flow control devices must be used to regulate the velocity of the entering liquid. The water entering the interceptor at the controlled rate is free of turbulence. It passes over a series of separator baffles which cause the grease, fats, and oils to separate from the waste water. Once released, these contaminants rise to the top of the interceptor by natural flotation and accumulate until removed. The waste water continues to flow through the unit into the plant liquid waste system. Flow control valves should be located as close to the grease source as possible.

INDUSTRIAL WASTES

Waste which is considered detrimental to the public sewer system or to the functioning of the treatment process must be treated and disposed of as directed by the local administrative authority. Prior approval of the local authority is advisable, and may avoid surcharges or other penalties which could be assessed.

UTILITY PIPING

Sanitation is consistently overlooked in the design and layout of most utility piping systems, regardless of their function. Emphasis is generally on major design criteria, such as pressure ratings, materials, sizing, and routing. The design engineer is reconciled to the prime importance of these major factors, but he must treat with equal importance the arrangement and location of pipes with respect to all sanitation requirements. Accessibility to facilitate cleaning, clearances, supports, and insulation are as important to the sanitation program as is a major piece of production equipment being installed. Both must meet sanitation needs.

Poor piping installations are the result of poor location planning. Pipe layout drawings that show location and installation details for all major utility piping should be provided. If pipe location and installation details are left to the field personnel, they should be coordinated by the resident engineer so that good sanitary design practices are followed.

PIPING ARRANGEMENT

Piping runs should be arranged in a neat, orderly manner. Where possible, main overhead piping runs should be grouped over aisles or other free floor space zones for cleaning access. If it becomes necessary to run multiple pipe groupings over floor-mounted equipment, which prevents access from the floor, the installation should include an access catwalk which parallels the pipe.

Pipe arrangements should not block off aisles or cleaning access under elevated equipment. Horizontal pipes

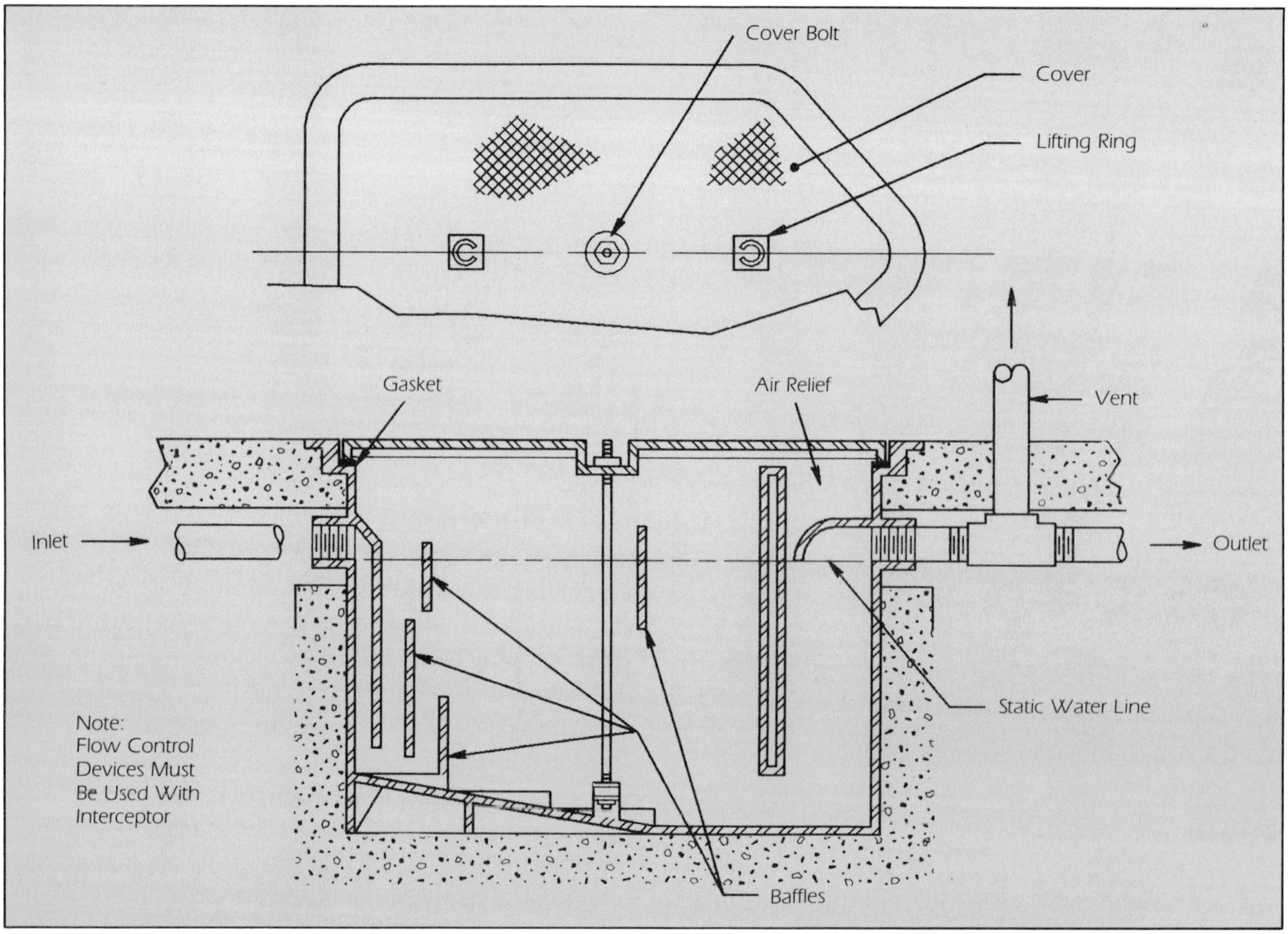

FIG. 3-61 Grease interceptors are needed to handle excessive grease globules from operations using oil, such as deep fat frying.

should be installed at an elevation equivalent to the bottom of the elevated machines they are servicing, but in no case less than 12 inches from the floor.

Vertical pipe groupings should not be bunched up, but should have at least four inches of space left around each pipe for cleaning.

Clearances. Adequate clearances above, below, and around pipe installations facilitate cleaning. Minimum requirements of six inches between horizontal pipe runs and the ceiling, and four inches between any parallel rows located below, should be considered. Horizontal pipes running in the same plane as adjacent pipes should have at least one inch between them. *(Figure 3-62)* When horizontal pipes parallel the wall or other long, flat surfaces, they should be held out from the wall or surface at least one inch. Vertical pipe runs that parallel a wall or another flat surface should be held out four inches or more. *(Figure 3-63)* Do not parallel a wall at a closer distance and then break from the wall at a point 18 inches or less from the floor to obtain the four-inch clearance. This creates a safety hazard.

Crevices. Crevices are formed when pipes are laid up against flat surfaces or adjacent pipes. These crevices trap solids and can result in insect, mold, or bacteria problems. Clearances should be maintained as suggested in the previous paragraph. The recommended clearances are in addition to allowances made for pipe insulation, when pipes are insulated.

WALL OPENINGS

Pipes subject to movement from expansion, contraction,

FIG. 3-62 Horizontal pipe runs should be adequately spaced.

FIG. 3-63 Vertical pipes mounted four inches from the wall allow for easier cleaning.

or vibration that pass through walls from one room to another should pass through sleeves or wall openings. Adequate clearance of two inches or more should be left between the outer surface of the pipe and the opening. Inadequate clearance results in inaccessible crevices or void areas. Pipe sleeves should be grouted into the wall with a nonshrinking cement grouting mixture. Wall openings should be framed with the frames well-secured and sealed to the wall.

Insulated pipes, in particular, must pass through wall openings constructed with the desired two-inch minimum clearance. Terminating the insulation at the wall and picking it up again on the other side is not a good practice. *(Figure 3-64)* In no case should you consider passing insulated pipes through the wall and attempt to seal the crevice with grout or caulking material. The soft surface of insulation is easily compressed and caulking or grouting seals are easily opened by the compression.

Grouted Pipes. Pipes not subject to expansion, contraction, or vibration can be successfully cement grouted into wall openings, or otherwise sealed with caulking material.

FLOOR OPENINGS FOR PIPES

Pipes passing from one floor to another should pass through a framed or sleeved opening which should have a watertight, four-inch-high curb or sleeve projecting above the floor. *(Figure 3-65)* If sleeved, the sleeve should be of a heavy gauge (12 gauge or larger) to withstand normal cleaning abuses. Where wet washing is anticipated, these openings should be capped off with sheet metal flashing to prevent water from reaching the floor below. *(Figure 3-65)*

Floor Opening Grouting. Pipes that are not likely to move because of vibration, expansion, or contraction can be sealed into the floor with grout. However, grouting is not advised in a wet process area where heavy water activity is contemplated.

FIG. 3-64 Insulated pipes should pass through a wall sleeve leaving adequate clearance in the sleeve for cleaning.

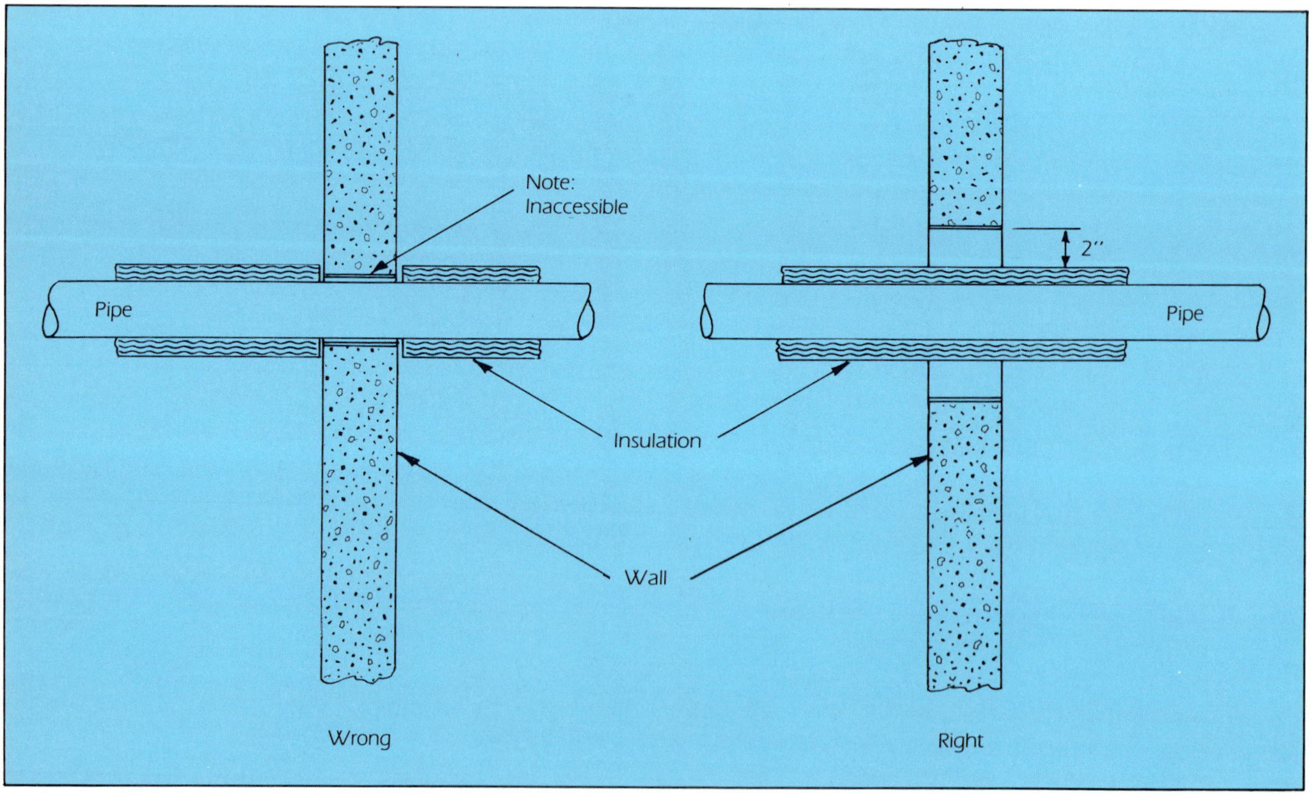

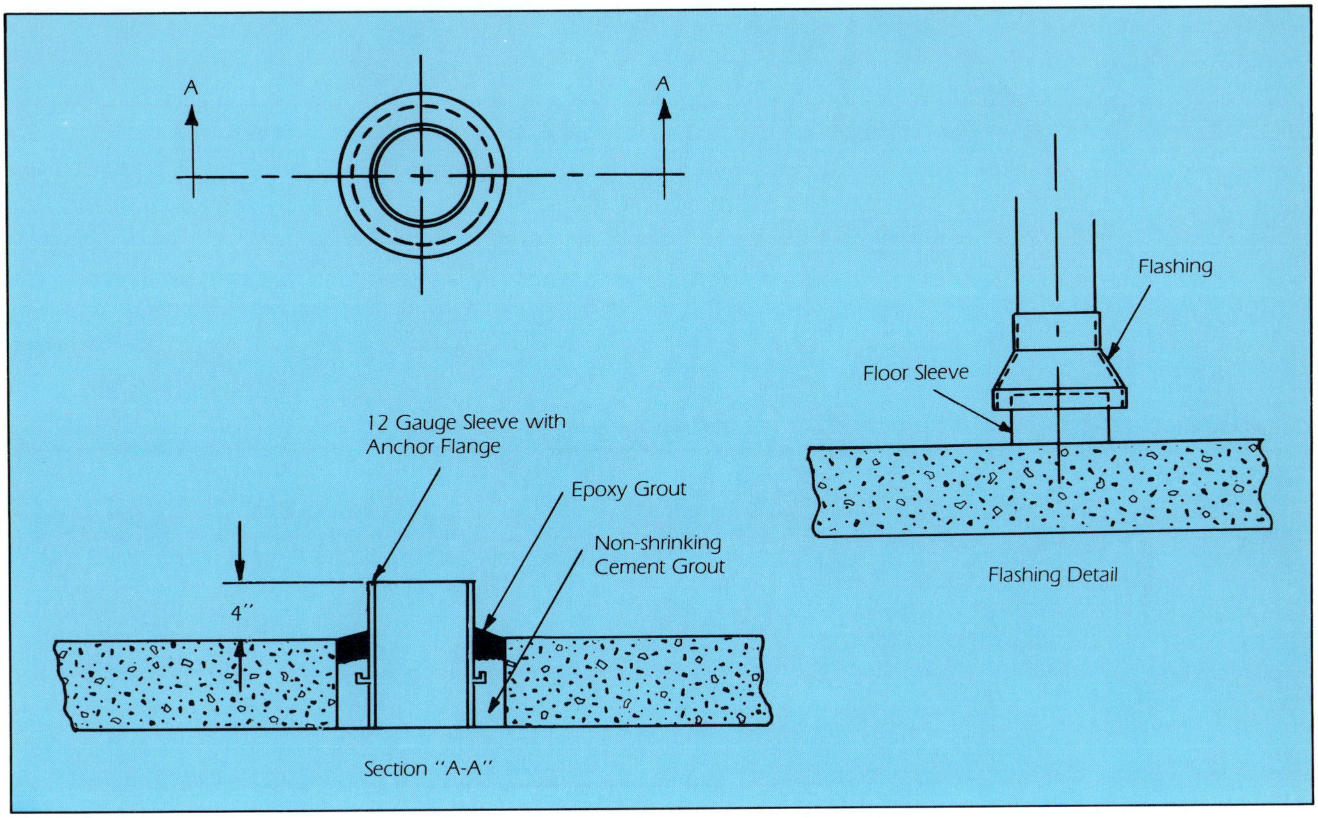

FIG. 3-65 Pipes passing through the floor should have properly designed floor openings, particularly in wet manufacturing areas.

PIPE SUPPORTS

Pipe Hangers. Pipe hangers should be simple in construction and arranged so they are easy to clean. *(Figures 3-66)* Structural shapes such as angle and tubing are acceptable as support members for trapeze hangers, as is flat bar stock. Channel is acceptable, but provides a second flat ledge for dust to lie, and should be used only where the load demands it. Formed channel with a flange crevice should never be used to support pipes. Formed channel is often constructed with longitudinal cuplike crevices at the edge of the flange. These crevices are difficult to clean. In all cases, angles should be used with the leg pointed down. If round or square tubing is used, all openings must be sealed.

All-thread hanger rods should not be considered since the threads are difficult and hazardous to clean.

Pipe Wall Brackets. Wall brackets should be constructed from standard structural shapes, such as angle or tubular members, and so arranged as to minimize crevice development at the junctions. Wall supports should be kept simple. The objective is to eliminate as much flat surface as possible. This is a basic criterion for any and all pipe or equipment supports.

Pipe Floor Supports. Pipe floor supports should be held to a minimum since they are difficult to clean around. Where necessary, use a single leg support to the floor. Consider supporting pipe from the existing equipment, work platforms, walls, and building steel.

Riser clamps consisting of two flanged bands, each being half the diameter of the pipe, are sometimes used at the floor junction to support the weight of a vertical pipe as it passes through the floor. Used in this manner, a void area is formed between the flanged bands and the floor, which is difficult to clean. This is not a good application of the riser clamp.

INSULATION

Poorly insulated pipes are also a sanitation concern. As with duct insulation and building insulation, pipe insulation can (and readily does) support insect

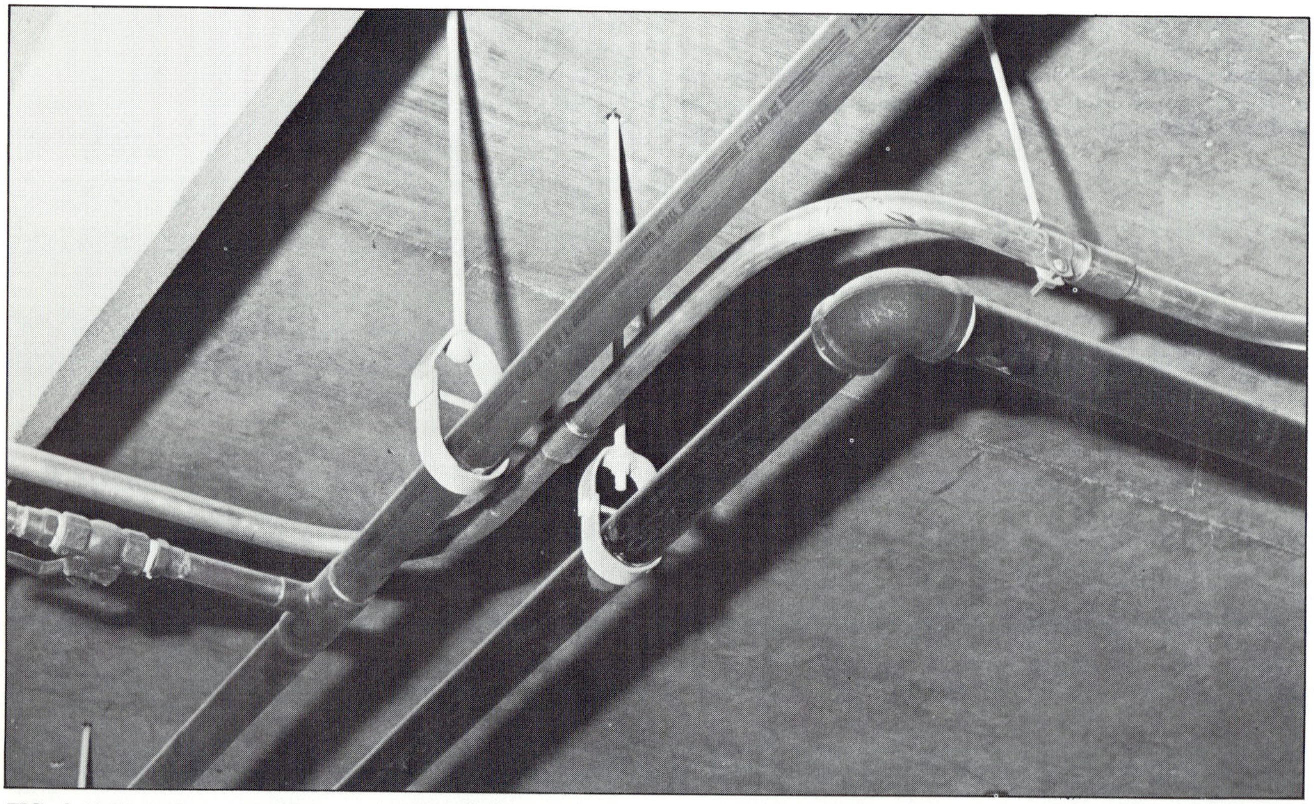

FIG. 3-66 Through the right choice of structural shapes . . .

. . . pipe supports can be kept simple to minimize dust collecting ledges.

populations in the presence of desirable food ingredients. Mold can grow on the insulation jacket, and the insulation itself is a contamination threat, particularly when it becomes damaged. It is commonplace to find numerous torn or frayed pieces of insulation dangling over open product zones or product zone surfaces in some food plants.

You might have heard that insects cannot live in fiberglass insulation as the glass fibers would scratch the waxlike surface of their bodies and they would dehydrate. This is not true. Most insects of the stored-product type get along well in fiberglass insulation.

There are many good pipe insulating materials available for use in and around food manufacturing areas. Of course, they are all adulterates and do not belong in foods. Some, such as asbestos, should not be used, as asbestos fibers are thought to be carcinogenic.

Density/Porosity of Pipe Insulation. The success of a satisfactory pipe insulation installation begins with the density and porosity of the material. Denser materials withstand more abuse than lighter, more porous materials. Denser materials are tighter and not as easily penetrated.

These are important sanitation features, and insulation materials should be selected accordingly.

Pipe Insulation Jackets. Standard insulation practices call for very porous canvas jackets to be placed over the installed insulation. Insects can easily crawl through the canvas and live in the insulation. Good, thick, or very viscous sealers that will seal these pores should be applied over the canvas. These sealers harden to a smooth, plaster-like surface which insects cannot penetrate. *(Figure 3-67)*

In wet processing areas, where wet airborne particulate matter is present, cloth jackets should not be used regardless of the sealer. Even the best sealers will usually stain from airborne particulate matter settling out on the sealed canvas jacket. Insulation that has a hard plastic exterior jacket should be used. This plastic should be laminated to the insulation, and the longitudinal and transverse joints should be glued or chemically sealed. This hard surface jacket is easy to clean. *(Figure 3-68)*

Protection for Insulated Pipes. In some cases, insulated pipes are located in areas where they will be exposed to cleaning or other forms of abuse. These pipes should be protected with a light gauge metal overwrap which should be sealed to the insulation with a pliable caulking material. The caulking material should be laid on the sheet edges prior to wrapping.

Another way to provide protection for insulated pipes is to select a round tubing approximately two inches larger in diameter than the insulated pipe. Split it in half lengthwise and place it around the pipe; then weld it back together. Fill in the void area with polyurethane foam.

FIG. 3-67 Canvas jacket pipe insulation can be sealed effectively with sealers designed for this purpose.

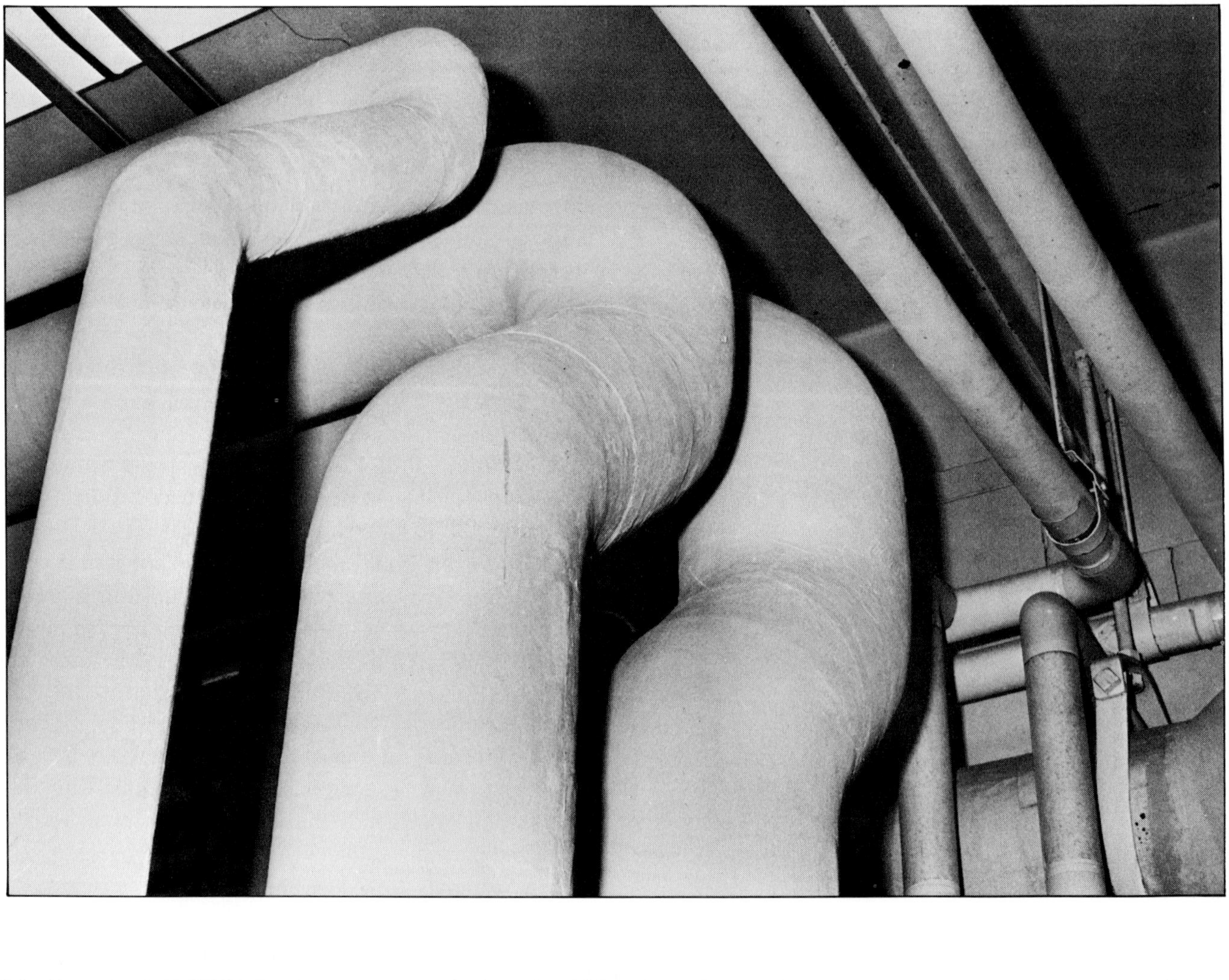

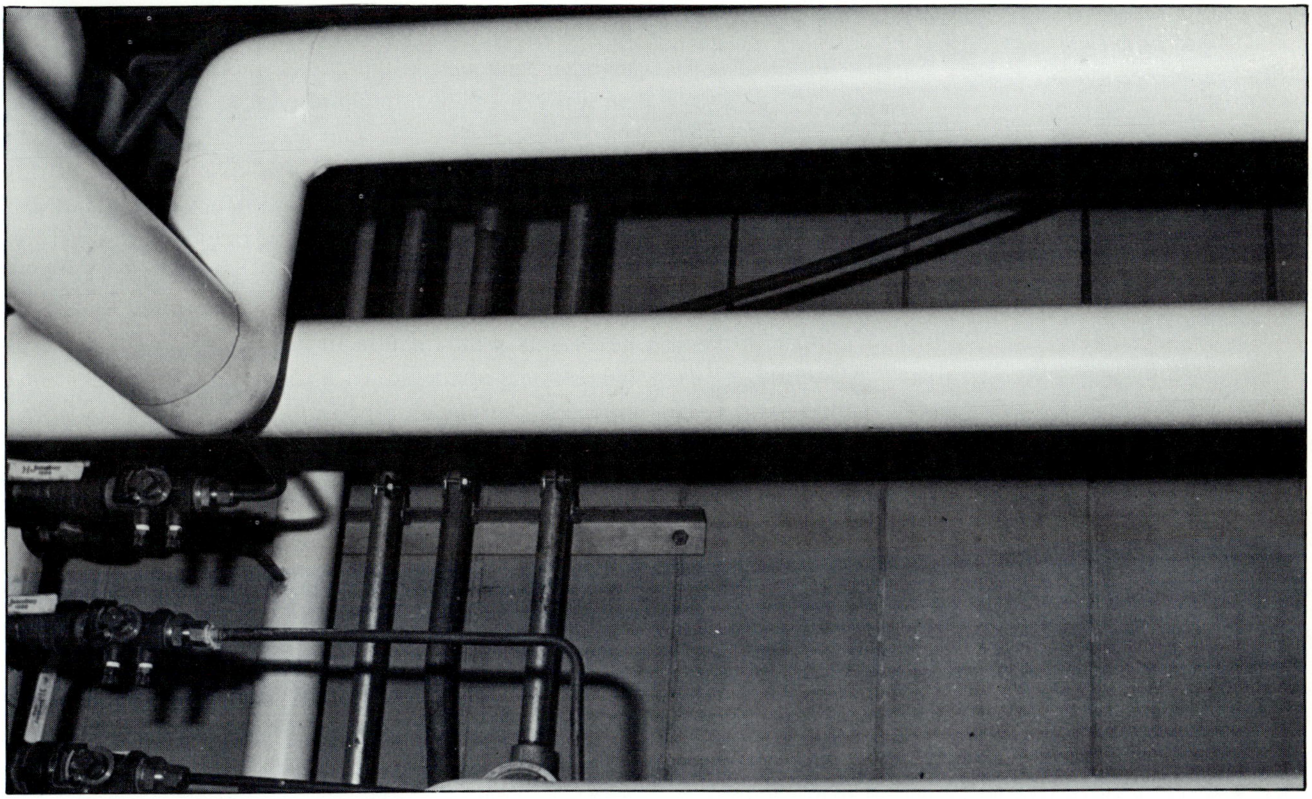

FIG. 3-68 Hard plastic pipe insulation jackets that clean easily are a good choice for wet manufacturing areas.

Polyurethane foam kits consist of two liquid parts that, when blended together in equal amounts, solidify and take a permanent set. Although the rate differs with each foam product, foam materials expand at a rate of approximately 35 to 1. When completed, the metal protective tube is fixed in place and provides the needed protection. The exposed set polyurethane end and/or ends should be sealed with epoxy paint. *(Figure 3-69)*

EMPLOYEE FACILITIES

When we talk about employee facilities, we are talking about the portion of the building in which the employee's personal needs are served. There is no more important area than this one, for it is here that the employee's work attitude is formed and pride is developed. Money spent here is money well spent, for all jobs are accomplished through people, the most valuable asset a modern food manufacturing plant has.

LOCKER ROOMS

The work day begins in the locker room so this facility needs to be bright and cheerful to the eye. It must be maintained in an exceptional condition and, therefore, must be constructed from easy-to-maintain materials. Wall, floor, and ceiling materials should conform to requirements outlined earlier. This facility must be well-lit and provided with the appropriate amount of air changes to maintain a pleasing atmosphere. Temperatures must be maintained at a comfortable level. Locker rooms should not be located directly in the manufacturing areas. Instead, they should be outside of, but close to, work areas and should be of sufficient size to avoid congestion and crowding. Adequate space should be provided for expansion.

Lockers. Each employee should have a locker (which can be locked) for personal belongings. The locker should have a master key for emergency purposes. The top should be sloped to prevent accumulation of debris. *(Figure 3-70)* Lockers should be raised off the floor four to six inches on concrete pads. Solid steel bottoms should be cut away

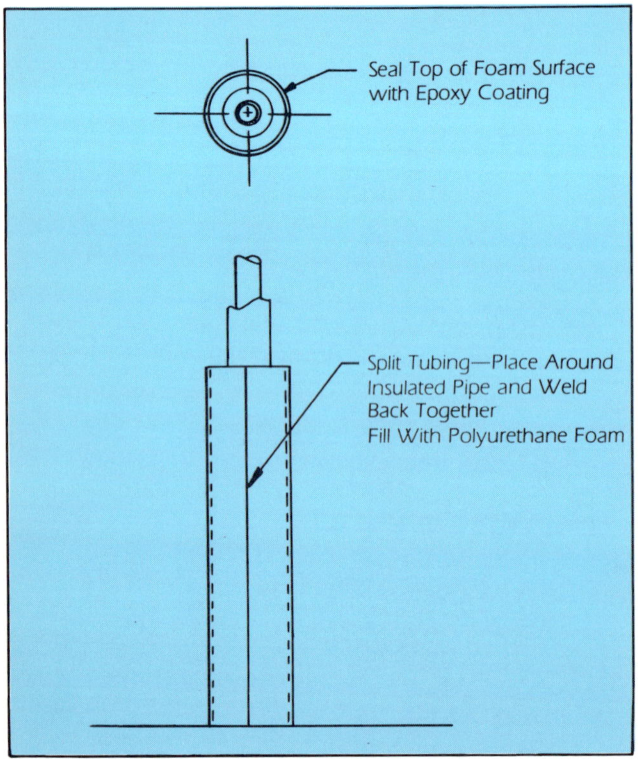

FIG. 3-69 Some insulated pipes need additional protection from various forms of operating abuses, including cleaning and maintenance.

to eliminate an insect crevice. Other desirable locker arrangements or configurations, such as mounting them on walls or ceilings, might also be considered.

Hand-Wash Basins. Each facility should be equipped with adequate hand-washing facilities with foot-operated valves. *(Figure 3-71)* These units should also include hand-sanitizing solutions where they are needed.

Hand-washing and sanitizing stations should also be located in the manufacturing areas where employees' hands come in contact with food or food product zone surfaces. Product inspection points, sorting tables, premium drop stations, and some packaging points are just a few of the areas where hand-washing and sanitizing stations might be needed.

Showers. Each locker room should provide a sufficient number of showers. Showers and shower rooms should include an adequate hot water supply, water mixing valves, soap trays, and ceramic-tiled floors, walls, and ceiling. The floor should be pitched to a floor drain.

FIG. 3-70 The employees' locker room is one of the most important parts of the plant. Lockers should feature sloped tops and be elevated four inches from the floor.

FIG. 3-71 Hand wash basin with foot operated valve

TOILET ROOMS

Toilet rooms must be maintained in an immaculate condition. Floors, walls, and ceiling must be constructed from moisture-impervious materials which are easily cleaned, as previously defined. *(Figure 3-72)* Rest rooms must be adequately lit and properly ventilated to prevent escape of offensive odors to the manufacturing area or other areas of the plant. Toilet room doors should not open directly into the production, packaging, or storage areas, but should be designed with vestibules and double doors having self-closing devices. The facility should be complete with wash basins, hot and cold water mixing valves, soap, paper towels, towel dispensers, hand sanitizing solutions, and covered waste containers. Larger, multiple-stall toilet rooms should be equipped with floor drains unless prohibited by local codes.

Water Closets/Urinals. Water closets should be good quality, properly trapped, wall-mounted units. The seats should have open fronts and be moisture impervious. Closet partitions should be suspended from the ceiling for cleaning ease. Each cubicle should be equipped with a door latch,

FIG. 3-72 Rest rooms or toilet rooms should be built from materials that can easily be wet-cleaned and sanitized.

FIG. 3-73 Employee cafeterias should provide a pleasant and relaxing environment.

paper holder, and coat hook. Urinals should also be mounted on the wall with automatic flushing devices. All plumbing fixtures should be equipped with appropriate backflow prevention valves, along with vacuum breakers to prevent contamination of the water system.

Women's toilet rooms should be equipped with special containers in each stall for the disposal of sanitary napkins.

Hand-washing signs must be mounted in a suitable location to remind employees to wash and sanitize their hands before returning to their work stations.

CAFETERIAS

The cafeteria is an important area to the employee. Since good manufacturing practices prohibit eating, drinking, or smoking in areas where foods are manufactured, it makes good sense to provide a pleasant and relaxing environment in which lunches and work breaks can be enjoyed. Little additional money is needed for good color selections, an appropriate lighting system, and comfortable furniture. Cafeterias should be attractively designed and they should be built with materials which are durable, moisture impervious, and easy to clean. Resilient surface floors (vinyl, vinyl asbestos, or asphalt tiles), concrete block with tile-like surface coatings or ceramic tiles for walls and easily cleaned drop ceiling panels are satisfactory. Lunch tables, chairs, and covered waste containers should also be easy to clean. Vending equipment in food serving areas should be constructed in accordance with acceptable food service guidelines. This equipment should be installed so it is easy to clean. Large machines should be equipped with casters so they can be moved in and out for cleaning. The ventilation system should provide adequate air exchanges as established by the appropriate building codes. Again, a well-designed cafeteria more than pays for itself in terms of good employee relations. *(Figure 3-73)*

CHAPTER 4
EQUIPMENT

CHAPTER CONTENTS

GENERAL EQUIPMENT DESIGN CONSIDERATIONS

EQUIPMENT CONSTRUCTION MATERIALS

CONSTRUCTION FEATURES

PROCESS CONTROLS

CONVEYING SYSTEMS

BULK SOLID STORAGE BINS AND USE BINS

LIQUID INGREDIENT STORAGE AND TRANSFER SYSTEMS

DUST COLLECTION SYSTEMS

As previously stated, the food industry is comprised of several segments. Each of these segments produces a variety of products which probably number well into the hundreds. With this understanding it is easy to see that many different types of equipment are required to produce the numerous varieties of food products. Each type of manufacturing machine will have its own peculiar characteristics and specific sanitary design requirements. Just as there is a lot of commonality in the industry's building or structural needs, there are many differences in equipment needs. These needs relate back to understanding the risks or sensitivities of the products being produced, as discussed in Chapter One. This understanding is going to establish much of the final equipment design criteria.

Because of the vast special requirements in equipment design, this chapter will deal largely with some of the fundamentals which will be required of nearly all food

manufacturing equipment. It will be necessary to select the fundamentals that apply to your specific equipment and determine to what extent they should be considered. These decisions, like those relating to special needs, should also be made with the product risk in mind.

We will also take a look at some specific areas where there may be some common equipment interests. These areas are: process controls, common conveying equipment, bulk solid storage and use bins, liquid ingredient storage and transfer systems, and dust collection systems.

It takes a design engineer about two years to develop his sanitary design skills. The engineering criteria are not particularly complex, and they can be mastered fairly quickly. Nonetheless, more time is needed to acquire a better understanding of the different manufacturing processes and their particular design needs.

In addressing the sanitary fundamentals of equipment design, we will discuss:

- General equipment design considerations
- Construction materials
- Construction features

GENERAL EQUIPMENT DESIGN CONSIDERATIONS

General design considerations are the basic sanitation needs that apply to every piece of equipment in all manufacturing processes. The attention paid to these general design considerations pays off in the ease with which sanitation problems can be solved; following them prevents accumulations of static food material in the equipment that impair product quality and safety. Taking these design considerations into account also makes equipment cleaning easier.

FOOD CONTACT SURFACES

The condition of the food contact surface has more effect on cleanability than any other single factor. Rough, porous surfaces support product buildup, bacterial development, and insect activity. They are also difficult to clean, and surfaces that are difficult to clean are costly to clean.

All food contact surfaces should be inert. Surfaces should be smooth, nonporous, and nonabsorbing. Product zone surfaces should be free of pits, checks, scale, cracks, crevices, and other imperfections. Contact surfaces should also be corrosion-free when in contact with food. Surfaces must also withstand the actions of cleaning and sanitizing chemicals as well as any pesticides that might be required.

ACCESSIBILITY

All food contact surfaces should be available for inspection and cleaning, either by access doors, removable covers, or complete disassembly. In cases where contact surfaces are not accessible, provision must be made for substantiating that current cleaning practices are effective and have eliminated potential bacterial or insect problems. This can be accomplished with inspection ports.

Access Doors. The access doors provided should not require the use of tools to gain entry. *(Figure 4-1)* Access doors should be large enough to permit thorough inspection, and for equipment where disassembly is not possible, the cleaning of all product zone surfaces.

Make sure the access doors are designed to accom-

FIG. 4-1 Access door should not require the use of tools. Note the quick-release fasteners on the screw conveyor drop bottoms shown in the lower section.

FIG. 4-2 Fasteners that do not require the use of tools save a great deal of time.

modate pressure conditions as required; they should be fitted with food-grade gaskets and should not leak. Door hinge assemblies should be durable and capable of withstanding normal abuse.

Removable Covers. Where applicable, provide removable covers, panels, or lids preferably with quick-release fasteners that do not require the use of tools. *(Figure 4-2)* The use of tools sometimes requires that a maintenance person be called, which slows the cleaning process.

Disassembly. To the extent possible, equipment should be designed for disassembly without the use of tools. Disassembly should be considered for smaller pieces of equipment such as pumps, valves, or extruders. Normally, equipment of this type cannot be adequately cleaned by access port or with automatic clean-in-place (C. I. P.) systems.

INSTALLATION

Elevated Equipment. Equipment should be designed to be installed elevated from the floor a distance of six inches or more. Large equipment such as dryers, toasters, cookers, and storage belts that occupy substantial floor space should be elevated 24 inches or more to facilitate cleaning. *(Figure 4-3)*

Floor-sealed Equipment. Equipment can be sealed to the floor if it is not practical to elevate it. The practice of sealing should be limited as it is very difficult to maintain a caulked seal on equipment that vibrates or is subject to expansion and contraction. The seal will ultimately break under these conditions.

NONCONTACT-ZONE SURFACES

Make sure all exterior or noncontact-zone surfaces are relatively smooth and nonabsorbent, and arranged to minimize soil accumulations. These surfaces must be easy to clean. *(Figure 4-4)* Noncontact-zone surfaces should be constructed of materials appropriate for the surrounding environment, so the materials themselves do not become a source of contamination, from peeling paint or corrosion.

EQUIPMENT FRAMING AND SUPPORTS

Keep equipment frames and supports simple. Your objective is to eliminate as many flat surfaces as possible through the careful selection and arrangement of structural shapes. Where possible, eliminate cracks and crevices. In some cases, support members can be eliminated by selecting heavier gauge sheet metals for equipment walls or panels. Stronger structural members can also eliminate the need for bracing or extra reinforcing.

DRIVES AND DRIVE MOTORS

Drives and drive motors should not be located over open product zones or in any other places where they can contaminate the product with leaking lubricants. *(Figure 4-5)* In addition, direct drives offer both a maintenance and a cleaning advantage, because they have fewer components.

Drip Pans. Drip pans should be used sparingly, although they do become necessary sometimes in dealing with an immediate oil contamination threat. Rather than using drip pans, the ultimate solution lies in relocating these drives and implementing a consistent seal replacement program.

BEARINGS

Use either sealed or self-lubricating bearings, and locate them out of the product zone. When you must install bearings in the product zone, consider food-grade materials such as nylon bearings that require no lubrication.

SEALS

Shaft seals that come in contact with the product should be nonabsorbent, nontoxic, nonexuding, and odorless.

FIG. 4-3 Equipment should be built so it is elevated from the floor at least six inches, with large equipment being elevated 24 inches or more.

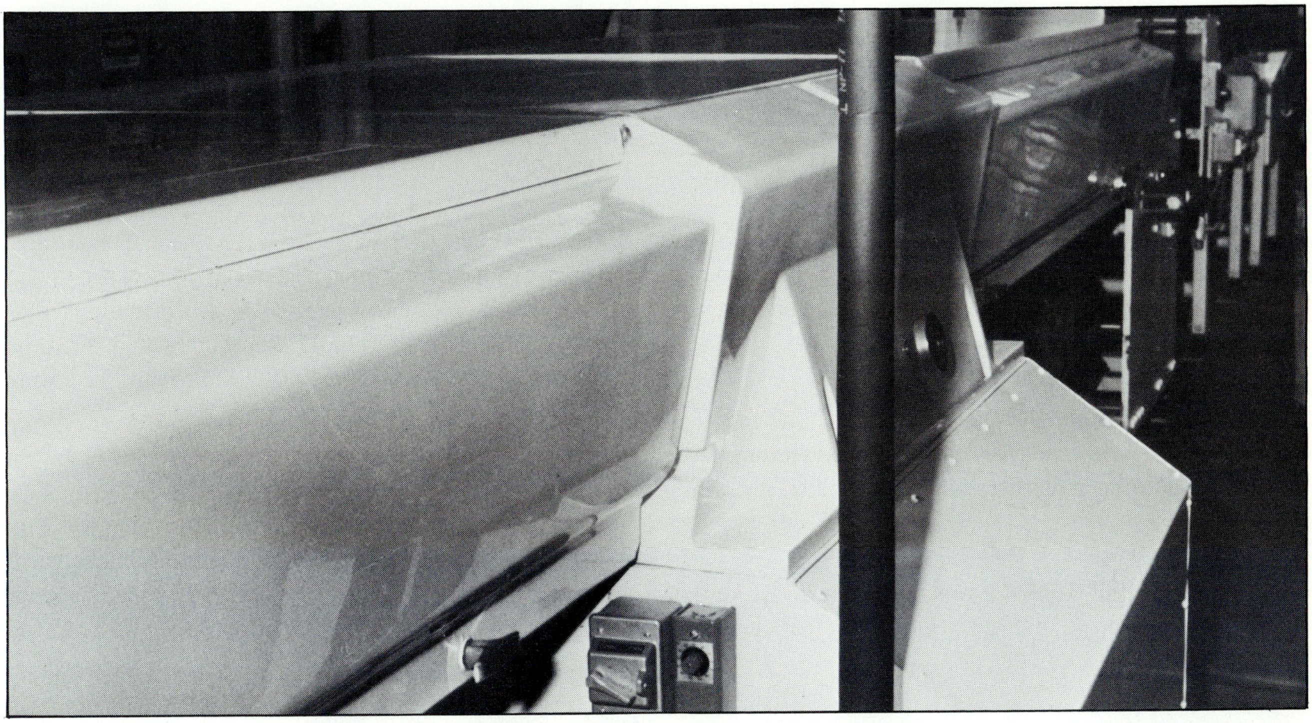

FIG. 4-4 Non-contact zone surfaces should be smooth and easy to clean.

FIG. 4-5 Equipment drives should not be located over open product zones.

Mechanical seals are preferred. Install all seals so they can be easily removed for cleaning and sanitizing. All seal materials should be unaffected by cleaning and sanitizing compounds.

HOODS

Provide dust collection or wet collection systems to remove dust or steam discharges. *(Figure 4-6)* These systems should be designed so they can be easily cleaned, sanitized, or fumigated, if needed.

SELF-DRAINING LIQUID KETTLES

Liquid handling equipment or equipment that requires wet cleaning should be constructed and installed so it drains freely. *(Figure 4-7)*

Mixing or Holding Kettle Covers. Make sure the kettle covers are designed so that mixer lubricants from leaking bearings cannot enter the kettles. A mixer shaft slinger and a raised lip around the mixer shaft opening in the cover will eliminate this problem. Split covers should not be hinged. Split covers should feature a raised seam and a vertical flange formed at the cover perimeter. *(Figure 4-8)*

EQUIPMENT COVERS

All open product zones must be covered. Where necessary, elevate covers to relieve excess vapor formations that otherwise might result in mold accumulations against the underside of the cover. *(Figure 4-9)* Make sure the covers can be easily removed for cleaning.

DIRECT STEAM COOKING

Direct steam cooking is not considered a good practice by many sanitation authorities. However, in some cases, there are no alternatives. To reduce concern about product safety, make certain that only high quality steam is used. Equip

FIG. 4-6 Dust collection hood

FIG. 4-7 Self-draining liquid kettle

the steam lines with check valves to prevent product backup and also with strainers and steam traps. Most importantly, make sure all boiler feed water additives are food grade. Many feed water additives contain harmful toxic substances.

COMPRESSED AIR

Compressed air which comes in contact with food or the food contact zone surface must be clean and oil-free. This includes applications of compressed air such as blowing open pouch-type packages, blowing packages off the forming mandrels of package-making equipment, and cleaning the cloth filter socks in dust collectors. Many lubricating oils contain toxic substances like lead or arsenic. Lubricants that have higher cooling properties, such as those recommended for compressors, are of particular concern. All lubricants are considered to be food contaminants, regardless of their toxic nature.

In addition to oil, compressed air can contain dust, pollen, water vapors, and other contaminants commonly found in atmospheric air. Air filters on the inlet side of the air compressors do not stop all of this material, so additional air filtering and drying systems on the discharge side of the air compressor might be desirable.

Early compressed air filtering and drying systems rarely removed all of the oil from the air. Unfortunately, many basic or simple systems currently available still fail to remove oil completely. Much of this equipment is considered to be 99.9% efficient and though .1% does not seem to be a large amount, it should be remembered that large quantities of air could be used daily; while .1% of oil per 1,000 cubic feet of air in a system might not represent a major concern, .1% of oil per 300,000 cubic feet is a serious potential source of contamination. In addition, oil can accumulate in low sagging areas of the air manifolds and eventually be pushed along by a surge in air pressure. Drip legs are helpful in removing oil and moisture from the compressed air flow and should be installed in all piping systems. However, drip legs do not remove all of the oil in the pipelines. Drip legs are typically provided with a blow-down valve to drain oil from the leg. This valve is a good place to check to get a feel for the quantity of oil and water in the compressed air system.

Over the years, there have been a number of technological improvements in compressed air filtering and drying systems. These improvements have led to even more sophisticated systems that use filter media and desiccant materials that are significantly more efficient. These systems commonly use individual filters in combination to remove particulate matter, liquids, and odors from the air. The better systems include indicators which sense the quantity of trapped lubricants and automatically drain it off. Alarms are also provided that warn of pressure differential increases which are caused by accumulations of contaminants or clogged traps.

These more sophisticated filtering and drying systems should be carefully evaluated on their own individual merits, claims, and test results. It should be remembered that any system is subject to malfunction and certain maintenance

FIG. 4-8 Mixing or holding kettle cover design

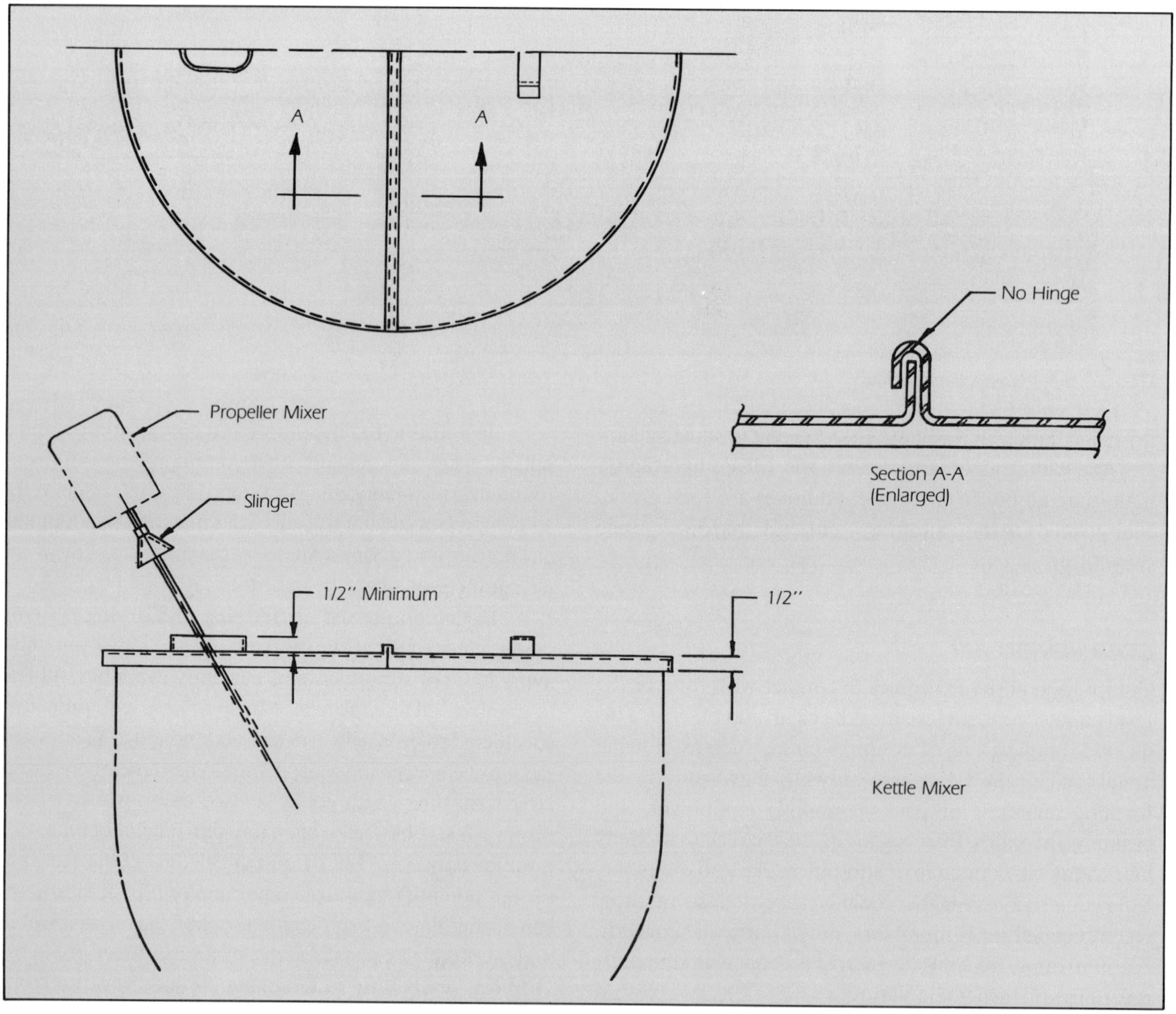

FIG. 4-9 Elevated equipment covers

needs. Monitoring equipment which has alarms would be of benefit. If properly considered, designed, and maintained, these more sophisticated systems probably are capable of performing better than the older and simpler systems used on oil-lubricated air compressors.

Teflon, carbon ring reciprocating air compressors, centrifugal, and dry screw machines do not require the use of oil lubricants. All of these are very good choices for food contact and food contact zone applications. However, they too should be equipped with filtering and drying systems to cope with contaminants in atmospheric air.

EQUIPMENT CONSTRUCTION MATERIALS

The materials selected to fabricate food-manufacturing equipment have an impact on the food safety and sanitation program in many different ways. If the materials selected are not compatible with the food in process, the food itself could be affected. Odor, taste, or color problems could occur, as well as direct contamination through a breakdown of the metal. The use of inert construction materials for food manufacturing equipment is mandatory. Food products can react to metals and form metallic salts, which often have a bitter taste that can easily be imparted to the food. Iron, nickel, and copper are oxidizing metals and in some circumstances can break down certain vitamins.

Improper selection of materials can also mean difficulty in using cleaning compounds, sanitizing solutions, and pesticides. Many of these products react adversely when in contact with certain metals, causing rusting and discoloration. It is important for the design engineer to ask questions, not only about the process, but also about how the equipment will be cleaned or the necessity of treatment with fumigants. Answers to these and similar questions will avert many potential sanitation problems and equipment failures. Emphasis is, and perhaps always will be, placed on stainless steel, aluminum, and other rust-resistant metals. But rust or corrosion-resistant properties should not be the only concern. In many cases the magnetic properties and/or hardness of the metal is important. Where there are fears or concerns for a metal screen disintegrating or a metal screw falling out, the magnetic property should be an important characteristic of the metal chosen. Where there are fears for metal working against metal, then hardness or durability must be an important consideration. And of course there are still many varied applications for mild steels despite these preferences. It simply boils down to a thorough assessment of the need, which necessarily includes food product safety.

METALS

Stainless Steel. Type 304 stainless steel is one of the most widely used materials for fabrication of food-handling equipment and it is suitable for most manufacturing applications. Stainless steel is a trade name given to corrosion-resistant steel. Normally it contains no more than .7% carbon and as much as 20% chromium. An 18-8 grade stainless, for example, contains 18% chromium and 8% nickel.

Type 316 is a better choice for food with a high acid content than 304, particularly if the food is processed at higher temperatures. Since type 316 has a lower carbon content than 304, it is somewhat more corrosion-resistant.

Stainless steel is available in many different finishes, from a flat 2B finish (standard from the steel mills) to a 6 or 8 highly polished finish. Some food manufacturers require the highly polished finishes with welds ground and blended. Others will specify a flat 2B finish on the inside and a number 3 or 4 finish on the outside. For the most part this is a matter of preference, but if you plan to grind welds to a certain finish, you should specify a certain sheet finish. The 2B flat mill run finish is not a ground finish. It comes off the mills with that flat, milky look. There is no grit or ground finish to polish or grind to.

Titanium. Titanium is used in its pure state and also in alloy form. It is light, very strong, and has good corrosion-resistant properties. Titanium weighs approximately 44% less than stainless steel and is resistant to the corroding effects of salt water, many acids, and alkalis. It is a particularly good choice for applications requiring a lightweight, high-strength metal.

Inconel. Inconel is a nickel-chromium alloy containing approximately 77% nickel and 18% chromium with the balance primarily iron. It has excellent resistance to many corrosive materials and to oxidation. It has very good mechanical properties and is considered to be more ductile than stainless steel. Inconel is a very suitable material for food equipment fabrication.

Mild Steel/Iron. Mild steel and iron will always have a place in food plants. Steel and iron do corrode and can cause some foods to discolor. Their use on contact zone surfaces should be limited to dry ingredient applications, some edible oils, syrups, and other similar materials.

Mild steel and iron can be used as a food contact material in storage tanks and equipment that handles dry products and some liquids. It can also be used as equipment framework or for supports in many areas.

Aluminum. Aluminum is resistant to most atmospheric influences, but is readily dissolved by hydrochloric and hydrofluoric acids and caustic solutions. Aluminum is similar to many other corrosion-resistant metals but contact with some commonly used metals in the presence of a corrosion medium can result in deterioration of the aluminum. Bright surfaces produced by rolling or polishing tend to darken and roughen through exposure to weather elements unless frequently cleaned. Pure aluminum is very soft. It can be hardened by cold working. Many aluminum alloys can be hardened by heat treating. Although this improves durability, it is still substantially softer than steel and will gouge and scratch much more easily, which is an important product safety consideration. Aluminum can be used as a contact zone material for many food products such as fruit juices, milk, pop, and other beverages.

Brass/Copper/Bronze. Copper and copper-bearing alloys are commonly known as the *brasses*. If a new copper or copper-alloy application is being considered, it might be advisable to conduct field tests for effects, or contact a metallurgist for an opinion.

The brasses do not corrode due to atmospheric conditions, but when they are exposed to various chemical agents, including cleaning chemicals, they may. The primary concern is for their ultimate effect on the food product due to various reactions that may result in offensive taste or discoloration.

Monel. Monel is a nickel-copper alloy containing about two-thirds nickel and one-third copper. It is strong, has excellent ductility, and good corrosion-resistant properties to such things as salt water, caustic soap, and many organic acids. However, because of the copper content it may discolor some vegetable products such as peas, corn, and beans. Monel is excellent for salt water applications.

Lead. Lead is considered to be very toxic.The potential applications of lead in a food plant are in paint products or solder. Lead-based paints should not be used, and solder should not contain more than 5% lead.

PLATED METALS

The plated metals, of which there are many, have various applications around the plant. However, these applications are generally provisional, as certain vegetable acids can break down plating materials, and other products can cause abrasive wear as well. Some of the plating materials contain toxic properties and for this reason cadmium, in particular, should not be used in the food plant.

Galvanized Iron. Galvanizing is a plating method of coating steel or iron with zinc. The finished product is commonly called galvanized iron.

Galvanized iron has been used for many years in all types of food plants. Although still being used, applications of galvanized iron in a food plant are being limited to ductwork and a few select food contact surfaces. Galvanized metal should not be used in contact with green vegetables as the zinc could be absorbed, causing the vegetables to darken. Fruit acids can dissolve zinc. Some cleaning solutions will also cause a zinc breakdown. Galvanized or zinc coatings are being used satisfactorily on some dryer belt and numerous equipment framework applications.

Tin. Tin is extremely resistant to corrosion and for years has been used to coat cans. In addition to coating cans, tin is also used in many food-grade solders. It is one of the better plating metals for contact surfaces, but the fact that it is

FIG. 4-10 An important product safety feature is metal scalper or sifter screens which use wire having a magnetic property.

soft and easily scratched or worn limits it to certain applications.

Nickel. Nickel has very good corrosion-resistant properties as well as excellent mechanical properties. The electroplated process of applying nickel to steel can produce adherent bonds which approach the strength of the metal. Nickel can be used satisfactorily as a food-contact surface material in a variety of applications.

Chrome/Nickel. Chrome plated nickel surfaces have been used on some food manufacturing equipment handling nonacid products. Food products containing acid or acid cleaners in particular can affect the bond between chrome and nickel, causing the chrome to loosen. Because of its hard surface, chrome/nickel plating has been used in applications requiring the use of a scraper device.

Cadmium. Cadmium is considered to be toxic. Cadmium-plated surfaces, including nuts, bolts, and other fasteners should not be used in the plant.

OTHER CONSTRUCTION MATERIALS

Wire. Wire used in sifter, scalper, classifier, and strainer screens should be made from steel that has a magnetic property. Most screens are made from small diameter wires that can, after a period of time, disintegrate from natural wear. *(Figure 4-10)* The wire fragments could be rather small and, if nonmagnetic, extremely difficult to sense with a metal detector. The magnetic characteristics would enhance the metal detector capability as ferrous contaminants are more easily detected, but more importantly, a magnet could also be used. Small, needle-like wires can be very injurious. A precaution of this type is an excellent safeguard. If wire screens must be corrosion-resistant, consider a plated steel wire in order to maintain the magnetic property. A series 400 stainless steel could also be considered. Series 400 does have magnetic properties.

Wood. Generally speaking, wood is not a satisfactory material for most food-producing equipment. It readily absorbs liquids, produces odors, is porous, and is subject to cracking as it ages. Wood surfaces often harbor insects and in itself represents a contamination threat. Wood can also splinter, producing slivers in the product. These wood fragments are of concern because they frequently cause infection to body tissue. For this reason wood is considered to be one of the most feared foreign material contaminants.

Glass. Glass is also one of the more feared food contaminants, as it can cause a great deal of damage to the body if accidentally ingested. Wherever possible, glass components such as light bulbs, dial gauges and product sight glass should be replaced with polycarbonate plastic materials or be protected by plastic enclosures. It may be argued that glass is a commonly used packaging material and if it is such a threat, then why is it used? It is true that glass is used as a packaging material and it is also true that there is a great deal of concern. For that reason the control programs must be impeccable. There are a few other justifiable applications for glass and the same impeccable controls are imperative.

Plastic. Plastic materials have a variety of applications in the food plant. Plastics made from polycarbonate resins are preferred, particularly as a glass substitute or glass protector. Polycarbonate plastics are specially suited for the many needs of the food industry. They are nontoxic and meet many regulatory requirements. Polycarbonate plastics also have outstanding impact strength, high heat resistance, and stain resistance.

Fiberglass. Like plastic, fiberglass materials have a variety of applications in and about the food plant. Fiberglass materials should be free of formaldehydes, phenolics, and should be formulated from materials meeting the requirements of the regulatory agencies. Some fiberglass resins give off toxic fumes when burned and make fire fighting exceptionally hazardous.

Rubber. Only rubber and rubber synthetics approved for food contact applications should be considered. These materials should be nonporous, nonabsorbent, and nontoxic. They should not be affected by ordinary cleaning methods or cleaning materials.

Cloth Fabric. A number of applications exist for cotton and wool, and also for nylon and other synthetic cloth materials. Materials selected should be approved for a specific application, such as sifting, filtering, or flexible connections. Vinyl-impregnated cloth materials should be

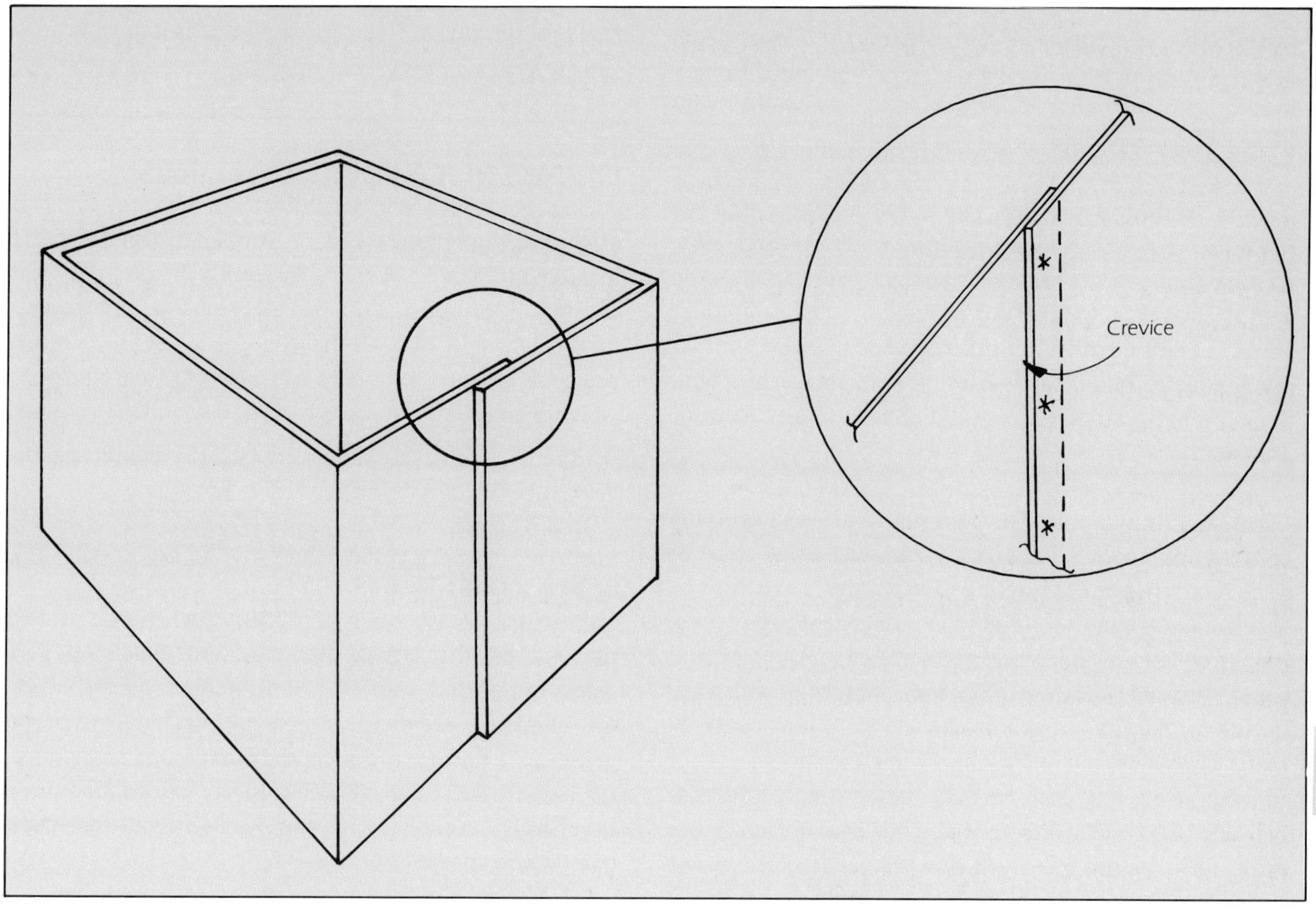

FIG. 4-11 Lap seams which are tack- or spot-welded result in cracks and crevices that can support insect life and bacteria.

FIG. 4-12 A major design objective is to eliminate product zone ledges in equipment.

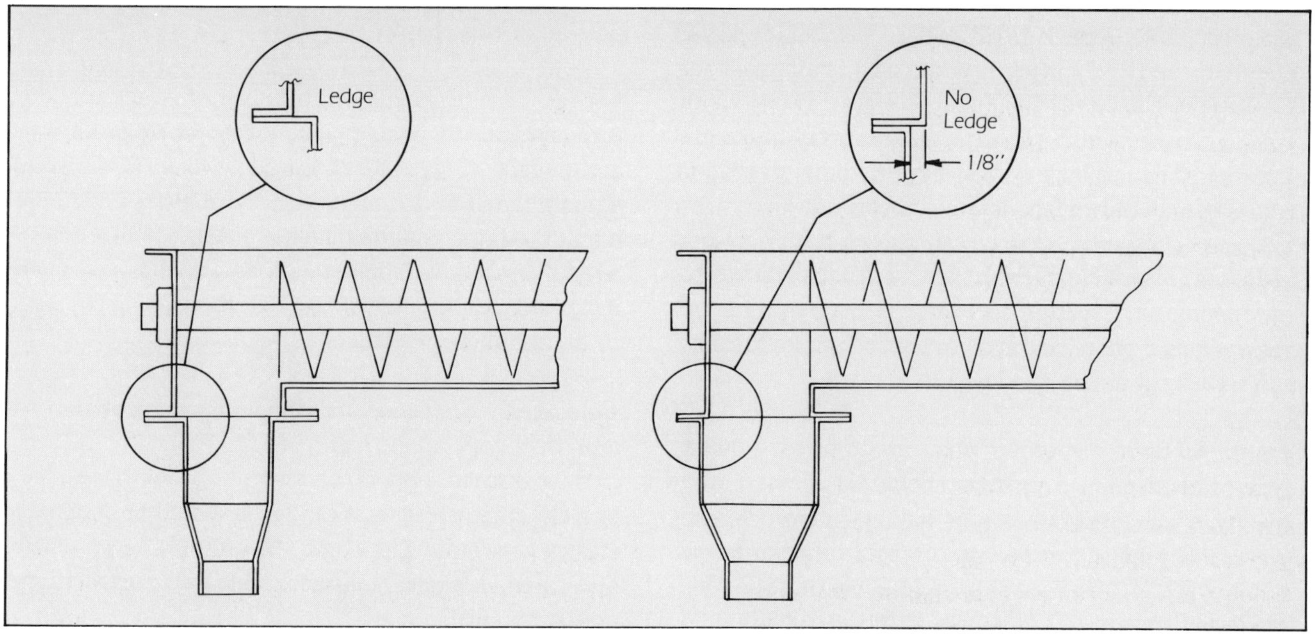

considered where applicable. Vinyl makes cloth impervious to moisture and tends to extend its life. Since most fabrics wear quickly, they should be installed so they can be easily inspected for wear and damage.

Teflon. Teflon is a trade name for a waxy, opaque material—polytetrafluoroethylene, which is used as a coating material to prevent sticking. Only food-grade Teflon should be used. It can be used to coat equipment surfaces to make them smoother and to create better flow conditions. Although this is effective, it might not last long. Teflon is soft and does not stand up well to any form of abrasion.

Caulking Compounds. There are many different caulking compounds; acrylics, latex, butyl, and silicone are some of the common ones. Caulking compounds have limited application around the food plant. In general, they must not be used in a product zone under any circumstances, and should be used cautiously on nonproduct zones which are in close proximity to open product zones. Many caulking materials will dry out or loosen with age and sooner or later fall out. Caulking materials can be considered for use between two bolting flanges where one end of each flange might lie in the product zone, providing that all the excess or squeezed-out caulking material is completely wiped off the product zone surface. However, this application should be limited to specific situations where welding of this seam is not possible because of dissimilar metal flanges, i.e. aluminum and steel and where gaskets have been ruled out. Gaskets can bulge and buckle under pressure. This situation creates crevices which can attract and support insect life. Because of their pliable consistency, the butyl or silicone caulking materials might suit this application best.

Asbestos. Asbestos materials should not be considered for food plant use. Asbestos fibers have been associated with various lung diseases, and in the interest of employee safety, substitutes should be considered.

Paint. All paint or primers which are to be used on any type of food plant equipment should be of food-grade quality, which essentially means that they contain no toxic properties. Many epoxy base paints meet this requirement. Paint which is to be used in product zones must be food-grade and approved for that use. Paints used on building surfaces should be free of heavy metals and other toxic substances.

CONSTRUCTION FEATURES

Construction features can be broken down further to include fabrication details, assembly details, and paint finishes. It is hoped that this summary of construction features, while not all-inclusive, will be helpful. As previously mentioned, there are many pieces of equipment which manufacture hundreds of different food products; each piece has specific design needs, but each also requires certain fundamental construction features.

FABRICATION DETAIL

Lap Seams. A lap seam is produced when one piece of metal is placed or lapped over another. *(Figure 4-11)* Lap seams are normally spot- or tack-welded, which leaves a tiny crack and a crevice. Insects and bacteria can harbor in the crack and the product easily hangs up in the crevice. Even where lap seams are continuously welded they are not necessarily satisfactory, as the product tends to hang or compact in the weld area.

Metal ends should be butted together and continuously welded. The weld should be ground. Welds and weld finishing will be discussed further in the assembly portion of this chapter.

Ledges. One of the major objectives of a sanitary design is to eliminate all contact zone ledges where product can accumulate. *(Figure 4-12)* Product which is allowed to accumulate or lie dormant attracts problems and in itself might spoil and contaminate the product in process. It is not only essential to eliminate product zone ledges, but the design engineer should strive to reduce or minimize exterior or nonproduct zone ledges as well.

Void Areas. Void areas are found adjacent to or alongside a product zone area. *(Figure 4-13)* Frequently the access to these areas is not good and various problems result due to a cleaning oversight. Void areas should be eliminated. If they cannot be eliminated, they should be completely sealed and/or made totally accessible by the installation of doors, removable panels, or the like.

Dead Ends. Dead ends are normally associated with product zones. They consist of dead-ended pipes which form pockets for product to flow into and become trapped, or in the case of screw conveyor troughs, they dead-end upstream of the product inlets or downstream from the discharges. *(Figure 4-14)* Stock accumulating in these dead-ended areas is certain to cause food safety problems.

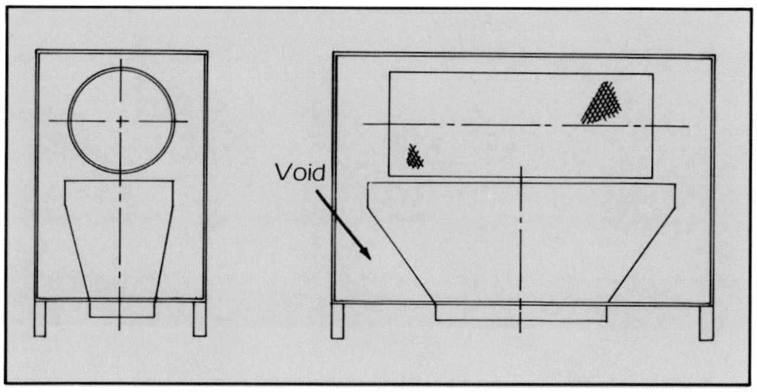

FIG. 4-13 Void areas are often unknown until a food safety problem surfaces such as an insect infestation or a food spoilage problem.

FIG. 4-14 Products can easily accumulate in dead-end areas.

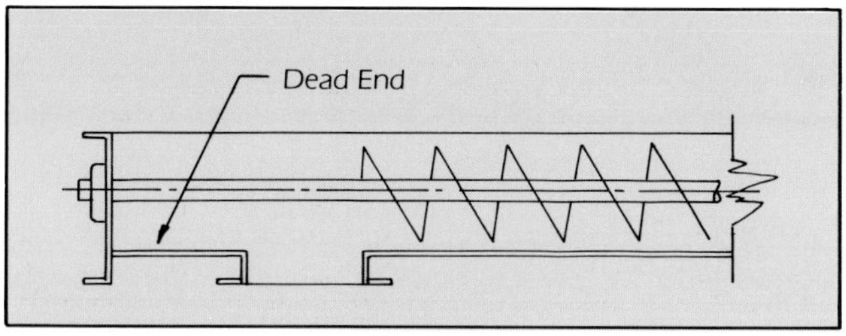

FIG. 4-15 Rolled edges should be sealed.

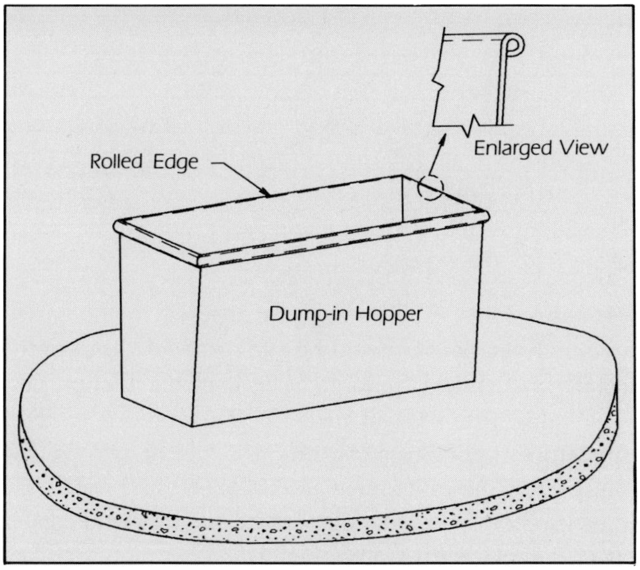

Rolled Edges. Often it is necessary to strengthen a sheet metal transition or hopper by forming a ridge or a roll in it. Rolled edges which are not sealed produce natural harbors for static stocks, insects, and bacteria. *(Figure 4-15)* Rolled edges should be sealed by welding, or the roll should not exceed 180° (half roll).

Rounded Corners. All product zone corners should be rounded. *(Figure 4-16)* The radius should be one half inch to one inch. Square corners cause the product to hang up and they are very difficult to clean.

Cove or Fillets. Corner welds or junctions should be coved or filleted. Coved or filleted surfaces should be ground smooth. *(Figure 4-17)*

FIG. 4-16 Rounded corners are easier to clean.

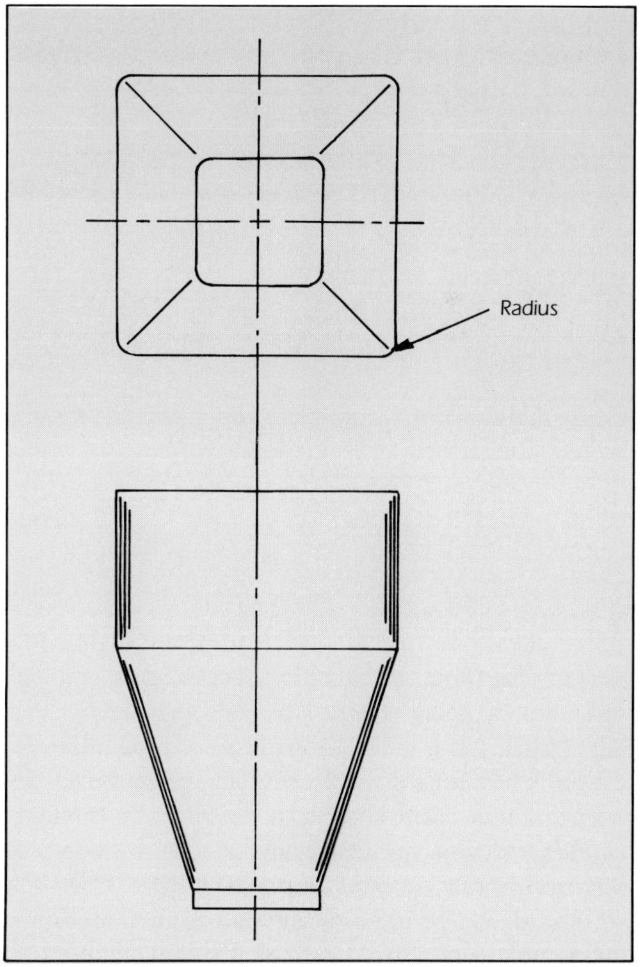

FIG. 4-17 Corner welds should be coved.

Pittsburgh Seams. Pittsburgh and similar sheet metal seams should be limited to ventilation ductwork. *(Figure 4-18)* Process ductwork should be continuously welded.

Cracks and Crevices. In product zones, cracks and crevices should be eliminated by continuously welding, soldering, or brazing. In nonproduct zones, caulking material may be used where it is needed to seal these areas. Caulking material must not be used in the product zones and should be used sparingly in nonproduct zones, particularly when in close proximity to a product zone.

Smooth Surfaces. Product zone surfaces should be smooth, free of pits, checks, scale, or rough deposits that might catch and trap food ingredients or food products. Cast iron product zone surfaces should be machined to a smooth finish. Rough surfaces are difficult to clean. Nonproduct zone surfaces should also be as smooth as possible, as this enhances the cleaning task. Cast iron surfaces should be free of blow holes, sand impressions, and be no rougher than American Standard number 125.

Framing. The sanitary design of equipment requires that frames have as few ledges as possible. Framing should include the use of round, tubular shapes to minimize flat surfaces where product can accumulate. In addition, a great deal can be done to eliminate ledges or minimize flat surfaces by the arrangement of the various structural shapes. When framing includes legs or floor supports, any and all horizontal framing members should be located a minimum of 12 inches from the floor. *(Figure 4-19)* In the case of equipment which will occupy large floor areas, the horizontal frame member might have to be 18 inches or more from the floor.

Mounting Pads. There are several different methods used to design equipment mounting pads or feet. Although the details of this might appear to be insignificant, they are not. The equipment used in most food operations requires numerous mounting pads, not just one or two. In the illustration, one mounting pad eliminates the crevice at the pipe leg/floor pad junction because a sheet metal cone is welded in place after mounting. The other mounting pad does not have this feature. *(Figure 4-20)* An alternate method for mounting pads when using tubular legs is also illustrated. *(Figure 4-21)*

Leveling Pads. Leveling pads should be free of as many exposed threads as possible. Threads are both difficult and hazardous to clean. Threads can be minimized by threading only a portion of the male adjusting bolt, allowing for a quarter to a half inch in elevation. If additional adjustment is required, this could be accomplished with full-sized shim plates or concrete grout.

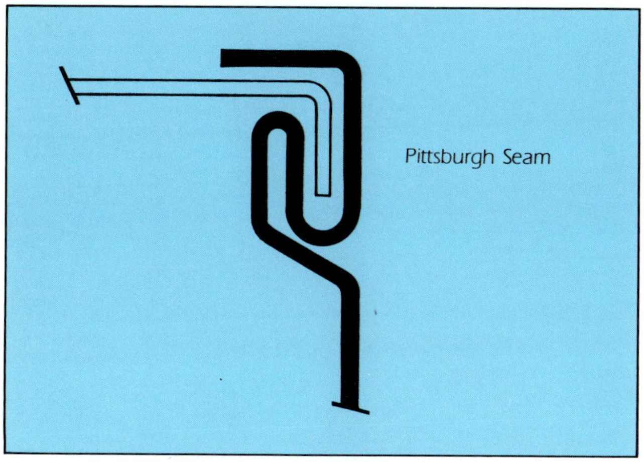

FIG. 4-18 Pittsburgh seams are not suitable for process ducts or spouts.

FIG. 4-19 Horizontal frame members should be raised 12 inches or more from the floor.

Cabinets or Enclosures. Cabinets or equipment enclosures should be constructed of materials heavy enough so they will not require reinforcement. If reinforcement is necessary, enclosure panels should feature sheet metal formed breaks or ribbing, arranged so that they do not trap material or become difficult to clean. *(Figure 4-22)*

Motor Mounts. Using solid plates for motor mounts results in extra flat surfaces that will require cleaning. *(Figure 4-23)* Angles can be used for this purpose just as easily and they reduce the cleaning surface accordingly. *(Figure 4-24)*

Motor Take-up Rails. Motor take-up rails can create void areas that are very difficult to clean. This occurs when the rails are fastened directly to a flat surface such as a solid plate or floor. *(Figure 4-25)* To eliminate this problem, take-up rails should be mounted on angles. *(Figure 4-26)* Take-ups can also be fabricated by placing two pairs of angles back to back. *(Figure 4-27)*

FIG. 4-20 Equipment mounting pads

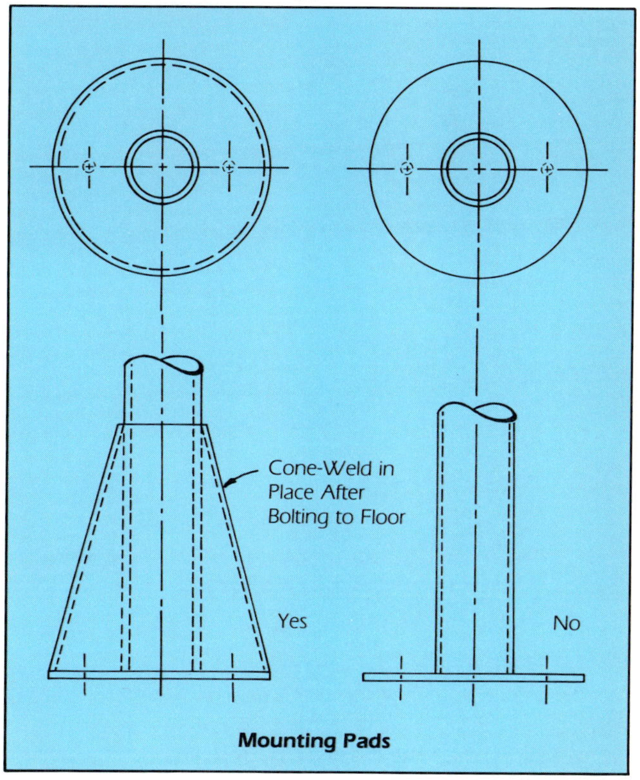

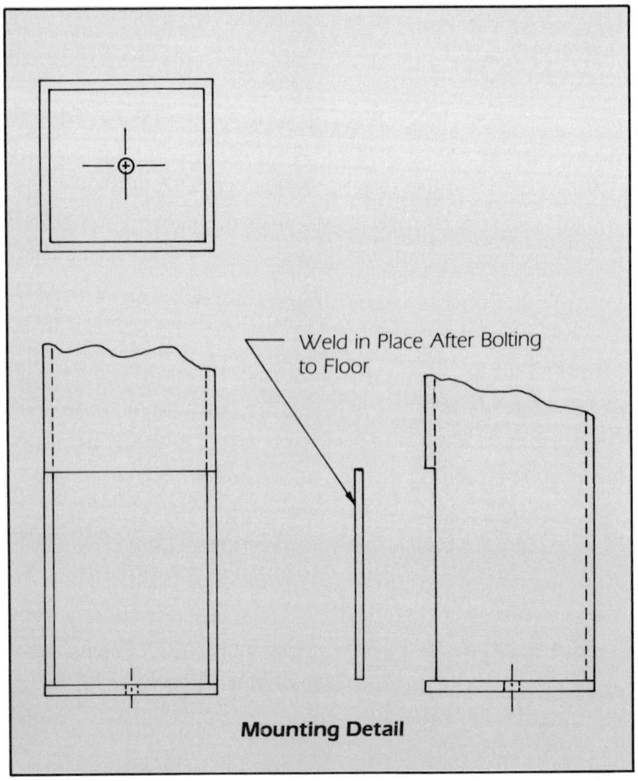

FIG. 4-21 Alternate mounting pad for tubing legs

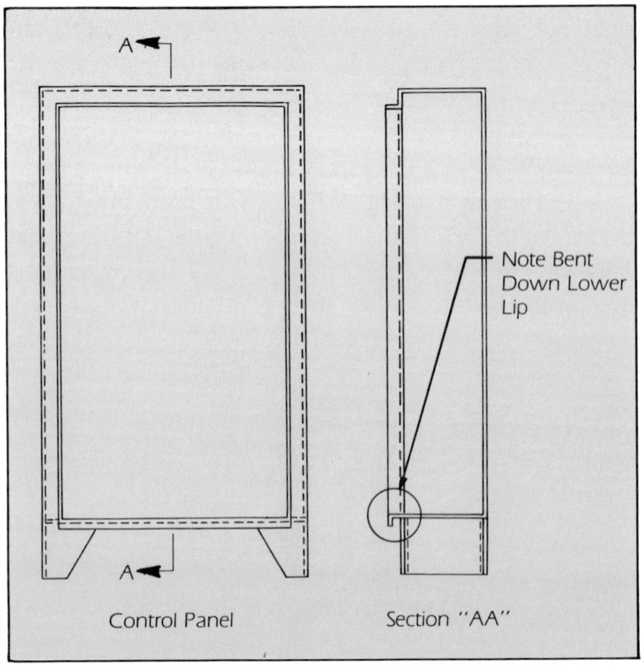

FIG. 4-22 Equipment enclosures and cabinets should also be designed for cleaning ease.

FIG. 4-23 Flat plate motor mounts are more difficult to clean.

FIG. 4-24 Angle iron motor mounts can facilitate cleaning.

Drive Guards. Drive guards should be constructed with semi-open bottoms. Solid bottoms collect dust and dirt and are difficult to clean. Semi-open bottoms should be constructed from 5/16-inch round mild steel bars on approximately 3/4-inch to one-inch centers. *(Figure 4-28)* The distance between the bars should be determined by the distance between the nip point and the bottom of the guard. The objective is to prevent an injury. The employee's fingers should not reach the nip point. Expanded metal, which is sometimes used to form semi-open bottoms, is difficult and sometimes hazardous to clean. Occasionally expanded metal sheets come through with razor-like burrs that can easily cut fingers or hands.

Drive Coupling Guards. Most direct drive coupling guards are fabricated from solid sheet metal in an inverted U shape and are mounted directly on the base plate, creating a void area which will be very difficult to clean. U-shaped drive coupling guards which are mounted 2 inches up from the base plate are more easily cleaned. *(Figure 4-29)* It is sometimes thought that drive coupling guards which are

FIG. 4-25 When mounted on flat plates, conventional motor take-ups create void areas which are very hard to clean.

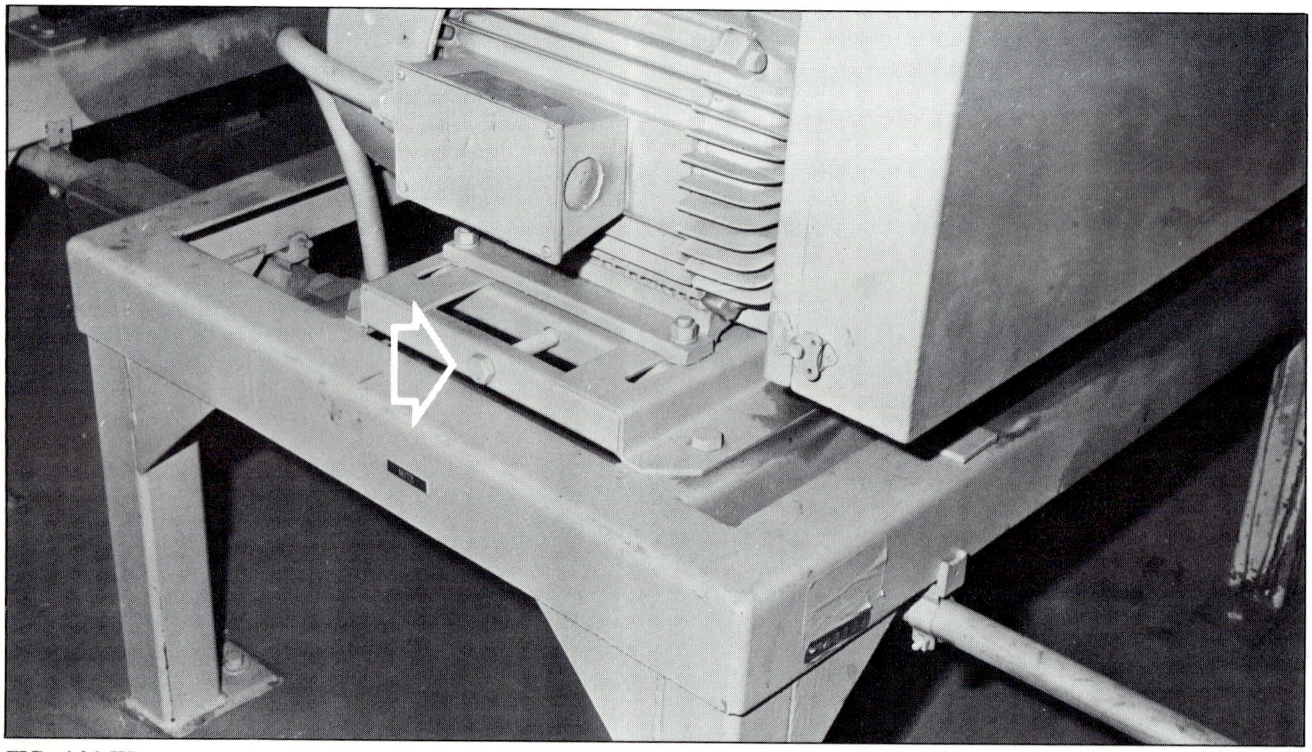

FIG. 4-26 When motor take-ups are mounted on angle irons, the void areas are virtually eliminated.

FIG. 4-27 Motor rails designed with back-to-back angle irons are also a satisfactory sanitary design.

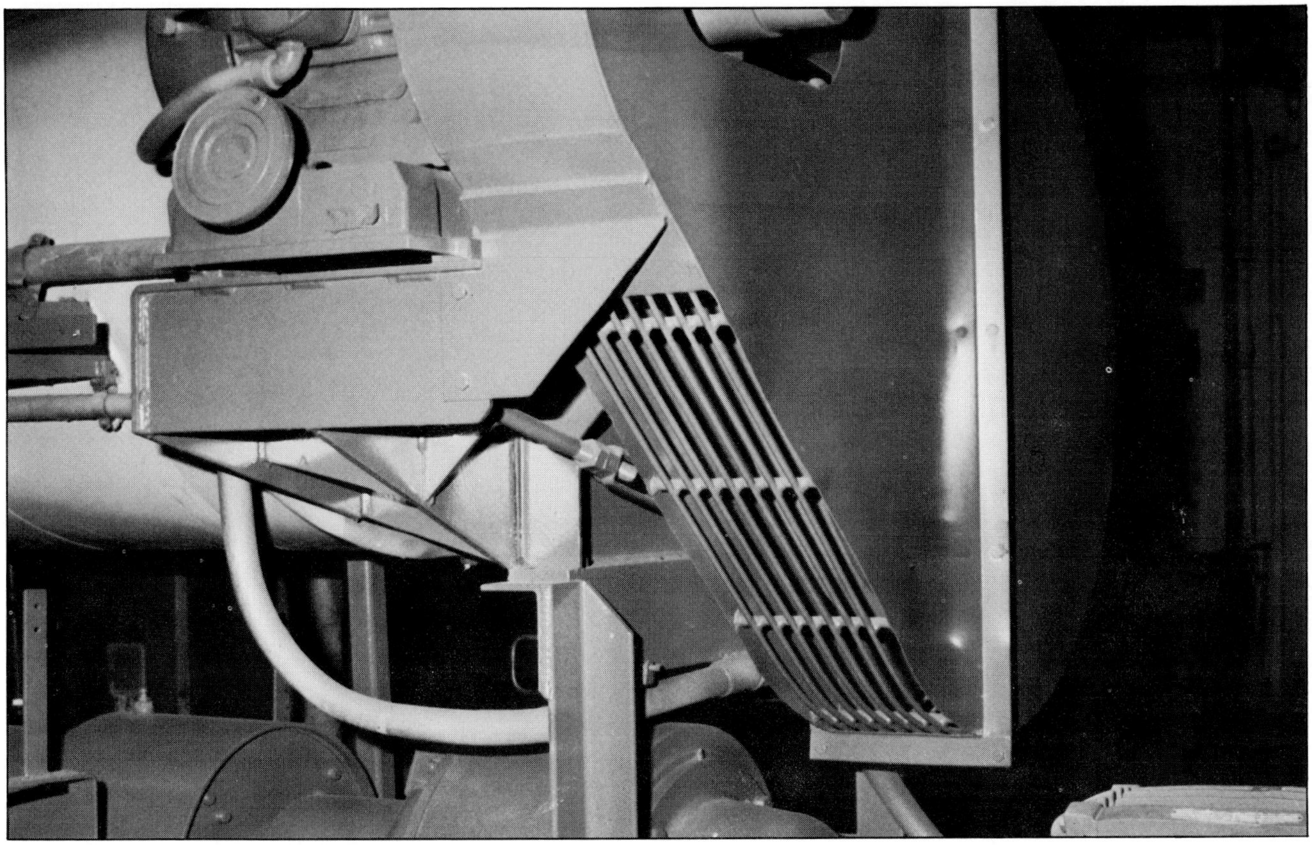

FIG. 4-28 A good sanitary drive guard feature has a semi-open bottom constructed of round steel rods.

FIG. 4-29 U-shaped drive coupling guard

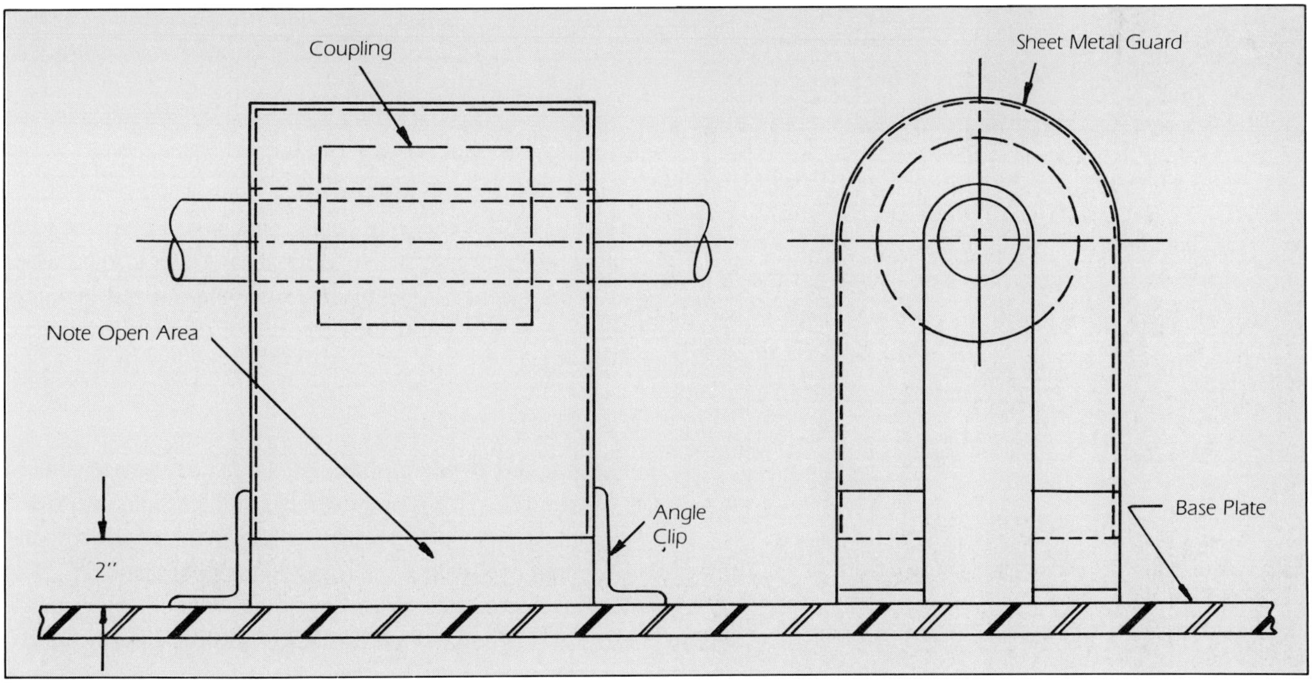

constructed from expanded or perforated metals can be easily cleaned, even when they are mounted directly on the base plate. This is not the case. Expanded or perforated metals must also be raised up from the base plate.

Pulleys. Belt or bucket elevator head and tail pulleys should not be of the deep well type. Shallow well or flush ends are preferred. *(Figure 4-30)*

Belt Conveyor Take-ups. Many of the standard or off-the-shelf belt conveyor take-up assemblies are far from sanitary,

FIG. 4-30 Belt or bucket elevator pulleys

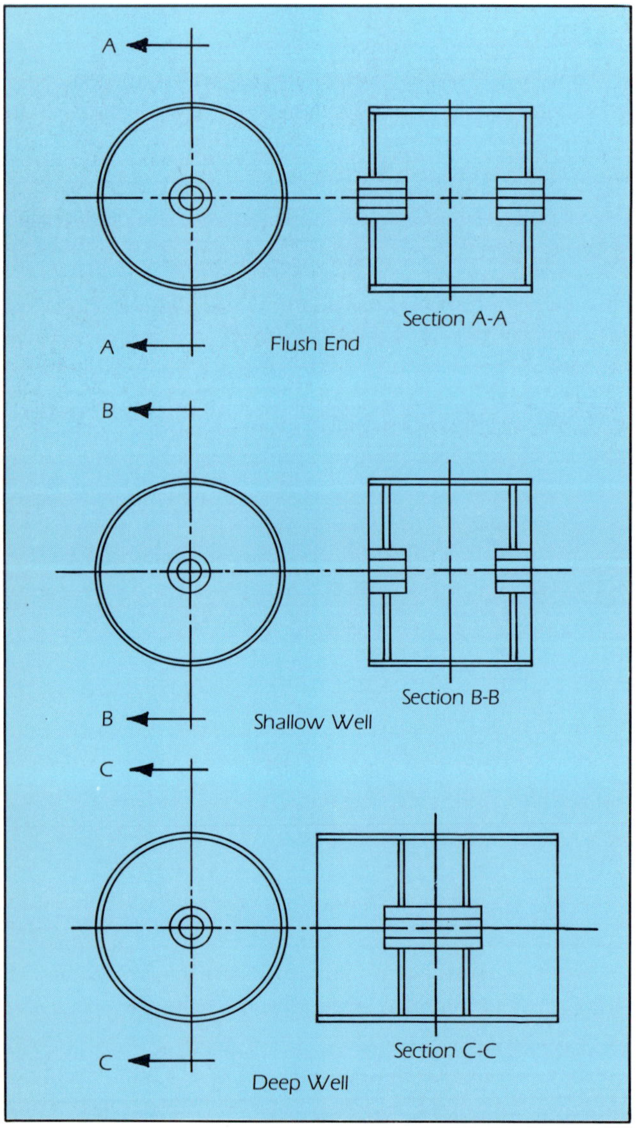

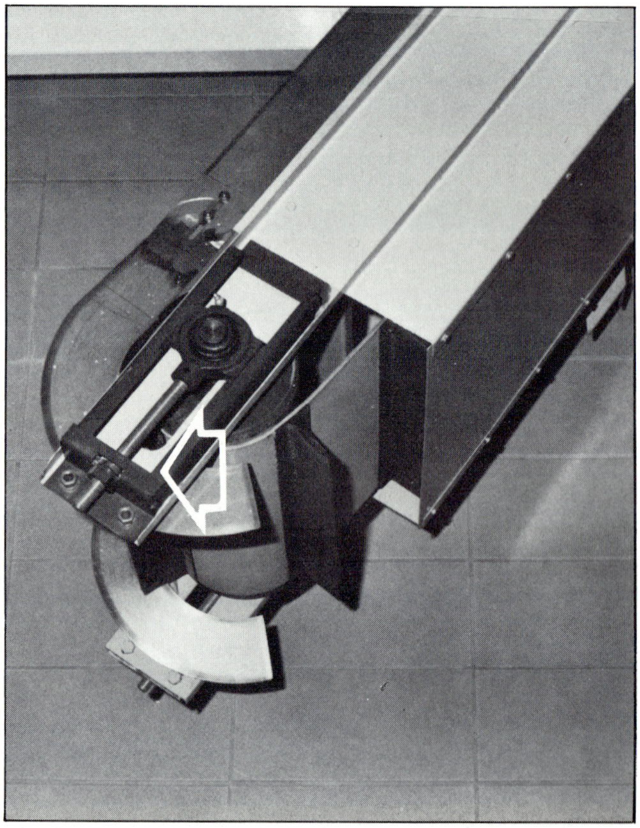

FIG.. 4-31 Standard belt conveyor take-up

having many crevices and void areas to trap material. *(Figure 4-31)* Take-up assemblies should be as crevice-free as possible. *(Figure 4-32)*

Endless Belt/Belt Lacing. Endless or laminated conveyor belt splices are still preferred for many food plant applications. However, metal lacing does have a place, so long as the potential product contamination is kept in mind when selecting applications. In all cases where metal lacing is selected, it is a good idea to choose one with magnetic properties.

ASSEMBLY DETAIL

Welding and Weld Finishes. All product zone welds should be continuous to eliminate objectionable cracks. All welds should be chipped and brushed. The extent to which a weld is ground or finished will depend on the product risk. The flow characteristics of some products which hang up easily may influence how the weld should be finished. Unfortunately there are no universal standards that call for certain

FIG. 4-32 Crevice-free belt take-up

weld finishes for certain conditions. This is compounded by the fact that grinding and polishing welds is expensive. Furthermore, it is hard to obtain a meaningful fabrication bid unless there is some understanding of what is meant by a narrative specification. The specification is likely to have a different meaning to each bidder. For that reason it is a good idea to ask each fabricator to submit weld samples showing several different grinding and polishing degrees. *(Figure 4-33)* In this way you will know exactly what to expect from a fabricator.

As mentioned in the preceding paragraph, the product risk or flow characteristics of the product will determine the extent of the finish; for example: for a product with a high degree of microbial risk such as milk or eggs, the welds should be ground flush, blended and polished to match the existing surface finish. When done correctly, it is extremely difficult to find the weld.

For a product with a high degree of insect infestation risk, such as solid or granular products, welds should be ground flush. The objective here is to eliminate the herringbone bead crevices of the weld. A product readily collects at this point, and insects could lie in the crevice or product accumulation.

Many products or ingredients do not present a serious product safety risk and present few, if any, flow problems. In such situations it might be necessary only to chip and brush the weld (depending on its quality) or perhaps it will just require being ground flat, not flush. Spot or stitch welds are not acceptable for product contact surfaces for obvious reasons. These welds should be restricted to nonproduct zone areas. But even in these cases stitch welds should be chipped and brushed to remove the weld scale and slag. In some cases it may be a good idea to eliminate the cracks between welds by caulking them with a good pliable caulking material, but only where necessary. Excessive caulking material is a contamination concern also.

Weld Studs. Weld studs are an excellent means of eliminating the need for nuts, bolts, or screws in the product zone. *(Figure 4-34)* One of the basic design objectives should be to eliminate all fasteners or other devices from the product zone. By doing so, the contamination threat posed by these devices coming loose is eliminated. Also eliminated are the many cracks and crevices created by the fastener heads.

FIG. 4-33 Weld samples; weld standards can have different meanings to different people.

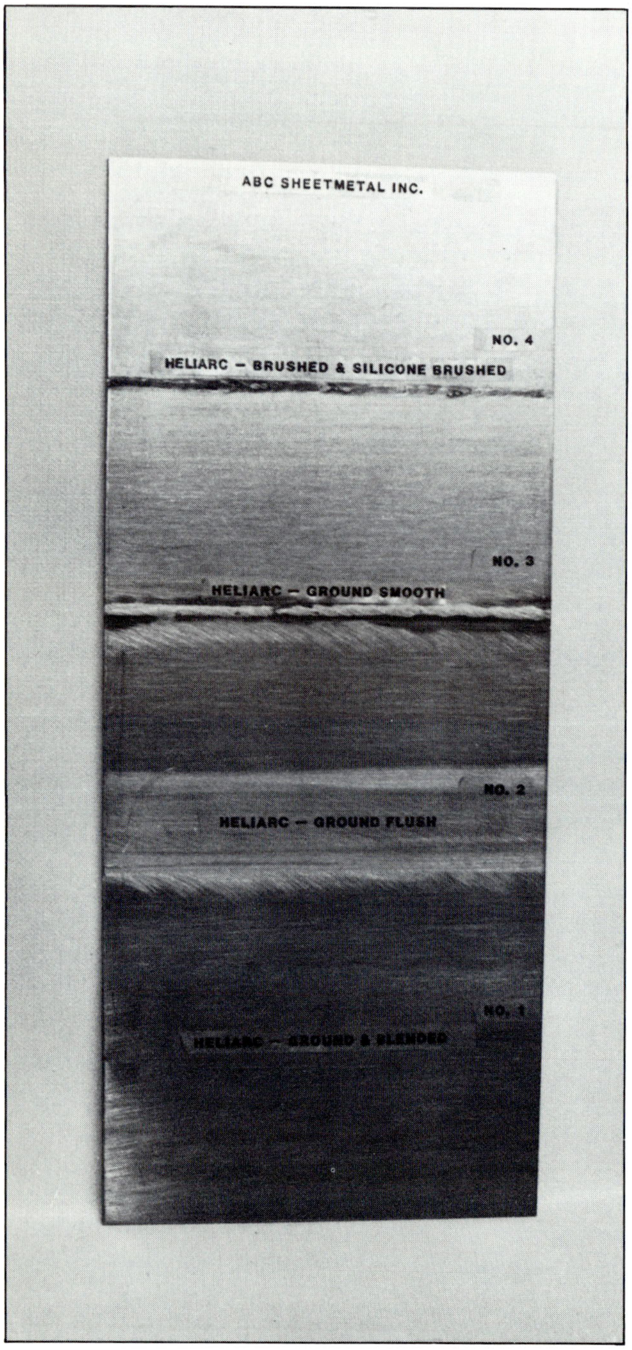

Nuts and Bolts. Despite efforts to eliminate nuts, bolts, and machine screws, there will always be some need for them. When they are needed, those having a magnetic property should be selected. In this way, if they do accidentally become loose and fall in the product, downstream magnets might remove them.

Lockwashers and Locknuts. Magnetic lockwashers of the split or tooth type should be used on all bolting applications in product zones or in close proximity to product zones. Where possible, the bolting arrangement should locate the washer and nut outside the product zone. Locknuts can be considered an alternative to lockwashers, but must be used within the recommended temperature limits. Locknuts of the elastic type (usually featuring a collar of nylon or similar material) will normally have temperature limits between 100°F to 400°F, with many having a 250°F maximum. The elastic material used as the locking element is normally nylon or a similar material that must be approved for food plant use. For higher temperature applications, locknuts of the split or tooth type should be considered.

In many cases, corrosion-resistant fasteners will be needed. When this is required, series 400 stainless steel or plated mild steel should be specified. Series 400 stainless is magnetic. Cadmium-plated fasteners should not be used.

Sheet Metal/Self-tapping Screws. Sheet metal and self-tapping screws should not be used in food contact zones. These fasteners are easily worked loose by vibration. If, for some reason, there is no choice but to use sheet metal or self-tapping screws, they should be of a metal of a magnetic type and tack-welded in place.

Rivet/Solder. Riveting and soldering as an assembly method is not widely accepted today. If it is necessary, only standard rivets should be considered. Solder should not contain more than 5% lead. Pop rivets are frequently over-stressed and should not be used.

Quick-release Fasteners. Quick-release fasteners do not require the use of a tool to tighten or loosen them. Many

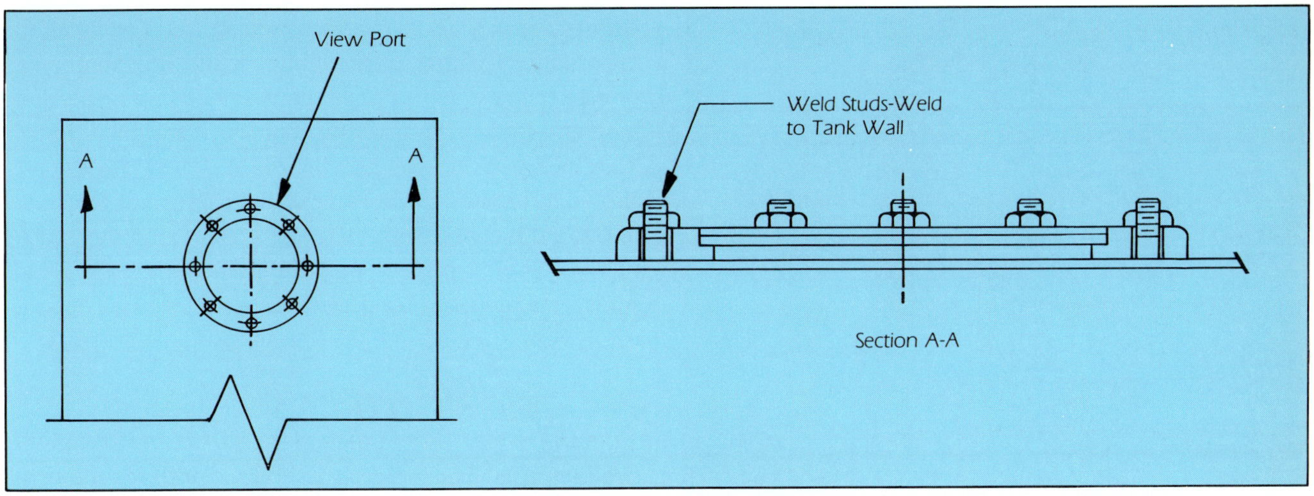

FIG. 4-34 Weld studs can be used to eliminate nuts and bolts from the product zone.

such fasteners are available such as thumb screws, knurled knobs, quarter turns, clamps, and many varieties of wedging lever units, to identify just a few. Quick-release fasteners are important to both the manufacturing process and the sanitation program. Durability and strength should be the prime requisites when selecting these fasteners. Weak or undersized fasteners fail or break easily; the results are product leaks and subsequent replacement with standard nut and bolt fasteners. The easy access feature is thus eliminated.

Hinges. Hinges should be the take-apart type so they can be cleaned. Hinges selected should be strong enough to withstand normal stress which can cause the hinge to be sprung or distorted. Continuous hinges are preferred.

PAINT FINISHES

As with other basic equipment construction materials, paint must be selected with product safety in mind. However, there is a significant difference. Paint is substantially softer and exceptionally sensitive to surface preparation and proper application. Surfaces must be properly prepared, including sandblasting when necessary to obtain satisfactory results. Paint does not stand up well to adversities, particularly when it has not been properly selected or when the surface has been improperly prepared. This results in chipping and peeling paint. The problem is compounded if the paint contains toxic substances such as lead. Only lead-free paint should be considered for use on food manufacturing equipment.

Paint. Only paint suited for incidental contact with food or classified as food grade should be used on product zone surfaces. Several epoxy-based paints are considered food grade. Food-grade epoxy enamels for nonfood contact zones are also worthwhile, although not absolutely necessary. Other epoxy-based paints, or paints that do not contain toxic components are acceptable for nonfood contact zones.

Paint should be selected with the application in mind. If it is to be used in a product zone it must stand up to the characteristics of the food. Generally, paint does not stand up well to abrasive materials and to many liquids. Moist, steamy, salty, or acid atmospheres call for very special corrosion-resistant paint. Paint must stand up to cleaning chemicals and cleaning methods as well as pesticides and sanitizing solutions. Many modern cleaning systems use high pressure/low volume hot water pumping equipment. Pressure to 750 psi or better is often produced at the nozzle of this equipment and offers a challenge to the very best painted surface. Paint manufacturers should be well informed of special needs so that proper materials can be recommended.

In some instances the actual food contact surface itself should not be painted as it would not stand up well to the product conditions, but the area directly above or adjacent to this area should be painted to prevent corrosion. *(Figure 4-35)* Although not in direct contact with food, this area is considered to be part of the product zone, thus requiring food-grade paint.

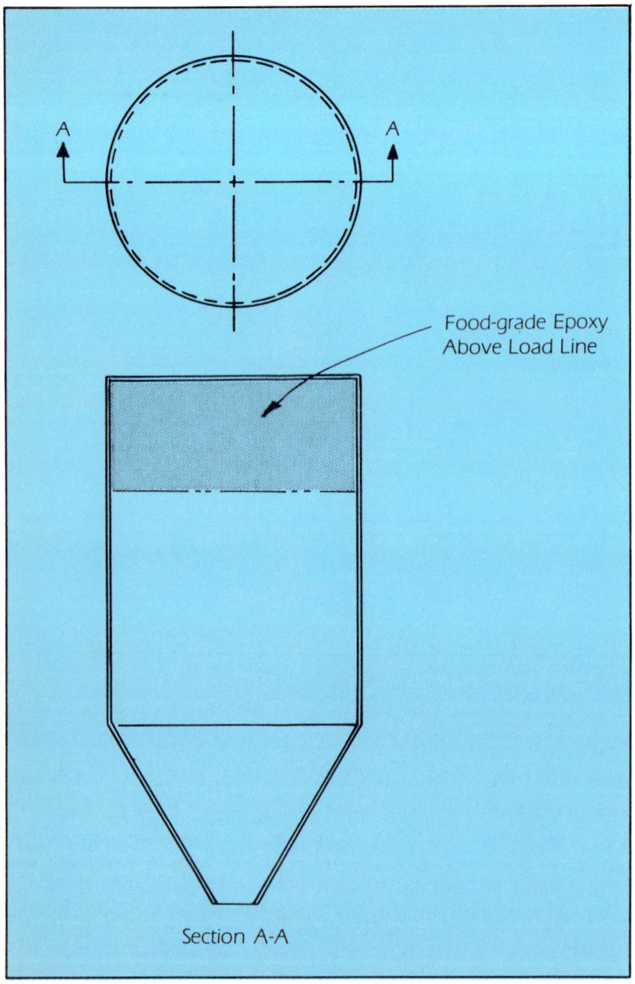

FIG. 4-35 When applied in mild steel storage or use bins above the load lines, food-grade epoxy paint can prevent corrosion problems.

Surface Preparation. As stressed in Chapter Three, surface preparation is the key to satisfactory results with paint. Certain metal surfaces and specific applications will require different surface preparation. In almost all cases product zone applications on mild steel surfaces will call for a recommendation to sandblast. Most food-grade epoxy paints stipulate that the surface should be sandblasted.

At the very least, nonproduct-zone surfaces will require removal of oil, grease, wax, chemicals, rust, and mill scale with solvents and brushes. Actual preparation will be directly linked to the condition of the steel prior to application. Special metal surfaces such as aluminum, copper, and galvanized iron will require special cleaning techniques, according to the paint manufacturer's instructions, to assure adhesion.

Primers. Just as proper surface preparation is essential, so are primers. Finish coats should not be substituted for base or primer coats. Primer paints contain bonding agents that have an adhesive characteristic that provides a good, stable bond between the steel and the finish coat. Primer coats should be applied on the same day the surface is prepared to prevent recontamination of the surface with dust, dirt, or rust. Special primers must be used for special metal surfaces (aluminum, copper, stainless steel, and galvanized iron). Primer paints should contain no toxic properties.

PROCESS CONTROLS

Product safety is not merely concerned with maintaining equipment and utensils in a sanitary condition through a well-designed and implemented sanitation program (including good sanitary design features from the beginning). It is also concerned with the product itself as it is in process. Many foods support the rapid development of bacteria. Such bacteria can be a public health concern in that they could result in decomposition of the food or could even result in one of the food-borne illnesses discussed in Chapter One. Microbiologically sensitive conditions will require the monitoring of such physical factors as time, temperature, moisture levels, pressures, and flow rates of the food product in its various process phases. Such monitoring is particularly necessary for processes in which an equipment malfunction could result in health problems. These process phases would include, but are not limited to, blanching, cooking, pasteurizing, dehydration, drying, toasting, acidification, refrigerating, and freezing.

ENGINEERING THE PROCESS CONTROL SYSTEM

The HACCP (Hazard Analysis and Critical Control Point) system mentioned in Chapter One can be very useful in engineering the process control system. As explained in Chapter One, the HACCP system involves a detailed analysis of the process flow for potential food contamination hazards and critical control points. When engineering the control system, this same HACCP system can be used to identify the critical areas of the process, and the appropriate controls, including alarms and recording devices, can be selected accordingly.

Control Devices. Depending on the complexities and length of the process flows, consideration should be given to a central control room where all physical factors of the process can be monitored. *(Figure 4-36)* For those factors more critical in nature, it might also be advisable to provide direct readout indicators at the equipment itself. In some cases, alarms, recording devices, and automatic shutdown features should be considered. Some equipment that requires remote readout units are pasteurizers, sterilizers, cookers, dryers, refrigerators, and freezers.

Many modern food processes are fully automated and certainly that is the direction of food plants of tomorrow. Automation means that controls are no longer used just to monitor; with the aid of a computer they are also used to operate part or all of the process. The need for alarms and record keeping is all the more important. Controls with constant, vigilant recording capabilities are essential. *(Figure 4-37)* Again, in the case of very critical control points, an alarm is a must.

Controls and recording devices should be easy to read, and they should not be a product safety threat. Pressure gauges, thermometers, and flow control gauges should be equipped with plastic faces instead of glass. The control cabinet or enclosure should be dust-tight and built for easy cleaning. If located in a wet-cleaning area, the controls must withstand the chemicals and hot water that will be used.

When the plan is to install the control in the immediate area of a dusty process, the control should be pressurized.

- Fluids. Fluids used in such control devices as thermometers and manometers must be nontoxic. Mercury must not be used, since it is highly toxic. Oils such as those found in

FIG. 4-36 A central control room allows a central point for monitoring all critical process factors.

FIG. 4-37 Temperature recorders offer continuous temperature monitoring of the processes.

dash pots on some gravimetric-type weighing devices must be approved for use in food plants.

- Air-operated Controls. Compressed air provided to air-operated controls should be furnished from a nonlubricated compressor. Airborne lubricants represent a contamination risk by virtue of their proximity to the food process; they also represent an operational and maintenance threat to the control itself, which is often sensitive to oil.
- Mounting. Difficult-to-clean junctions, crevices, and void areas must not be created when mounting controls, particularly in manufacturing zones. Ledges where dust can accumulate should be minimized. Mounting brackets should be constructed of materials compatible with the cleaning plan.

Control Room. Control rooms should be designed with easy, cleanable sanitation features in mind, as outlined in Chapter Three. In addition, control rooms should be pressurized with clean, filtered, and conditioned air. The room should be well lighted to accommodate easy readout of the various controls.

Although the control room is not classified as a wet-process area, it should include such easy-to-clean features as smooth walls and ceilings and coved floor/wall junctions. Flooring materials should be easy to clean and comfortable for walking and standing. Concrete in combination with resilient floor overlays is adequate.

CONVEYING SYSTEMS

Although there are many dissimilarities in food manufacturing equipment from one segment of the industry to another, there are enough similarities in conveying equipment for us to identify some of the key requirements of the sanitary design. There will, of course, be exceptions, and the equipment design should allow for such exceptions.

The common conveyance methods used by the industry are:

- Belt conveyors
- Screw conveyors
- Bucket elevators
- Bucket lifts
- Pneumatic conveyors

Each of these conveyors has some very specific sanitary design needs. These needs can best be determined by identifying product risks associated with the equipment. The cleaning methods and materials which are proposed to maintain the equipment at the accepted sanitation level should also be considered.

Conveying systems have one thing in common: their primary purpose is to move product from point 'A' to point 'B' while completely containing the product within the conveying mechanism through the entire moving process. There should be no leaks or spills associated with the equipment. The conveying systems must be built in such a way that they do not harbor pests or trap product or ingredients and they must be built to be cleaned! Lastly they must be constructed and maintained so they in themselves do not contaminate product in process through malfunction of one type or another.

The guidelines found elsewhere in this chapter and the chapters on layout, buildings, and electrical systems will have some application to the sanitary design of the five conveying methods described below. In some cases, these sections and chapters are directly referenced in the guidelines for each conveyor method.

BELT CONVEYORS

Belt conveyors are built in many sizes and shapes. They move food products of various shapes and forms, to various points within the manufacturing process. They also move raw commodities, process ingredients, and packaged finished product. The discussion of the belt conveyor design will concentrate on the movement of bulk raw commodities, bulk ingredients, and bulk (unpackaged) finished product.

Belt conveyors used to move bulk products are primarily trough-type, as opposed to flat-type. Trough type belt conveyors are built to contain the product on the belt. The standard design for troughs is 20° or 45°. They are usually formed by idler pulleys or slider beds. Adjustable trough idler pulleys are also available and could be considered. Trough angle selection should be associated with the flow characteristic of the product.

Belt conveyors that handle dry solids can be built with a mild steel superstructure. When wet materials are used or when wet cleaning of the conveyor superstructure or frame is required, the construction material should be stainless steel or other corrosion-resistant materials.

Slider Beds. Slider beds should be fabricated to form a trough at the desired 20° to 45° angle. Slider beds should be constructed from materials heavy enough so that continuous longitudinal reinforcing members will not be required. To minimize open area where product or dust can accumulate, the width of the slider bed should not exceed the belt width by more than two inches. Bends or breaks placed at the base of the vertical sidewalls of the slider bed should not form a right angle flange; this arrangement forms a flat ledge for product to accumulate. This break should be 45° to the vertical sidewall. Slider beds should be welded or bolted together. *(Figure 4-38)*

Troughed Idler Pulleys. When troughed idlers are used, solid rubber cushions or solid steel idler pulleys with flush end faces are preferred over the ends which are recessed. Recessed pulley ends collect dust and dirt and are difficult to clean.

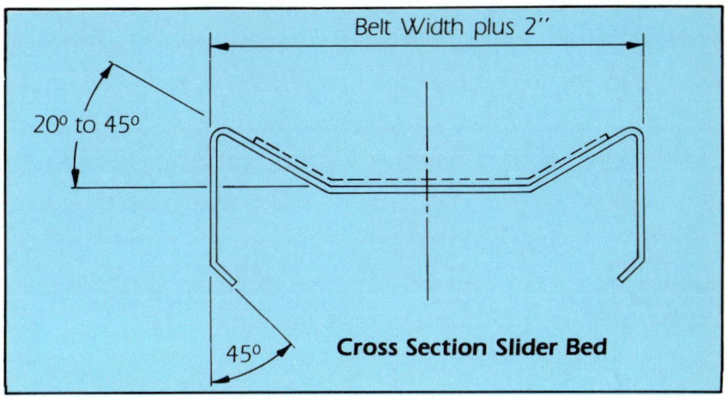

FIG. 4-38 Belt conveyor slider bed

Conveyor Frame. Conveyor frames should be constructed of round tubular shapes to reduce dust collecting surfaces. All open ends of the tubing should be closed. Holes drilled in the tubing should be drilled and tapped to effect a tight seal. Avoid building conveyor frames from channel. Channel has an extra flange to collect dust.

Return Belt Idlers. The sanitary design of return belt idlers should consist of flush or solid end faces, as opposed to recessed faces.

Head/Tail Pulleys. Welded steel head and tail pulleys with shallow wells or flush ends are preferred over the deep well type.

Bearings. Sealed roller or ball bearings should be used. Bearings should be located out of product zones.

Take-ups. Belt conveyor take-ups should be free of void areas in the take-up frame assembly.

Conveyor Belts. Conveyors should be provided with white food-grade rubber or synthetic rubber belts. The belt surface should be smooth and nonabsorbing. Belt ends should be continuously joined by splicing. If metal belt lacing is used, it should have a magnetic property.

Covers. Belt conveyors should be equipped with removable covers. Cover sections should be no longer than three to four feet, so they can be easily removed for cleaning.

Drives/Drive Guards. Drives should not be located directly over belts or adjacent product zones. Drive mounts should be designed to minimize ledges as outlined in this chapter under construction features. Motor take-ups should be arranged so they do not create void pockets between the take-up housing and the mounting surface. Drive guards should have semi-open bottoms similar to those shown under the construction features heading.

Paint. Food-grade epoxy enamel should be used on surfaces requiring paint.

There are many bulk product belt conveyor designs, some of which have good sanitary design features. A typical example is provided. *(Figure 4-39)*

SCREW CONVEYORS

Screw conveyors have a wide number of applications in the food industry. They are used to convey solid materials and can be used as volumetric feeders and as mixing conveyors. Screw conveyors are made in a variety of lengths and sizes. In outward appearance, they are all very similar. Nonetheless, screw augers take on a number of different appearances depending on the primary use. In general, sanitary design requirements are much alike, although allowances must be made for some specific design conditions.

Construction materials will vary with the products being handled. In areas and operations where dry commodities or ingredients are handled and stored, mild steel construction material for all screw conveyor components is adequate. In other instances where conveyors are operating in a wet processing area of the plant, but handling a dry product or ingredient, construction materials might call for stainless steel trough and end plates and mild steel screw flights and shafts. This combination is often preferred because it gives an attractive appearance to equipment. In other situations where the conveyor is handling moist solids or sticky ingredients requiring a wet clean-up, conveyor components made entirely of stainless steel might be necessary.

Troughs. Conveyor troughs should be constructed from materials heavy enough to eliminate the need for angle iron along the top edge of the trough. The top flange should be formed as part of the trough. Where extra strength is needed, a double break should be used to form this flange. The second break should point down in a vertical plane. Trough connecting flanges should be continuously welded in place. *(Figure 4-40)*

Drop Bottoms. Drop bottoms should be fabricated so they do not leak. Doors should be equipped with hinges and heavy-duty quick-release fasteners that do not require the use of tools.

End Plates. Standard screw conveyor end plates should be used. Need should determine the construction materials.

Discharges. Single-screw discharges should be arranged so that no portion of the trough section extends beyond the discharge opening. To meet this requirement, it is necessary to use the end plate as one wall of the discharge transition. Multiple discharges should be equipped with contour-fit slide gates. Standard slide gates used at the base of a discharge transition leave a pocket in the transition for product to accumulate.

Inlets. Screw conveyor inlet points should be located at the farthest position opposite the discharge end. Intermediate location of inlets can lead to product buildups in unused areas of the screw conveyor. These areas frequently go uncleaned.

Screw Flights. Screw flights should be continuously welded on each side of the flight. Welds should be ground to form a small 3/16- to 1/4-inch radius.

Coupling/Drive/End Shafts. Standard coupling, drive, or end shafts should be used. Shafts should be secured to screw conveyor flights with bolts and self-locking hex nuts. No shaft holes should be unused.

Bearings. Exterior conveyor bearings should be sealed flanged bearings or pillow blocks. Thrust bearings should be used where needed to prevent the screw flight from contacting the end plates. When such metal-to-metal contact occurs metal contamination of the product being conveyed results.

Hanger Bearings. Hanger bearings should be made of nylon or polycarbonate thermoplastics. Any bearings using or requiring lubricants should not be used in product zones.

FIG. 4-39 Belt conveyor frames fabricated with pipe or tubing structural members offer sanitation advantages and are easily cleaned.

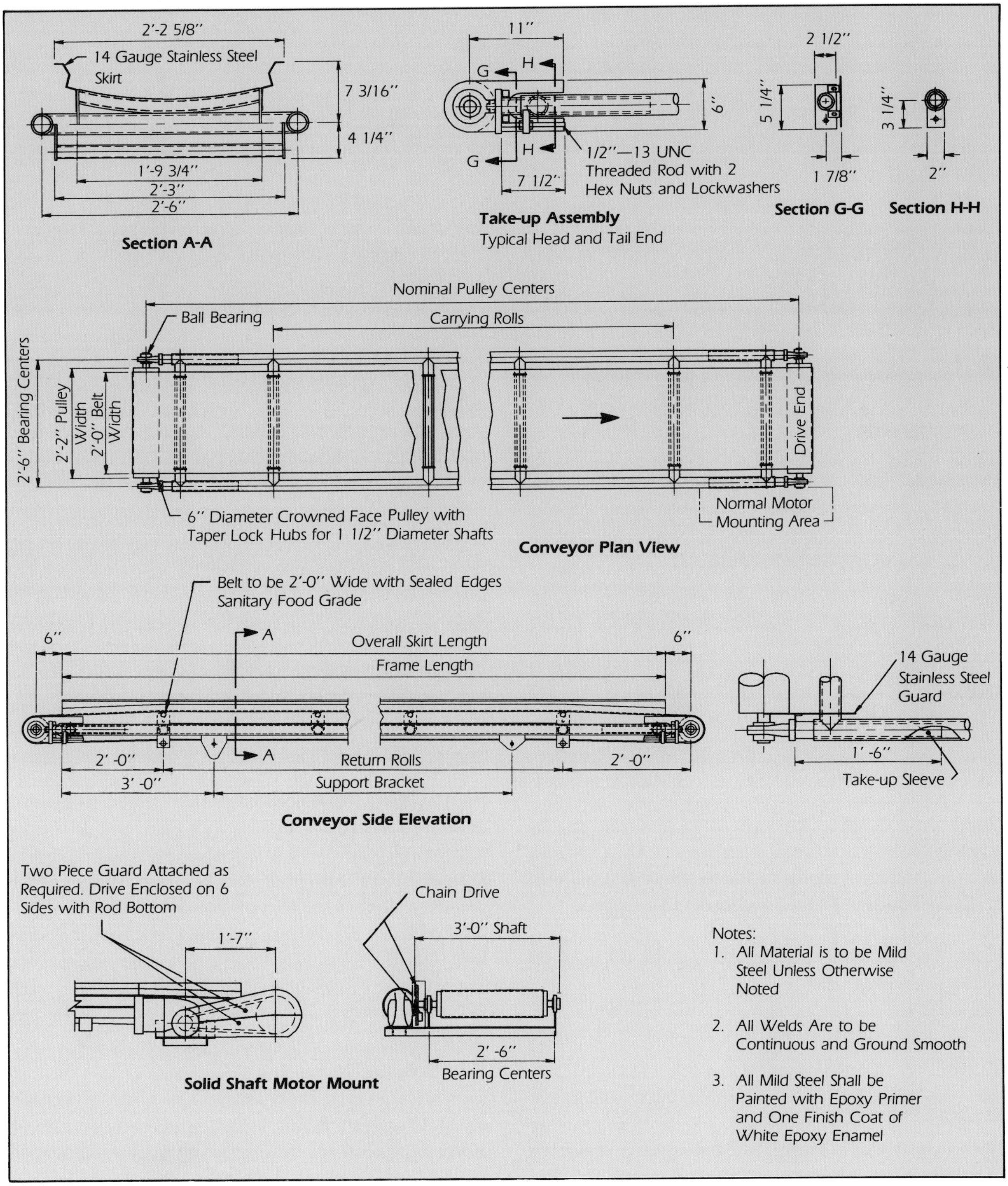

FIG. 4-40 Formed flange screw conveyor trough section

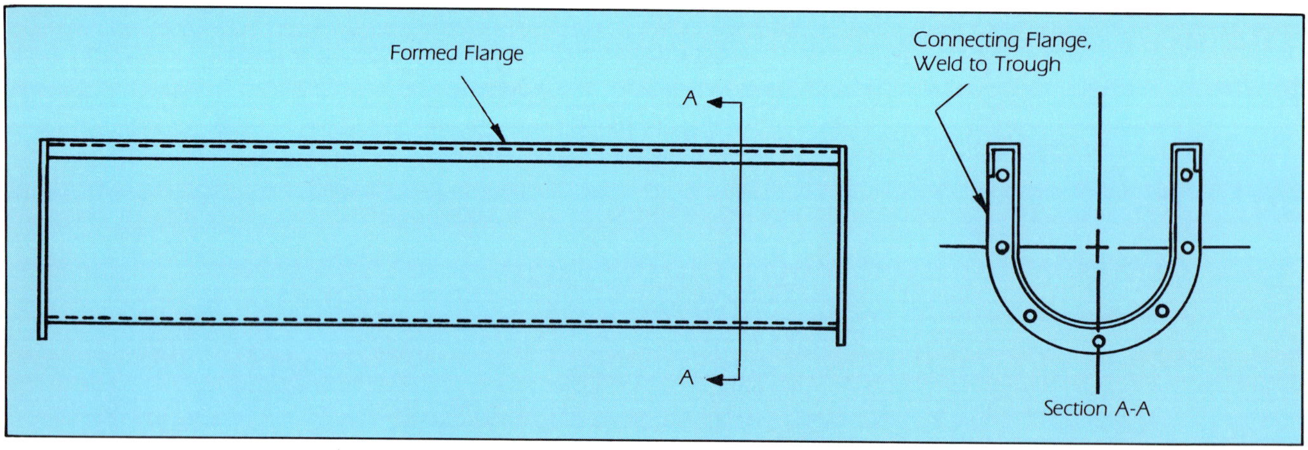

Hanger bearing brackets should be constructed of bar stock. They should not form pockets where food could become trapped.

Shaft Seals. Seals are designed to keep the product from reaching the bearing and bearing lubricants from reaching the product. Lip seals rather than waste packing seals should be used where dust seals are required.

Covers. Removable covers which do not require the use of tools should be provided.

Gaskets. Screw conveyor gaskets should be constructed from food-grade rubber or food-grade synthetic rubber. Sponge rubber should not be used. The gaskets should be nonporous and nonabsorbing.

Drives/Drive Guards. Drives should not be located directly over the conveyor or adjacent product zones. Drive mounts, take-ups, and drive-guards should be designed as outlined in the construction features section of this chapter.

Paint. Food-grade epoxy enamel should be used on surfaces requiring paint.

An illustration of a typical screw conveyor design is provided. *(Figure 4-41)*

BUCKET ELEVATORS

Bucket elevators are capable of handling large quantities of grain and other raw commodities per hour of operation. This equipment can be designed to operate indoors or outdoors. The outdoor type requires special weather-tight design features. Smaller capacity bucket elevators are used to elevate processed ingredients and finished food products in the manufacturing process. This discussion of bucket elevators will concentrate on the smaller size, although many of the design conditions and sanitation requirements are applicable to large bucket elevators.

Bucket elevators designed for handling processed ingredients or finished food products should be of a sanitary design. These units usually feature total access to the bucket and belt area. Stainless steel is used for units which operate in a wet environment or which require wet cleaning because they handle sticky or microbe-sensitive ingredients. Bucket elevators should be designed with dust-tight features.

Boot Sections. Boot sections should be designed with a bottom radius corresponding with the travel arc of the moving buckets. A removable side on the boot housing provides access to the boot. Quick-release fasteners which do not require the use of tools should be installed on this removable side. Construction details of the boot, including inlet transition, should include continuous welding in the contact zones. Welds should be ground flush and corner welds should be ground to form a 3/16- to 1/4-inch corner radius.

Head Sections. Removable side panels on head sections permit total access to the head area. Quick-release fasteners which do not require the use of tools should be used to secure the removable side. Construction details of the head section should feature continuous welds in the contact

FIG. 4-41 Screw conveyor with formed trough

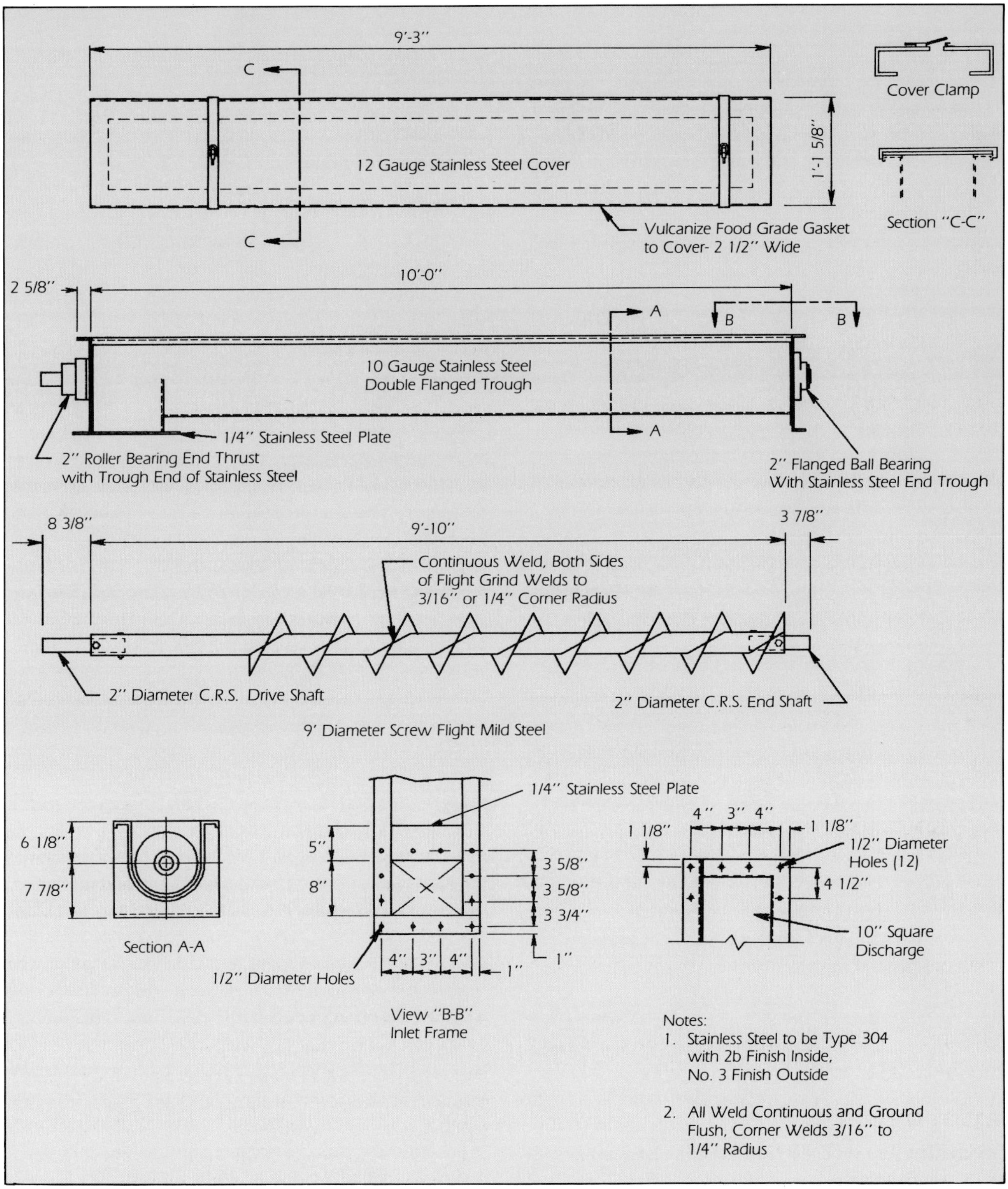

zones, including discharge transition. Welds should be ground flush and corner welds should be ground to form a 3/16- to 1/4-inch corner radius.

Leg Sections. Leg sections for smaller bucket elevators should be constructed as a single leg enclosure. One side of the enclosure should be designed to be removable. Quick-release fasteners which do not require the use of tools should be used to secure the removable side. Welds required in the production zone should be continuous and ground flush. Corner welds should be ground to form a 3/16- to 1/4-inch corner radius.

- Vertical Companion Flanges. Boot, head, and leg sections should be constructed with formed vertical companion flanges on the removable panel side.
- Horizontal Angle Iron Companion Flanges. Horizontal angle iron companion flanges can be used to join elevator sections together. These flanges can be stitch-welded to the leg section.
- Gaskets. Gaskets should be constructed from food-grade rubber or synthetic rubber which is nonporous and nonabsorbent. Sponge rubber should not be used.

Pulleys/Head and Boot Pulleys. Wing-type self-cleaning head and boot pulleys should be considered. If conventional crown-faced drum pulleys are used, the end facing the nonremovable side of the elevator boot and head section housing should be brought out flush with the edge of the pulley face. A recessed pulley end on this side is difficult to clean.

Belt. The bucket belt should be constructed of smooth, light colored, food-grade rubber or synthetic rubber which is nonporous and nonabsorbent.

Buckets and Bucket Bolts. Light colored food-grade polyethylene or seamless steel buckets are recommended. Buckets with lap seams are difficult to clean. Standard elevator bolts having a magnetic property should be used to fasten buckets to the belt. Series 400 stainless steel or plated mild steel should be used when corrosion-resistant bolts are required.

Bearings. Bearings should be mounted on the outside of the leg. Only sealed ball or roller bearings should be used.

Shaft Seals. Mechanical shaft seals should be used. Do not use packing.

Take-ups. Void areas where product can be trapped should be eliminated from pulley take-ups.

Drive/Drive Guard. The drive and drive guard should conform to the description in the construction features section of this chapter.

Paint. Food-grade epoxy enamel should be used on surfaces requiring paint.

CONVEYING ELEVATORS

With conveying elevators the product to be elevated is deposited in the bucket. In contrast, bucket elevators are designed so that the product flows into the boot and is scooped up by a moving bucket. Conveying elevators are used to convey horizontally as well as vertically. This equipment is used on a variety of food products because it conveys and elevates product with little agitation or damage. Bucket lifts are commonly used to move fragile products.

The easy accessibility of buckets, chains, and other components for cleaning enhances the sanitation aspects of this equipment. Although conveying elevators are available without covers, for product safety reasons, covers should be included. The product risk and proposed cleaning plan determines construction materials.

While mild steel satisfies many product needs, the processing of some foods, particularly fruits and vegetables, requires stainless steel or other corrosion-resistant materials.

Frame/Side Panels. Frame and side panels should be constructed with materials appropriate for the food being handled. These components should be made of stainless steel when wet cleaning is anticipated. Horizontal bracing between the side panel should be constructed from round shapes or from angular members with the heel of the angle pointing up. The frame should be free of as many ledges as possible. The entire conveying elevator frame should be designed and built so that it can be exposed for cleaning.

Any required side panel longitudinal flanges should be fabricated as part of the panel wall. Angle iron companion flanges placed where sections are joined should be stitch-welded to the exterior of the side panels. The frame and side panel construction detail should include design features that elevate the bottom horizontal conveying section off the floor a minimum of six to nine inches.

Chain Track and Chain. To minimize product contamination risk from metal to metal contact of moving components, either construct the chain track of hardened steel or equip chains with nylon rollers.

Sprockets. Chain sprockets should be used at each turn. A formed track curve is not recommended as this kind of construction can cause excessive wear at the contact points.

Buckets. Polycarbonate, polystyrene, or stainless steel buckets should be used for most conveying and elevating applications. Buckets with lap seams are considered unsanitary.

Inlet/Outlet Transitions. Inlet and outlet transitions should be constructed with continuous welds. Interior welds should be ground to a smooth finish and corner welds should be ground to form a small 3/16- to 1/4-inch corner radius. Transition construction materials should be similar to those used for the side panel.

Covers. Product zones should be covered to minimize contamination. Both vertical and horizontal sections should be equipped with removable covers. To allow for easy handling, cover sections should be no longer than three to four feet. Covers should be attached with quick-release fasteners that do not require the use of tools.

Bearings. Bearings should be sealed and mounted outside the side panels. Self-aligning bearings should be used where necessary.

Take-ups. Take-ups used to adjust chain tension should be free of void areas where product can be trapped.

Torque Limiters. Torque limiters should be considered to minimize the risk of product contamination caused by jamming buckets.

Drives/Drive Guards. Drives should not be located over a horizontal section of the equipment. Vertical motor mounts should be considered. Drives and drive guards should be designed as suggested in the construction features section of this chapter.

Paint. Food-grade epoxy enamel should be used on surfaces requiring paint.

PNEUMATIC CONVEYING

Pneumatic conveying has widespread application across the food industry for handling dry, solid materials. Finished products as well as raw commodities and processed ingredients can be pneumatically conveyed. *(Figure 4-42)* From a sanitation viewpoint pneumatic conveying is one of the best means of transporting a product from one point to another, if care is taken to use filter-clean air as the conveying agent.

Properly designed pneumatic conveying systems are virtually self-cleaning. A few key points in a typical system will require some manual cleaning.

Both vacuum and pressure systems are used to handle food products. Because the bulk of the products handled are dry solids, conveying pipes or tubes are generally manufactured from mild steel. In some special applications, aluminum and stainless steel are also used. Other conveyor components are fabricated with metals selected according to need.

Blowers. Both rotary positive blowers and high-pressure fans are used to provide air, the conveying agent. Rotary positive blowers are normally provided with blower stands. *(Figure 4-43)* Blower stands should be fabricated to elevate the blower 10 to 12 inches above the floor or mounting surface. Blower mounts and motor take-up rails should be mounted so that the take-up rails do not create any void areas between the contour of the take-up rail itself and the mounting surface. Drive guards should be easily removed and have semi-open bottoms as set forth in the construction features section of this chapter.

Dry filters should be used on blower intakes. Do not use oil-impregnated or oil-immersed filters. Filters should be capable of filtering particles of 50 microns or larger from the air.

Silencers should be used on air streams only. Products should not pass through silencers. Check valves should be

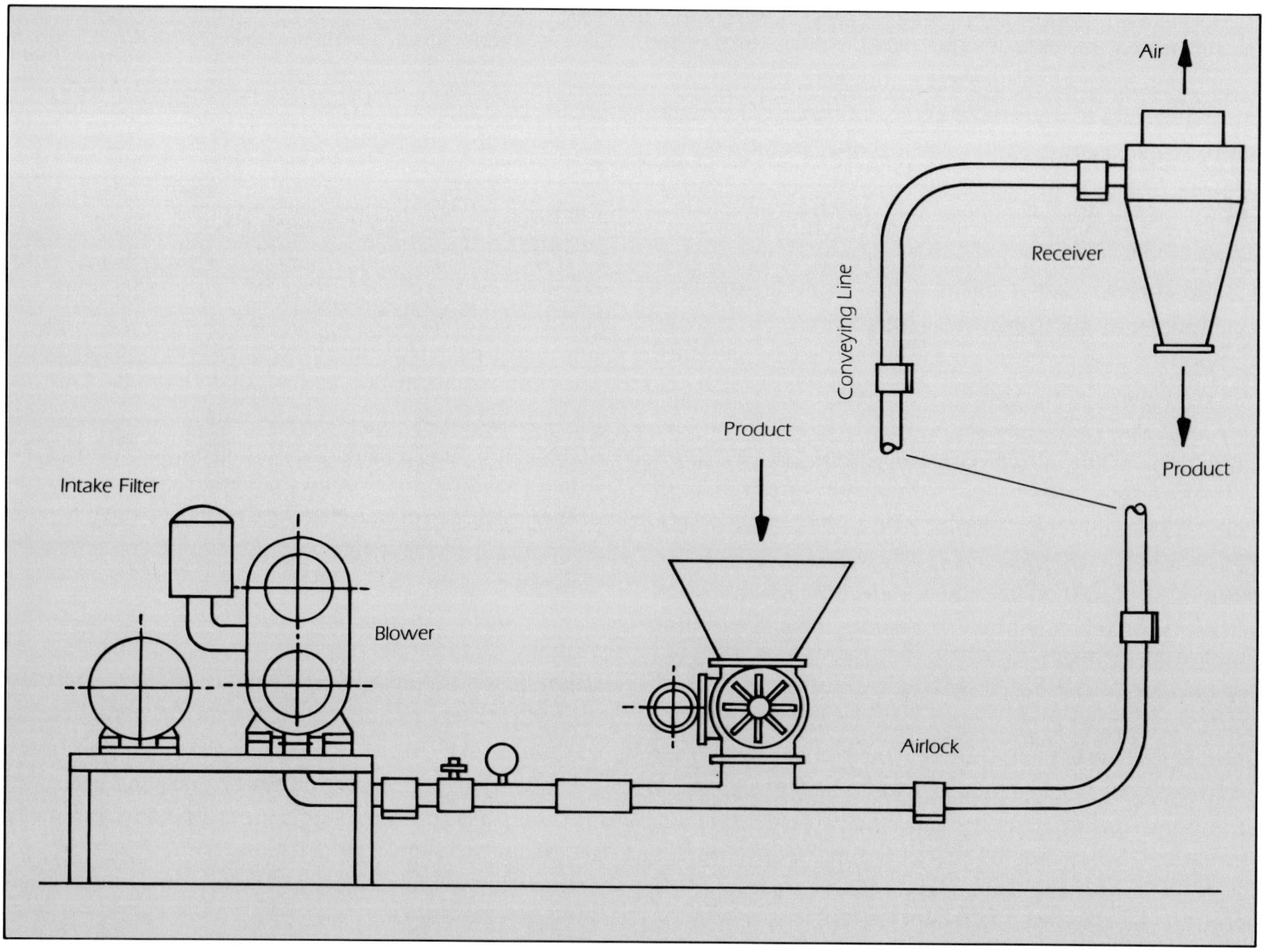

FIG. 4-42 Components typically found in a pneumatic conveying system

used between silencers and the point where the product is introduced into the conveying line. Failure to install silencers accordingly will result in product accumulation in the silencer media. When silencers are to be installed under blower stands, the stand should be high enough to allow a clear distance of six inches to nine inches between the bottom of the silencer and the floor.

Fans. High-pressure fans or exhaust fans should be mounted on stands to elevate them 10 to 12 inches from the floor. Fans mounted in the overheads should be placed on specially designed supports that minimize flat surfaces where dust can accumulate. Fan housings, including drive shaft and motor mounting housings, have been designed to minimize vibration problems. When fans are mounted directly on flat, solid surfaces, these housing construction features create void areas which are often difficult to clean. Fans should be mounted on angle iron or tubing members instead of solid flat plates.

Rotary Valves. To allow for cleaning, rotary valves should be installed with access doors in the inlet transitions. These access doors should be provided with proximity safety switches to shut down the rotary valve when the access door is open. Additional safety devices can be provided to protect the hands from the rotary should it start accidentally during a cleaning operation.

Rotary valve drives and drive guards should be designed as suggested in the construction features section of this chapter. Vertical arrangements of drive units should be considered to avoid the horizontal placement of motor mounting plates. Mounting them in a vertical position creates fewer dust-collecting flat surfaces.

Vacuum Line Air Intakes. Vacuum line air intakes normally found at the end of the intake line should be elevated above the floor or ground surface to assure that the intake velocity is not pulling dirt from the adjacent surfaces. When used outdoors, these intakes should be equipped with dry filters to remove particles of 50 microns or larger. The filters are generally not necessary when vacuum-line air intakes are used indoors.

Pneumatic Receivers. Pneumatic receivers in all product zone areas should be constructed with butt seams that are continuously welded. Welds should be ground flush. Avoid using lap seams. Mild steel product zone surfaces should be painted as required. Interior product zones should be free of ledges. Receivers using filtering media shall be provided with access doors to permit filter changes and receiver cleanouts. Filter media should be selected according to the manufacturer's recommendations for the products being handled. Access doors shall be provided in the clean air plenums for inspection and cleanout. These doors should be approximately 10 x 18 inches, which is a sufficient size to allow for cleaning.

When receivers are located in the overheads, work platforms should be provided for filter changes and cleanout. Larger receivers should be equipped with an interior round bar grate capable of supporting plant personnel during the cleaning and filter changing tasks.

Compressed Air. Compressed air used to pulse filter media should come from a nonlubricated air source; air from a lubricated compressor source must be dried and filtered. Only the nontoxic lubricants described in the chapter on maintenance should be used in these compressors.

FIG. 4-43 Blower stands should be elevated above the floor.

Primary and Secondary Receiving Systems. Primary and secondary receiver systems should be considered on difficult product/air separation applications. *(Figure 4-44)* Primary collectors remove the larger share of the solids; the filtering work of the secondary unit removes any remaining particles. Improved separation efficiency can result from this arrangement.

Breather Bags. Individual breather bags relieve air at the termination source; they also relieve the back pressure from a rotary airlock on a positive pressure conveying system. The use of breather bags should be limited to special applications, and they should be changed frequently.

Tubing and Tubing Fittings. Pneumatic lines should have smooth unobstructed interior surfaces. Cracks develop when tubing ends are joined by tubing clamps. These cracks can be minimized by cutting the tubing ends square. Tubing clamps should be installed securely and in a manner consistent with good sanitary design installation practices. Elbows should be large radius sweeping bends.

Conveying lines should be large enough to assure conveying velocities that will prevent product dropout or accumulation. There should be no dead ends in pipe sections.

- Flexible Hoses. Flexible hoses used to

FIG. 4-44 In this primary/secondary pneumatic receiver arrangement, the bulk of the solids are removed in the primary receiver.

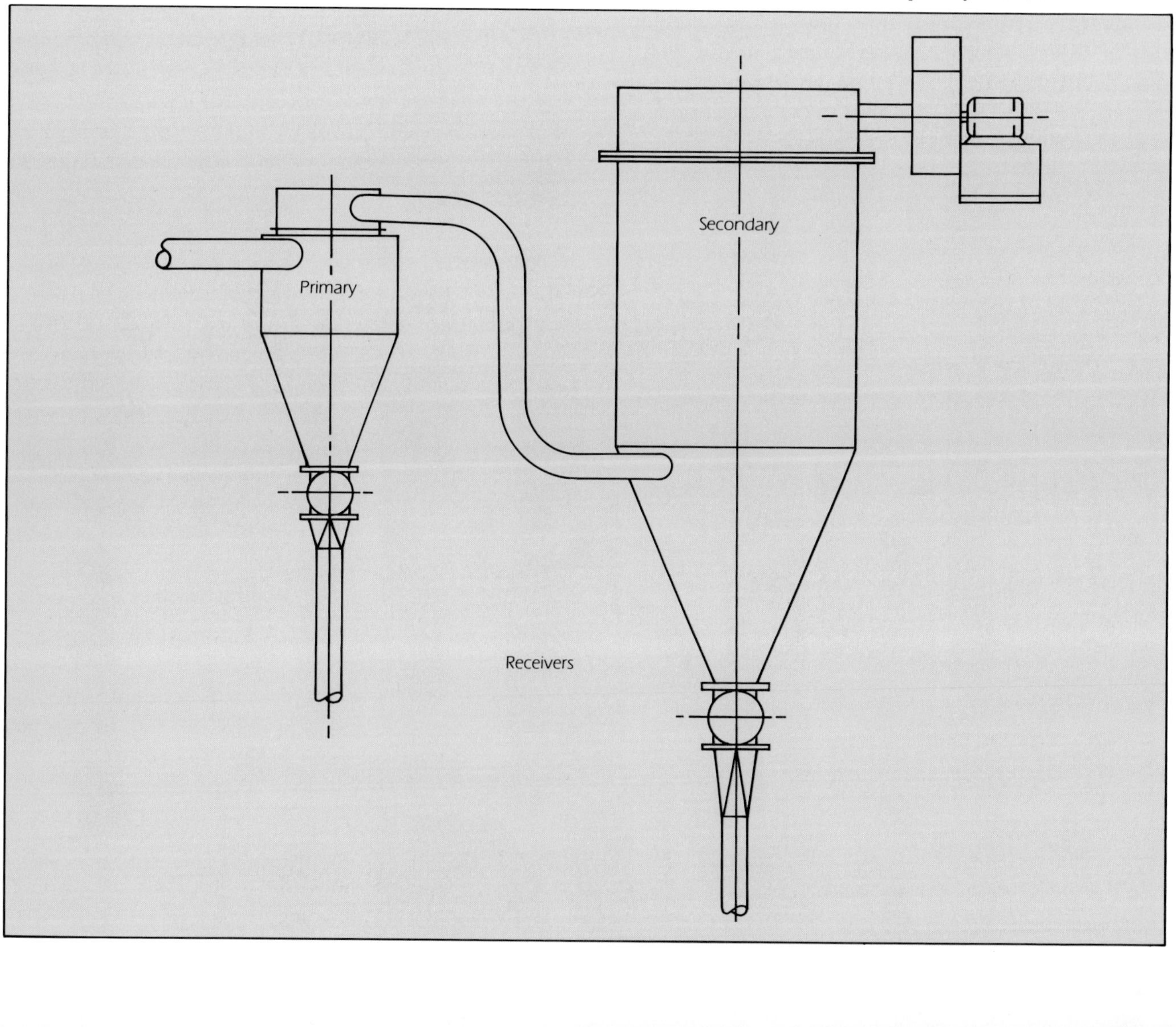

complete system connections at unloading points or transfer stations should be constructed of food-grade materials. Rubber hoses in particular should be white or off-white and of food-grade quality. Flexible metal hoses should be considered for products having high abrasion characteristics.

- Sight Glasses. Sight glasses should be made of polycarbonate plastic and should not be used on abrasive products.
- Diverting Valves. Diverting valves should be self-cleaning and free of void areas where product can accumulate. They should also be free of sharp surfaces that might damage the product.
- Pneumatic Line Supports. Pneumatic lines should be installed and supported in a manner consistent with the suggestions outlined in the utility piping section of Chapter Three. Pneumatic lines must be rigidly supported to prevent them from coming apart.

Gaskets. Gaskets used throughout the system should be constructed of food-grade, nonporous and nonabsorbent rubber or synthetic rubber. Do not use sponge rubber.

Bearings. Sealed bearings should be used on various system components.

Shaft Seals. Shaft seals should be nonabsorbent, nonexuding mechanical seals. Packing should not be used.

Electrical Features. Electrical features should include time delays to purge the pneumatic conveying system before the blower is shut down.

Paint. Food-grade epoxy enamel should be used on surfaces requiring paint.

BULK SOLID STORAGE BINS AND USE BINS

Bins for the storage and holding of bulk solid raw commodities and processed ingredients take on many shapes and sizes. Most bulk storage tanks are round in shape and have storage capacities ranging from several thousand pounds to well over a million bushels. In this section, we will be thinking in terms of small to intermediate, round-shaped bulk storage bins.

Sanitary design requirements change very little for square or rectangular bins. However, significant structural changes, such as the reinforcement of the flat surfaces of the sidewalls, create special sanitation needs. If reinforcing is required, structural members should be placed in a vertical position. Members placed in a horizontal position form ledges which collect dust.

Use bins and/or holding bins may be defined as smaller bins of many different shapes which hold the product for a short period of time before it moves on to another stage of the manufacturing process or before it is packaged. Use bins and holding bins can be built of mild or stainless steel. Stainless steel is often a better choice, particularly if the manufacturing environment requires frequent wet cleaning of the use bin. Furthermore, stainless steel use bins can serve a wide variety of purposes in the plant.

Bulk solid storage bins are almost always built from mild steel. These bins should not be located in wet manufacturing areas unless extreme care is taken to protect the mild steel surfaces from moisture. The fabrication of large storage bins from noncorrosive materials such as stainless steel should be considered when it is necessary to ensure the safety of the product. In instances where corrosion is a threat to the interior surface of mild steel bins, food-grade epoxy coatings might be useful, particularly if applied above the load line.

ROUND STORAGE BINS

Mild steel bins that are of welded construction design should include continuously welded butt seams. The corner weld should be ground and have a 3/16- to 1/4-inch radius. All butt seam interior welds should be ground flush, and exterior welds should be chipped and brushed. Interior tank surfaces should be smooth and free from pits, checks, and weld spatter. *(Figure 4-45)*

The interior zones of the bins should be free from any horizontal ledges where the product might collect. Discharge hoppers should be properly sloped or equipped with mechanical unloading devices to assure complete

FIG. 4-45 Typical bulk bin with several sanitary design features

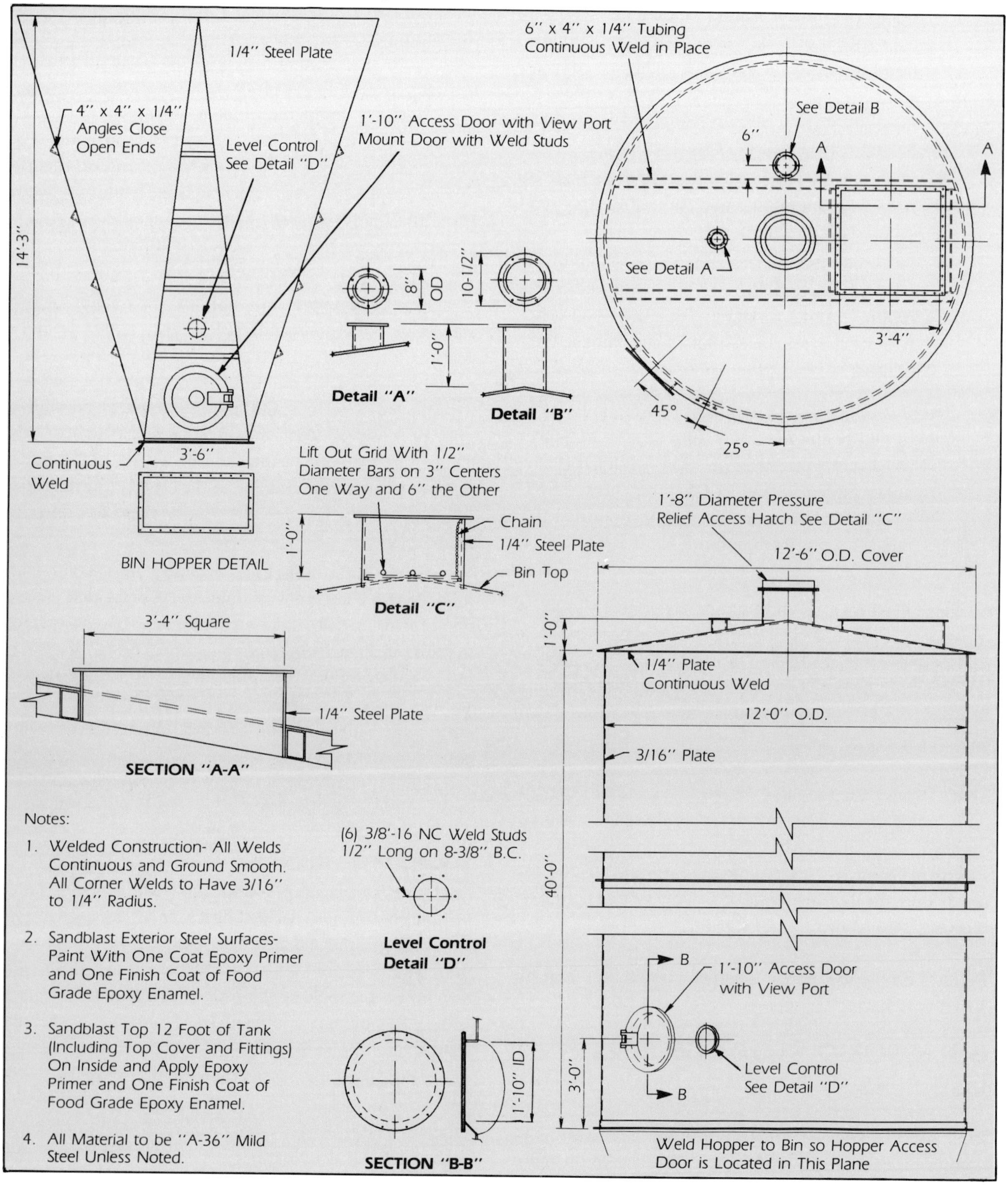

discharge of the product. Reinforcing members for flat bins tops should be externally mounted. Preferably, these tops should be constructed of material heavy enough to eliminate the need for reinforcement. If internal reinforcement is used on the tank top, it should be continuously welded to the bin top to eliminate all cracks and crevices.

When bin vents are required, they should provide protection against the entry of moisture and foreign material and should be equipped with easily removable filters to separate out particles of 50 microns or larger.

Storage bins should be equipped with access doors measuring a minimum of 18 inches. A fabricated door transition should be constructed to elevate the access doors two or more inches above the tank top of vertical bins. Access doors at the bin bottom should be positioned not more than four feet from the discharge opening. The doors should be fabricated with quick-release fasteners that do not require the use of tools. Additional doors should be considered as required by special needs.

Bin level indicators should be installed as required, and in such a way that they are accessible to service and cleaning. No mounting threads or bolts should project into the bin.

Companion flanges used at discharge openings, bin inlets, and access doors should be continuously welded to the bin walls when the crack formed by the companion flange and the bin wall is located inside the product zone. *(Figure 4-46)* When they are located outside the product zone, cracks formed by companion flanges and bin walls can be stitch-welded and caulked between welds.

FIG. 4-46 Bin companion flanges should be continuously welded.

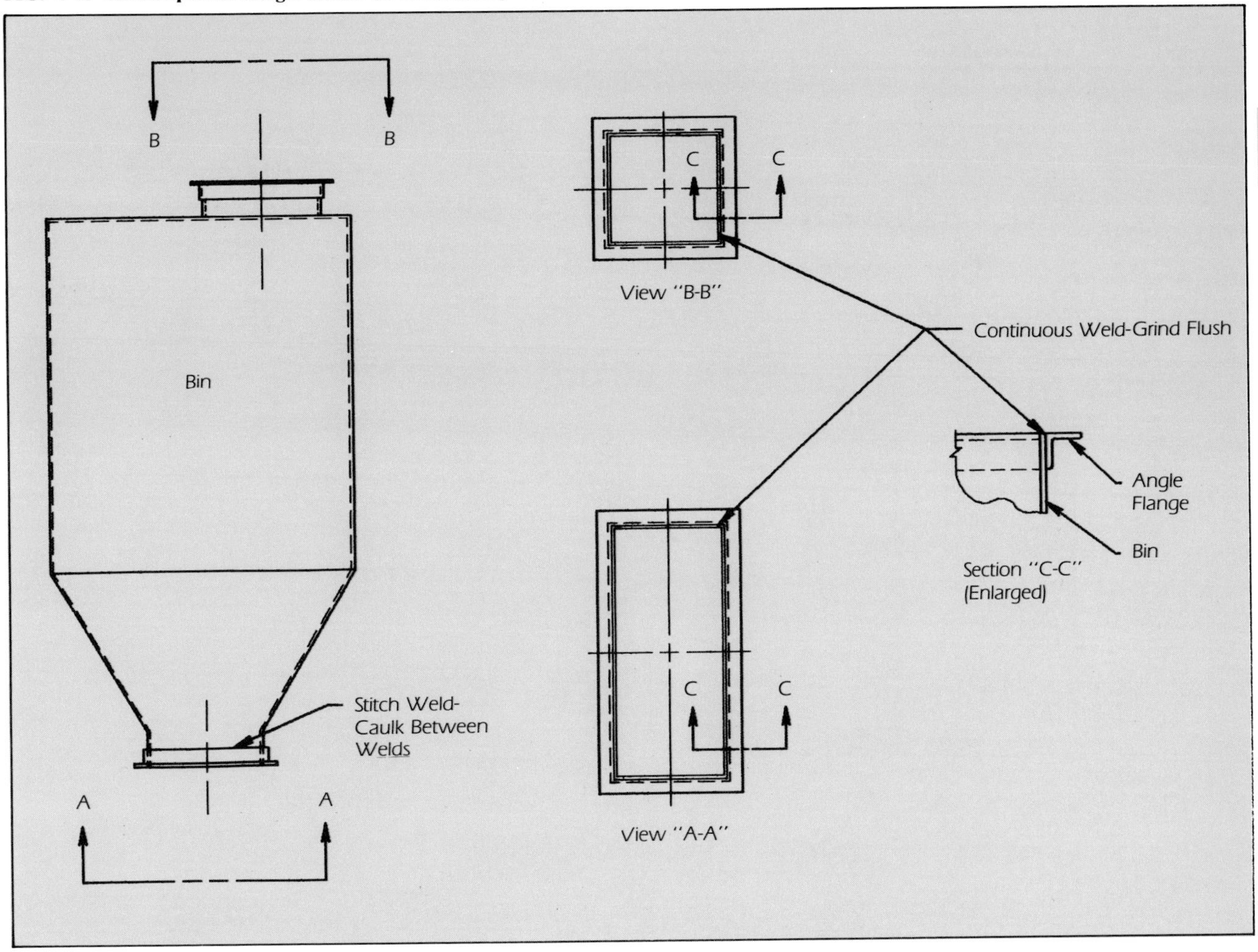

Bin interiors subject to moisture from condensation formations should be painted with food-grade epoxy enamel. Often it is only necessary to paint the bin top and side wall areas located over the product load line. The product movement down the walls of the tank will keep the remainder of the bin walls smooth and rust free.

Storage bin exteriors should be painted with food-grade epoxy enamel. Rust-proof primer should be applied to bins that are to be located outside.

USE/HOLDING BINS

Use/holding bins should be of welded construction. Butt-welded seams should be continuously welded and ground flush on the inside. Corner welds should be ground and have a 3/16- to 1/4-inch radius. Exterior welds can be chipped and brushed. Interior tank surfaces should be smooth and free of pits, checks, and weld spatter.

Type 304 stainless steel is generally adequate for bins handling dry solids. A number 4 exterior finish should be considered to provide a good appearance. A 2B mill finish is satisfactory for the inside. Neatly done exterior heliarc welds or automatic exterior machine welds might only require silicone brushing, while welds of lesser quality might require grinding. *(Figure 4-47)*

Interior zones of the bin should be free of horizontal ledges where product might collect. Discharge hoppers should be properly sloped or equipped with mechanical unloading devices to assure complete discharge of the product. Bin tops may be dished or sloped for reinforcing purposes. Flat bin tops should be constructed of materials heavy enough to eliminate the need for reinforcement.

Square or rectangular bins should have rounded corners. The corner radius should be 1/2 of an inch on small bins and 1 inch to 1-1/2 inches on larger bins.

FIG. 4-47 Typical use bin detail

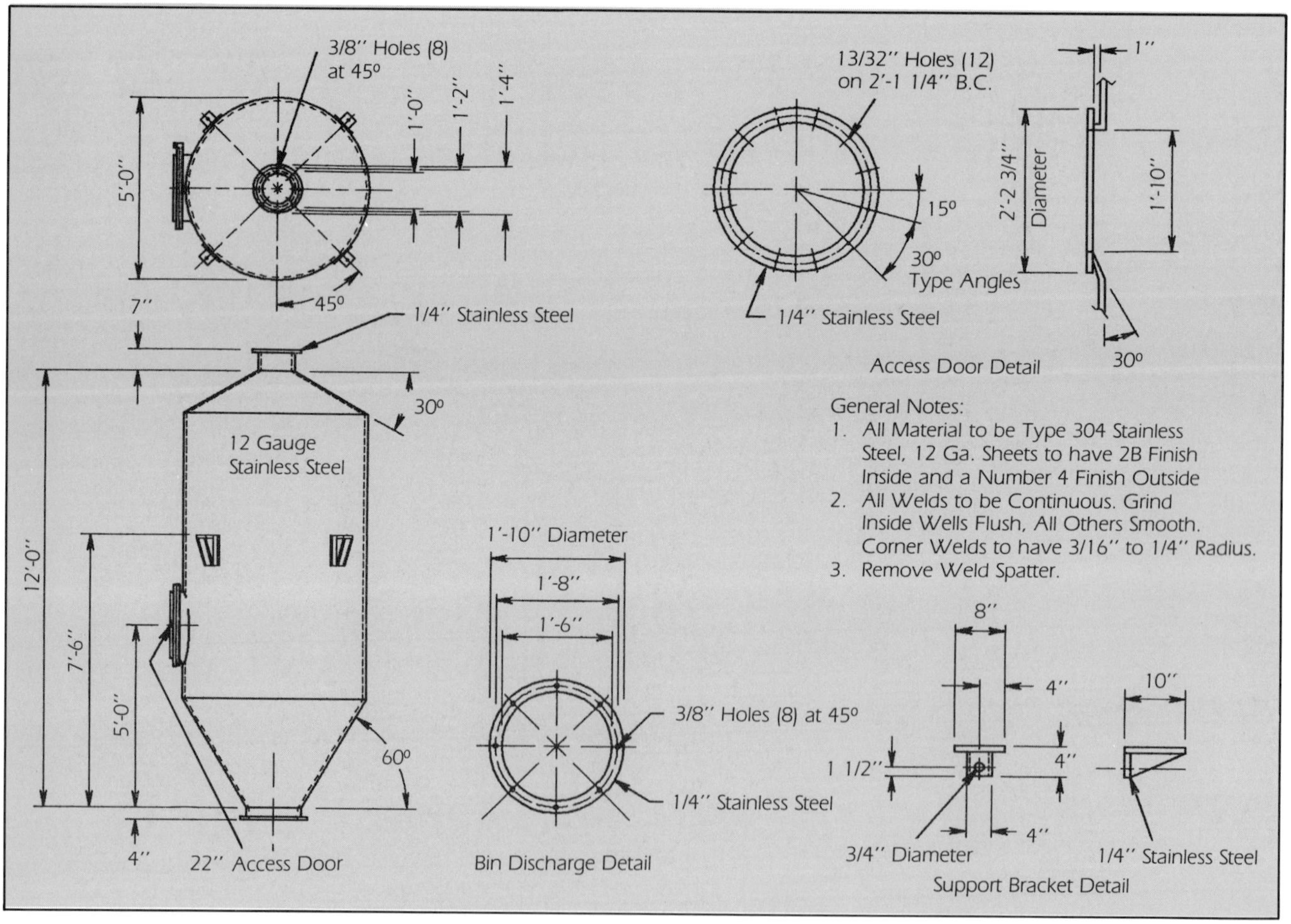

Access doors measuring a minimum of 18 inches should be provided at a convenient location. If the access door is placed in the top of the bin, it should be raised two or more inches with a fabricated door transition. Doors should be fabricated with quick-release fasteners that do not require the use of tools.

Bin level indicators should be provided as required and should be installed so they can be reached for service and cleaning. No mounting threads or bolts should project into the bin. Companion flanges at discharge openings, bin inlets, and access doors should be continuously welded whenever they form cracks or crevices in the product zone. Companion flanges located on the outside of the bin can be stitch-welded. Mild steel is satisfactory for these flanges.

Exterior reinforcing members used on square or rectangular bins should be arranged to minimize exterior ledges. These members can be mild steel.

Sight Glasses. If sight glasses are used, they should be fabricated from polycarbonate plastic.

Portable use or holding bins should be equipped with a good set of industrial-grade castors having ball or roller bearings.

Paint. Mild steel surfaces requiring paint should be painted with food-grade epoxy enamel.

LIQUID INGREDIENT STORAGE AND TRANSFER SYSTEM

Liquid piping systems can follow a simple plan consisting of a liquid storage tank, fill line, distribution pump, an on/off valve, and a pipeline that takes the liquid to a use point. Large dairy or edible oil plants, however, use a highly sophisticated liquid storage and distribution system. All systems, however complex, must be designed to enhance product safety and cleaning efficiency. Self-draining and take-apart or clean-in-place features are required to assure the cleanliness of the system.

Liquid storage and process system design begins with identifying the nature of the risk the product presents to the consumer. Systems considered to be sensitive to microbiological risk should be designed of stainless steel and be self-draining. They must also be free of pipeline sags, low spots, dead ends, and fitting wells where liquids can accumulate. Finally, they must be free of pits, checks, cracks, and crevices where bacteria can live and multiply.

Because of the high microbiological risk associated with their product, the dairy industry has long recognized the need for sanitary standards. In response, the International Association of Milk, Food and Environmental Sanitarians, the United States Public Health Service, and the dairy industry have developed the 3-A standards. These standards provide an excellent model for the processing of microbe-sensitive liquids. Reference to the 3-A standards will be made when applicable.

Not all liquid ingredients represent high microbiological risks. Liquid sugars, syrups, and honey are rarely troubled by bacteria; however, they do support mold and yeast growth if exposed to moisture and oxygen. Moisture can develop in storage tanks as a result of condensation formed when warm liquids are loaded into cold tanks.

Because of the low moisture level, sugar and syrup systems need not be constructed exclusively of stainless steel. Storage tanks constructed of mild steel with special interior food-grade epoxy coatings are suitable. Nonetheless, the system's piping and pipe fittings should be standard stainless steel. Liquid sugar and syrup spills and splatter are difficult to clean up and require the use of hot water and a mild detergent cleaning solution. If spills and splatter are anticipated at use points, holding vessels and associated handling equipment should be constructed of stainless steel.

Other liquids include vegetable and animal oils. These liquids also have low moisture levels and do not represent a serious microbial risk. Rancidity, however, can be a problem. Rancid oils have a disagreeable odor and a sour or stale taste sometimes described as *rank*. Rancidity generally results from aging or from exposure to high temperature. To control this problem, it is necessary to maintain the temperature necessary for easy flow. To prevent separation of the oil, inventories should be turned over frequently, and the system should be kept clean.

Oil systems do not normally require the use of corrosion-resistant metals. Mild steel is satisfactory for bulk storage tanks, but parts of the system will benefit from stainless steel. Do not lose sight of the fact that oil and fat storage systems require the use of strong alkali cleaners which can be corrosive. Copper-bearing metals should be avoided as they cause discoloration and oxidation of some oils.

STORAGE TANKS

Liquid tank design should include the selection of fabrication metals appropriate to the need or the product risk. Round storage tanks are preferred because rectangular tanks are harder to clean and more costly to fabricate. Tank seams should be continuously welded and ground flush on the inside product zone surface. The interior surface should be smooth and free of pits, checks, weld splatter, and weld slag. Access doors should be located in the bottom and at the ends of horizontal tanks, and they should be placed at a convenient height from the ground or floor. In addition, doors should be large enough to allow a worker to enter the tank.

Vertical Tanks. Vertical tanks should have access doors located in the top or side. Side doors should be located at the bottom of the tank to provide access from the ground. Doors to elevated tanks should be accessible from a work platform.

Doors that can be opened without tools should be provided for tanks requiring frequent cleaning. Larger tanks should have two or more access doors. Doors should be equipped with removable food-grade gaskets which are nonporous, nonabsorbent, and nonexuding. All corner junctions, such as those found at the ends of a horizontal tank and at the top and bottom of a vertical tank, should be rounded. The radius should be at least 3/4 of an inch on smaller tanks and one inch or larger on larger tanks. Vertical tanks should be constructed with slanted bottoms sloping towards the tank outlet. The slope should be 3/4 of an inch per foot. Horizontal tanks should feature a pitch toward the tank outlet that measures 1/8 of an inch per foot. To minimize sediment buildup, tank outlets should be installed in the tank bottom. Tank legs should be of a sanitary design. Vents, inlets, outlets, and special equipment fittings should not trap liquids and they should be easy to clean. There should be no exposed screw threads inside the tank. The inside radii of all welded tank fittings should be not less than 1/4 of an inch.

Milk Tanks. Milk storage tanks should be constructed of stainless steel and conform to the 3-A dairy standards. *(Figure 4-48)*

Liquid Sugar and Syrup Tanks. Liquid sugar and syrup tanks can be constructed of mild steel. Interior product zone

FIG. 4-48 Milk storage tank designed and built to meet the 3A dairy standards

surfaces should be coated with food-grade epoxy enamel. This application prevents pitting and checking of the mild steel surface caused by occasional exposure to air and by moisture formation in the tank.

Fiberglass tanks should also be considered for sugar and syrup storage applications. *(Figure 4-49)* Fiberglass tanks should be fabricated with food-grade resin materials, and they should be steam-cured. A smooth, nonporous and nonabsorbent product zone surface should result. Exterior surfaces should be constructed so they too can be easily cleaned. The construction details for easy cleaning outlined in the storage tank introductory paragraph apply to fiberglass as well.

Some syrups and liquid sugars are sensitive to mold formation caused by the presence of moisture in the tank. In such cases the tank should be equipped with ultraviolet lights and vent fans to circulate air through the tank. Circulating units are available that include a replaceable filter on the inlet side of the fan and an ultraviolet light installed in an enclosed box; this box is mounted to the discharge side of the fan. *(Figure 4-50)* The air is pulled in through the filter and discharged into the enclosed box. It then passes through the ultraviolet light area on its way into the tank. The tank is equipped with a vent pipe that features a replaceable filter. Filter media should be capable of separating out material 50 microns in size or larger.

Ultraviolet lamps have a life expectancy of 12,000 hours. However, they tend to lose some of their efficiency sooner than that. Consequently, they should be replaced approximately every eight months or 6,000 hours.

Edible Oil Tanks. Oil tanks can be constructed from mild steel materials. Special coatings are not normally required. Fiberglass tanks can be considered for the storage of some oils. Steam curing fiberglass is recommended to minimize odor problems. Heating or cooling coil loops installed inside the tank should be of a sanitary design and a material compatible with the liquid to be held in the tank. Coil loops must be sufficiently spaced to allow cleaning access to the tank bottom. Eight inches between the pipes is the most advisable distance. Coils should be pressure tested for leaks after installation. They should be installed six to twelve inches from the bottom of the tank so that they do not trap liquids or sediment. In addition, coils should be rigidly mounted to endure normal abuse from cleaning.

Fabricated hot water, steam, electric, or cooling water jackets should be considered for tanks requiring heating or cooling. These jackets surround the periphery of the tank and, in some cases, form part of its essential structure. Because they are mounted externally, water jackets will not contaminate the stored liquid should a leak occur in them. In addition, the tank interior is more easily cleaned. Tanks may also be installed in temperature-controlled rooms when heating or cooling is a requirement.

Agitators are used on some tanks containing liquids which require constant agitation to remain blended or in suspension. Agitators should be installed so they can be readily cleaned at the time the tank is cleaned. Agitator shafts and mixer propellers should be constructed of

FIG. 4-49 Fiberglass tanks fabricated with food-grade resins can safely store or hold many liquid ingredients.

FIG. 4-50 Ultra-violet light units can reduce mold development in liquid syrup tanks.

stainless steel and equipped with sealed bearings and mechanical seals. Mechanical seals should be nonabsorbent and nonporous food-grade materials. Stuffing boxes should not be used. Nitrogen blankets are a method of replacing oxygen in the tanks with nitrogen. Because mold does not grow in the absence of oxygen, they are effective in controlling mold in liquid tanks. The control of molds with nitrogen can be costly, depending on the size and the number of tanks involved. Tanks designed for nitrogen purging should be equipped with relief valves and vacuum breakers.

Nitrogen blankets may also be considered to stabilize oils that are in storage for prolonged periods of time. Oxygen accelerates the aging process of many oils; aging, in turn, results in the oils becoming rancid.

TRANSFER SYSTEMS

Liquid transfer systems should be of sufficient size to handle the designed flow rate. Microbe-sensitive liquids will require the use of sanitary pipe, fittings, and valves that conform to the 3-A standards.

Pipelines. Sanitary pipe installations should be installed so they are self-draining and have no sags where product or cleaning solutions can accumulate. If the system is to be cleaned in place, sanitary weld fittings with standard take-apart fittings provided at pumps, valves, tanks, and other points of equipment connection should be considered. Pipelines should be rigidly supported to assure their fixed position.

Liquid sugar and syrup piping systems should be

constructed of stainless steel pipe and stainless steel fittings. Aluminum might also be considered. Pipe connections can be screwed, flanged, or clamp-coupled. Piping systems should be installed to slope back toward the pump. This maintains a full liquid level in the pipeline.

Because of the viscosity of these liquids, piping sizes should be generously oversized. A three-inch diameter pipe is the minimum size. Oil handling systems can be constructed of mild steel pipe. The pipe connections can be screwed, flanged, or clamp-connected. The piping system should also be installed to pitch back toward the pump.

Liquid Pumps. Liquid pumps handling microbe-sensitive liquids should conform to the 3-A standards. The pumps should be located so that they can be easily cleaned.

Pumps handling liquid sugars and syrups may be cast iron, bronze trimmed units. However, serious consideration should be given to cast stainless steel, which is more durable. Cast iron without the brass trim is preferable for oil pumps.

Pipeline Strainers and Magnets. Pipeline strainers and magnets should be installed on the suction side of the pump to protect it. Stainless steel housed strainers and magnets should be used on the liquid sugar and syrup systems. Mild steel units can be used on the oil systems. Strainer screens should be constructed of metals with a magnetic property. Actual screen size will depend on the viscosity of the liquid. The rule of thumb is to provide the finest mesh that the handling system will allow.

DUST COLLECTION SYSTEMS

Clean, well-run plants are constructed to prevent the escape of even minute particles of a product from a manufacturing process. Product dust escaping from equipment often represents large ingredient losses which affect manufacturing costs noticeably. In addition, the dust strewn around the plant means additional cleaning costs. In some cases dust escapes as fast or faster than it can be cleaned up. The work attitudes of plant employees can be affected by this demoralizing situation.

There are many factors that affect our ability to control dust. One such major factor is air balance within the facility. In many cases, the atmosphere of some buildings is robbed or starved of air. This negative air atmosphere will have a major impact on a dust collection system designed for atmospheric conditions. Depending on the severity of the situation, a lack of air in the plant can reduce the handling capacity of dust collection by 50% or more. A well-designed dust system begins with the restoration of the air balance in the plant.

In addition to a good air balance, properly designed dust collection systems will require an adequate supply of air at collection points and sufficient intake velocities. There are many opinions about how much air is needed and some decisions are best determined by individual circumstance. However, good dust collection design engineers have a number of useful guidelines for the engineering process.

Along with air quantity and velocity requirements, there are a number of important sanitation considerations with respect to the air and cleaning needs of the dust collection system.

Often dust collection systems will collect dust which is very sensitive to insect infestation, and in such cases good insect-proof design details are very important. In other cases, the dust is very sticky and hard to handle. A design that emphasizes good access for cleaning and appropriate construction materials becomes important to the proposed cleaning method.

DESIGN REQUIREMENTS

Air Velocity Needs/Hood Design. Sufficient quantities of air must be available to provide needed capture velocities at the point of dust emission. Exact quantities will depend on the nature of the operation. Hood design should provide capture velocities at the point of emission. Where release of dust is both large and rapid, velocities ranging from 500 to 2000 FPM are needed. At a minimum, capture velocity should be 300 to 400 FPM.

High-velocity Pick-ups. Where small quantities of dust are emitted 8 to 12 inches from the collection point, high-velocity slotted pick-ups should be considered. *(Figure 4-51)* Slotted pick-ups can also be effective in containing large quantities of dust confined in a smaller area such as some ingredient dump-in hoppers or packaging scales.

Conveying Velocities. Conveying velocities should be sufficient to prevent particle dropout. Generally 3500 FPM is the minimum velocity required.

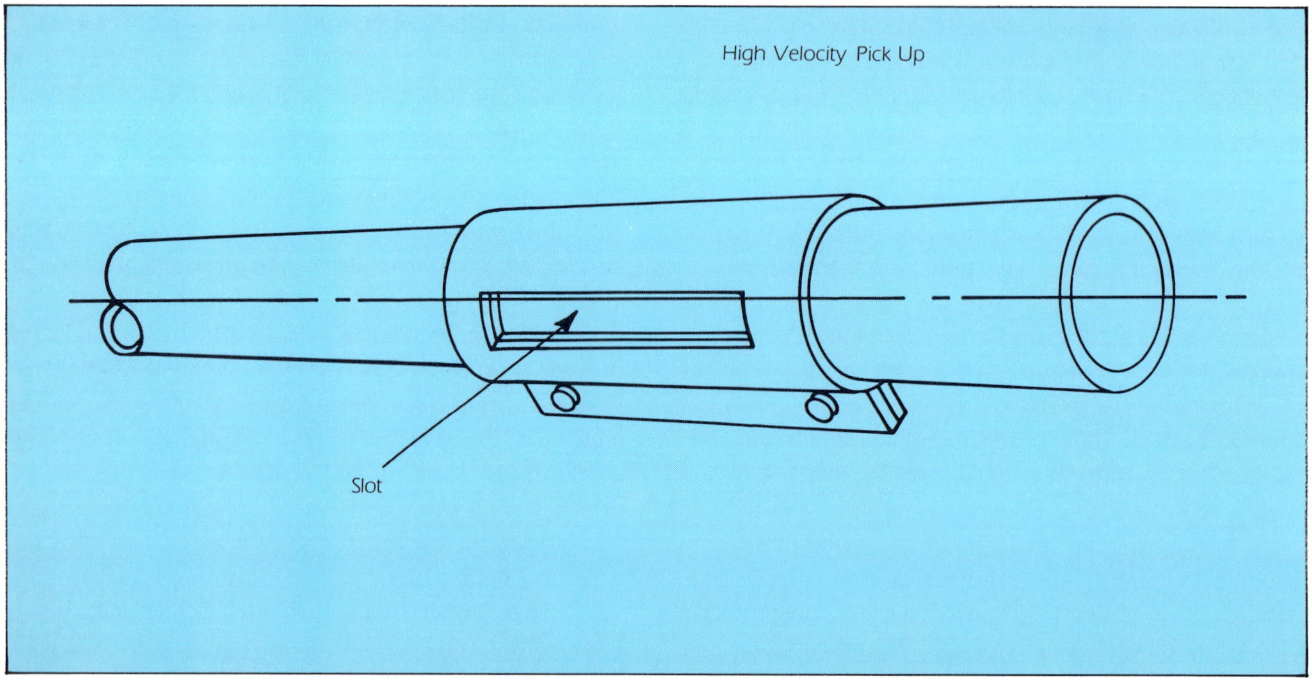

FIG. 4-51 High velocity dust collection pick-up

Blast Gates/Air Intakes. Blast gates and air intakes should be provided to regulate air quantities at collection points and to maintain balance in the collection system to avoid particle dropout. Accumulations of insect-sensitive stock in the bottom of a dust collection line are sure to become infested if left for a long period of time.

Duct Design. Duct design should feature seamless tubing whenever possible. Seamed tubing should be butt-jointed and continuously welded. Slip-fit joints or lap seams, which are riveted and soldered, and fabricated seams should not be used. Construction materials should accommodate the product need.

Cleanout doors featuring quick-release fasteners should be provided every 20 feet in pipes 18 inches and larger in diameter. Door openings should be large enough to allow for inspection and cleaning. Any gaskets used should be food-grade and should be nonporous and nonabsorbing. Sponge rubber should not be used.

Ducts smaller than 18 inches in diameter should feature a 24-inch removable section every 20 feet. *(Figure 4-52)* This section should be secured in place with clamps which can be opened without using tools.

Ducts should be supported as suggested in the utility piping section of the chapter on buildings.

Dust Collectors. Dust collectors with appropriate filtering media should be provided. This filtering media should be installed according to the manufacturer's recommendation, including recommendations for air to cloth ratios. Access doors should be of sufficient size to permit filter media changes and easy cleaning of the inside surfaces. Exterior platforms and interior grates should be provided as required for large dust collectors or dust collectors located in the overheads. Platforms should be constructed in accordance with the platform suggestions given in the section on layout in Chapter Two. Interior grates should be constructed from round shapes and be capable of supporting plant personnel during cleaning and filter media changes.

Access doors should be provided in the clean air plenum and be of sufficient size to allow for cleaning; approximately 10 to 18 inches is adequate. It is not uncommon to find accumulations of dust and insects in this area.

Primary/Secondary Collectors. Primary/secondary collection systems should be considered when the dust loading per cubic foot of air is excessively heavy. Primary collectors remove a large share of the dust particles, thus the filtering work of the secondary unit is reduced and an improved collection efficiency often results. Primary collectors are normally cyclone collectors and secondary collectors are filter media collectors.

Compressed Air. Compressed air used to pulse the filter media should be from a nonlubricated air source if the dust is to be recovered for reuse or used in animal feed. Air from a lubricated compressor should be dried and filtered. The approved lubricants described in Chapter Seven should be used in these compressors.

Fans. Fans should be mounted on stands to elevate them 10 to 12 inches from the floor. Fans in the overheads should be mounted on specially designed supports that minimize flat surfaces where dust can accumulate. Fan housings, including drive shaft and motor mounting housings, have been designed to minimize vibration problems. When fans are mounted directly on flat surfaces, these housing construction features create void areas which are often hard to clean. These void areas can be eliminated by elevating the fans and by using structural shapes such as angle iron and tubing instead of solid plates in the support and stand design.

FIG. 4-52 Removable sections of dust collection ducts permit better cleaning access.

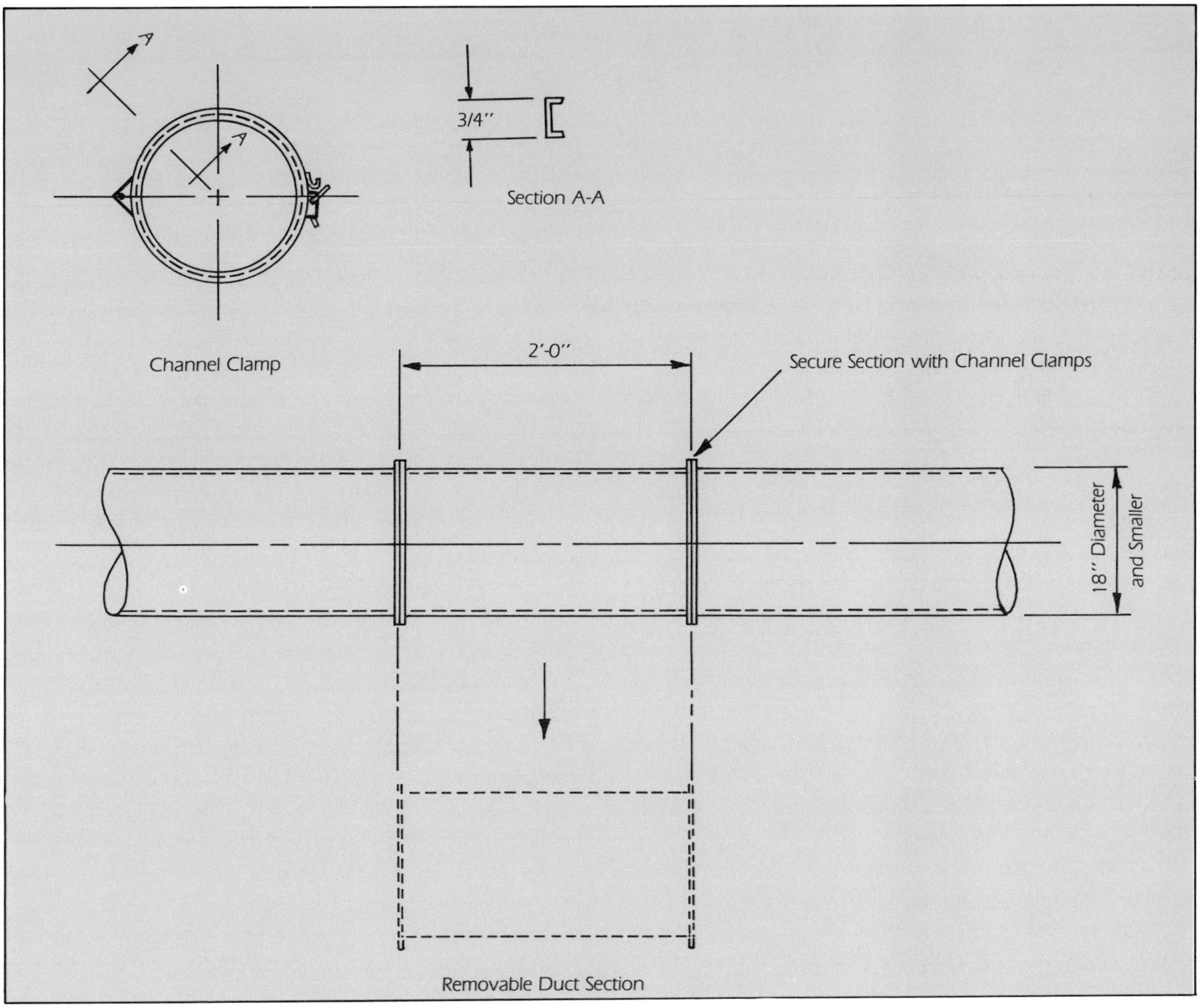

Rotary Valves. Rotary valves installed on the bottom of the dust collector should be installed so they can be cleaned if necessary. An access door installed over the valve in the dust collector hopper might be necessary to facilitate cleaning. Access doors must be provided with a proximity safety switch to shut down the rotary valve when the access door is open. Other safety devices can be considered to present possible injuries caused by a turning rotary.

Bearings. Only sealed bearings should be used on system components.

U-tubes/Pressure Gauges. Dial faces of pressure gauges and/or U-tubes used to measure system pressure conditions should be made of polycarbonate plastic.

Fumigation Plugs. Fumigation plugs should be provided in the dust collector if the system is handling insect-sensitive dust. They should be placed in the upper clean air plenum and the filter plenum. Fumigation plugs can be 3/8- or 1/2-inch spring-loaded oil caps. These are the same oil caps that are commonly used on some bearings.

Drive/Drive Guards. Drive and drive guards should be designed as suggested in the construction features section of this chapter.

Paint. Food-grade epoxy enamel should be used on surfaces requiring paint.

CHAPTER 5
ELECTRICAL SYSTEMS

CHAPTER CONTENTS

ESTABLISHING DESIGN REQUIREMENTS

ELECTRICAL EQUIPMENT

ELECTRICAL HARDWARE

Electrical systems represent special sanitary design challenges. The nature of these challenges varies according to the product safety risks of the particular plant. Most insect sensitive products benefit from an electrical system that is dust-tight or explosion-proof, while microbe-sensitive products dictate the need for watertight enclosures and sanitary motors. Transformers, control centers, motors, and electrical hardware are some of the electrical components that must be designed with product risks in mind, and certainly, the electrical system must be designed to accommodate the latest insect control and cleaning methods proposed by the plant sanitation staff.

ESTABLISHING DESIGN REQUIREMENTS

INSECT-SENSITIVE ELECTRICAL DESIGN

Electrical systems are extremely vulnerable to major insect infestation. Open-style enclosures or dust-tight enclosures that are not maintained in a dust-tight manner provide thousands of cracks, crevices, and void areas for the insect and its food. In a large food plant, there are literally miles upon miles of interconnecting conduits and electrical wireways that can turn into an endless infestation network. Because of the warmth they generate, light fixtures become natural incubators for insect eggs. In addition, insects are attracted to the ultraviolet rays and warmth that are produced by many light systems. The electrical design engineer must become familiar with insect control methods, because of the impact they have on the electrical system. If the use of lethal chemicals to control insects is prescribed, electrical components must be chosen that will not be damaged by the use of those chemicals. If the control plan calls for heat sterilization, the electrical components chosen must be compatible with the proposed sterilization temperatures of 140°F or more.

MICROBE-SENSITIVE ELECTRICAL DESIGN

Microbe-sensitive conditions call for extensive wet cleaning and sanitizing of all equipment, including electric motors, control panels, and controls. However, the cleaning and sanitizing materials used for this are generally very corrosive and can damage even the most carefully designed systems. To avoid this kind of damage, the design engineer responsible for a microbe-sensitive electrical system will want to use sanitary motors, stainless steel controls and control panels, with watertight construction features. *(Figure 5-1)*

ELECTRICAL EQUIPMENT

In addition to the sanitation considerations discussed in this chapter, the electrical equipment selected and its installation must meet appropriate electrical code requirements.

TRANSFORMERS

Dry Transformers. The best location for dry, air-cooled transformers is in specially designed, well-ventilated rooms that are dust-free. If all these conditions cannot be satisfied, these transformers should at least be mounted in dust-free areas as they usually have open windings which are very difficult to clean.

Oil-Insulated Transformers. Because of their enclosed nature, these transformers have sanitation advantages. Whenever possible, install oil-insulated transformers outdoors. For a roof installation, choose a location where there is no danger to finished food products or food ingredients, in case a transformer oil tank ruptures. Also, be sure to provide the necessary dikes to contain any oil spills. If you install the oil-insulated transformer indoors, place it in a well-ventilated, dust-free room. This room must be designed to be easily cleaned and to accommodate all fire code requirements including confinement of the cooling oil, in case of a tank rupture.

FIG. 5-1 Areas that are wet cleaned should be provided with watertight stainless steel electrical control panels.

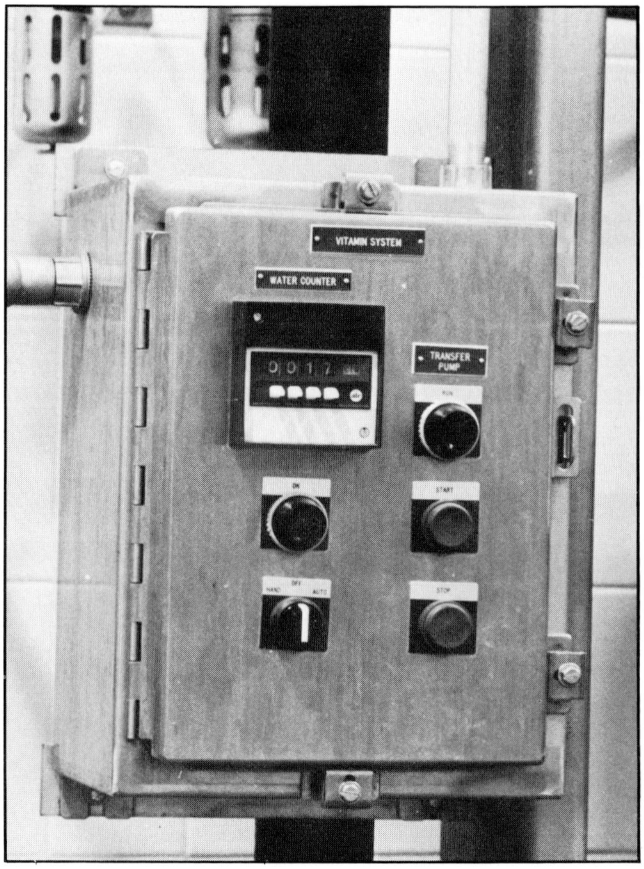

FIG. 5-2 Motor control centers should be mounted off the floor on concrete pads.

Polychlorinated Biphenyls (PCBs). In the past, polychlorinated biphenyls were a common cooling oil for transformers and capacitors. As is recognized today, PCBs are highly toxic substances. Recent federal laws prohibit the use of PCBs in and around food plants beginning in 1985 for some PCB-containing equipment and for all such equipment by the late 1980s.

SWITCH GEAR AND CONTROL CENTERS

Main switch gear and individual motor control centers should be installed in pressurized rooms that are air-conditioned, well-lit, and easy to clean. The installation of motor control centers in manufacturing areas should be discouraged. However, if it does become necessary, these centers should be located in dust-tight enclosures that are pressurized with a temperature-controlled, filtered air source.

Regardless of location, all motor control centers should be mounted off the floor on a four- to six-inch solid concrete pad so the floor can be wet cleaned without damaging internal electrical components. *(Figure 5-2)* Also, be sure that all unused machine screw holes or conduit knockout openings are properly sealed, to eliminate access for dust and insects. Check all reset buttons to be sure they will operate from outside the panel, so you are not required to open the cubicle door to reset the buttons. Cubicle doors that are opened frequently are often not closed and properly secured, exposing the interior to dust and insects.

CONTROL PANELS

Control panel enclosures should be used to contain relays, timers, and allied equipment including complete logic modules. Whenever possible, the enclosures should be mounted in a central control room. This room should be pressurized, air-conditioned, well-lit, and easy to clean.

Control panels installed in manufacturing areas should be designed to be dust-tight and in the case of wet manufacturing areas, watertight. Where appropriate, consideration should be given to pressurizing the panel with clean, filtered air. Make sure the bottom reinforcing flange of the control panel breaks down instead of up, to make it easier to clean out the bottom of the panel. (See Chapter Four, Figure 4-22.) Design the cabinet with clean, smooth exterior lines for easy cleaning.

Elevate floor-mounted control panels on legs six to eight inches high. *(Figure 5-3)* Full-length, larger control panels should be mounted directly on four-inch solid concrete pads.

Control panel light lenses, instrument faces, pressure gauges, and similar devices should be made of polycarbonate plastic. Under no circumstances should glass be used. Make sure all unused control panel openings are sealed.

FIG. 5-3 Floor-mounted control panels with legs

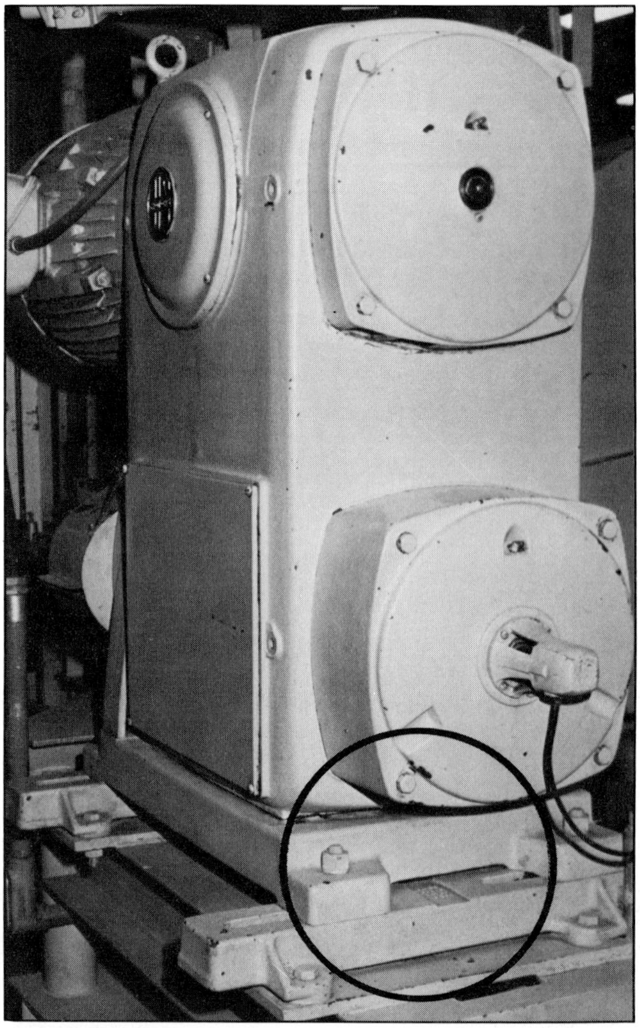

FIG. 5-4 Difficult-to-clean void areas, normally associated with rail and flat mounting surfaces, are formed when motor take-up rails are bolted down on a solid steel plate.

MOTORS

Sanitary motors should be selected for microbial-sensitive processes that will require wet cleaning with the use of corrosive cleaning and sanitizing chemicals. Most sanitary motors are equipped with smooth outer housings and specially designed enclosures that can withstand washing or hosing down with hot water and various cleaning compounds. Electrical codes require that special motor enclosures be used for specific atmospheric conditions. However, it is wise to minimize the use of open motors even when permitted. At some point in time, all motors require cleaning. Open motors are difficult to clean because their openness limits the cleaning method: wet cleaning is likely to create damage.

Motor supports should be designed to eliminate large flat surfaces that collect dust. If motor take-up rails are used, install them so difficult-to-clean void areas, normally associated with rail and flat mounting surfaces, are eliminated. *(Figure 5-4)* Whenever possible, motors should be mounted on specially designed supports, in the open, where they can be easily cleaned. In no case should motors be suspended in or over product zones or product zone surfaces. *(Figure 5-5)*

PUSH-BUTTON STATIONS

Select the push-button stations for individual motors with specific atmospheres and code requirements in mind. From a sanitation viewpoint, a dust-tight or explosion-proof enclosure installed in dusty atmospheres eliminates many insect harbors. Push-button lights should have polycarbonate plastic lenses.

When mounting push-button or any other start-stop stations on the wall, make sure the box is sealed to the wall with a good pliable caulking material. *(Figure 5-6)* If the station cannot be mounted tightly to the wall, extend it from the wall a distance of at least two inches to allow for cleaning. All unused holes in the switch enclosure should be sealed so that insects cannot enter.

GLASS

Precautions should be taken to eliminate glass on electrical equipment. Plastics should be substituted for glass on such electrical equipment as pilot light lenses, amp meter dial faces, and alarm lights. Polycarbonate is the preferred plastic composition, as it is considerably stronger than many other plastics.

FIG. 5-5 When motors are mounted over open product zones, product contamination from leaking lubricants is a possibility.

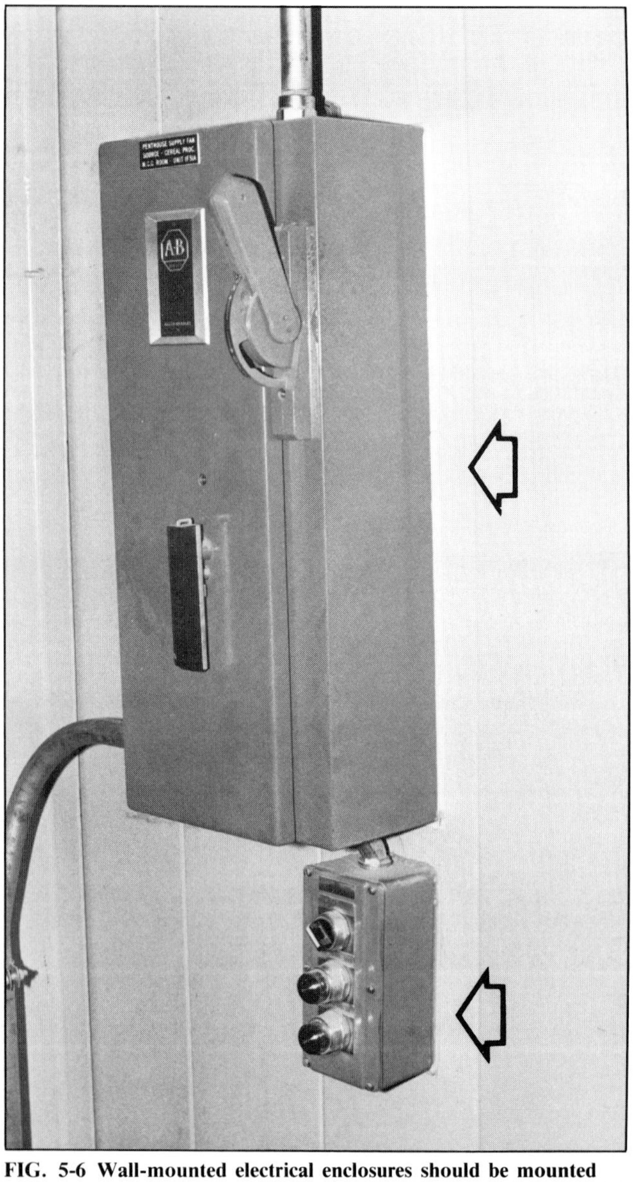

FIG. 5-6 Wall-mounted electrical enclosures should be mounted away from the wall or sealed to it with a pliable caulking material.

ELECTRICAL HARDWARE

CONDUIT

Conduit used in manufacturing areas should be rigid steel or heavy-duty aluminum. Conduit couplings should be standard threaded. When conduit is mounted along a wall, extend it from the wall approximately one inch, using stand-off clips. For conduit that is mounted in a horizontal plane at the floor level, elevate it 18 inches or more from the floor. Single and multiple conduit runs should be installed in accordance with recommendations outlined in the utility pipe section of Chapter Three.

Conduit systems must be well-sealed. In conditions of extreme insect-sensitivity, sealing the open ends of conduits at junction boxes or switch box junctions with a pliable sealing material may be advisable.

Conduit fittings should be constructed of high quality cast fittings complete with gaskets and should be capable of receiving threaded conduit without danger of breakage. When possible, "LBs" (elbows with bottom access plates) or "LRs" (elbows with right-side access plates) should be installed in a vertical position to minimize flat surfaces that collect dust.

Flexible Conduit. The use of flexible conduits should be limited to motor connections or similar applications and should be the smooth, vinyl-covered metal type. Use ring connectors which apply uniform pressure to the full circumference of the conduit. Uneven pressure applied to the connector will break the fitting, causing a sanitation and safety problem. *(Figure 5-7)* Do not use flexible conduit in place of rigid conduit.

JUNCTION BOXES

Junction boxes installed in manufacturing areas should be dust-tight at a minimum. If a specific atmospheric hazard has been identified, the design of junction boxes should accommodate it, i.e., Class II Group "G" or NEMA 4. All junction box lids should be provided with gaskets. Neoprene gaskets should be of the closed-cell type.

SWITCH BOXES

Switches should be mounted in dust-tight enclosures. When installed against a flat surface, such as a wall or flat mounting plate, they should be sealed to the flat surface.

As an alternative, switch boxes can be extended away from the flat surface at least two inches, but preferably more. All switch boxes should have gaskets or have a machined surface. All unused openings should be sealed.

DUCTWORK/WIREWAYS

The use of ductwork or wireways should be limited. Wireways are exceptionally difficult to keep dust- or watertight because of the inevitable abuse and gasket deterioration that occurs. Furthermore, the numerous openings for sheet metal screws and conduits are commonly left unsealed, increasing insect infestation risks. However, there are several situations where the use of wireways is acceptable. One such situation is where numerous small conduits are grouped together in several horizontal rows. Conduit arrangements of this type are difficult to clean and the choice of a wireway would eliminate the need for numerous small conduits and simplify the cleaning task. Wireways constructed from heavier gauge metals and limited to eight to nine inches in width are preferable as they are more rigid. Wireway covers should also be constructed of heavier gauge metals and formed with approximately one-half inch vertical flanges which break down to form a watertight lip. Cover gaskets should be closed-cell rubber or synthetic rubber vulcanized to the cover.

CONDUIT SUPPORTS

Conduit supports using formed channel with a J-like hook at the base of each flange, (commonly used by electricians to support conduits), should not be used in a horizontal position because of the cupped crevice that is formed. Food product accumulates in this crevice and attracts insects. *(Figure 5-8)*

When horizontal conduit runs are placed tightly against another vertical flat surface, a longitudinal crevice

FIG. 5-7 When damaged, flexible electrical conduit motor connections create both safety and sanitation problems.

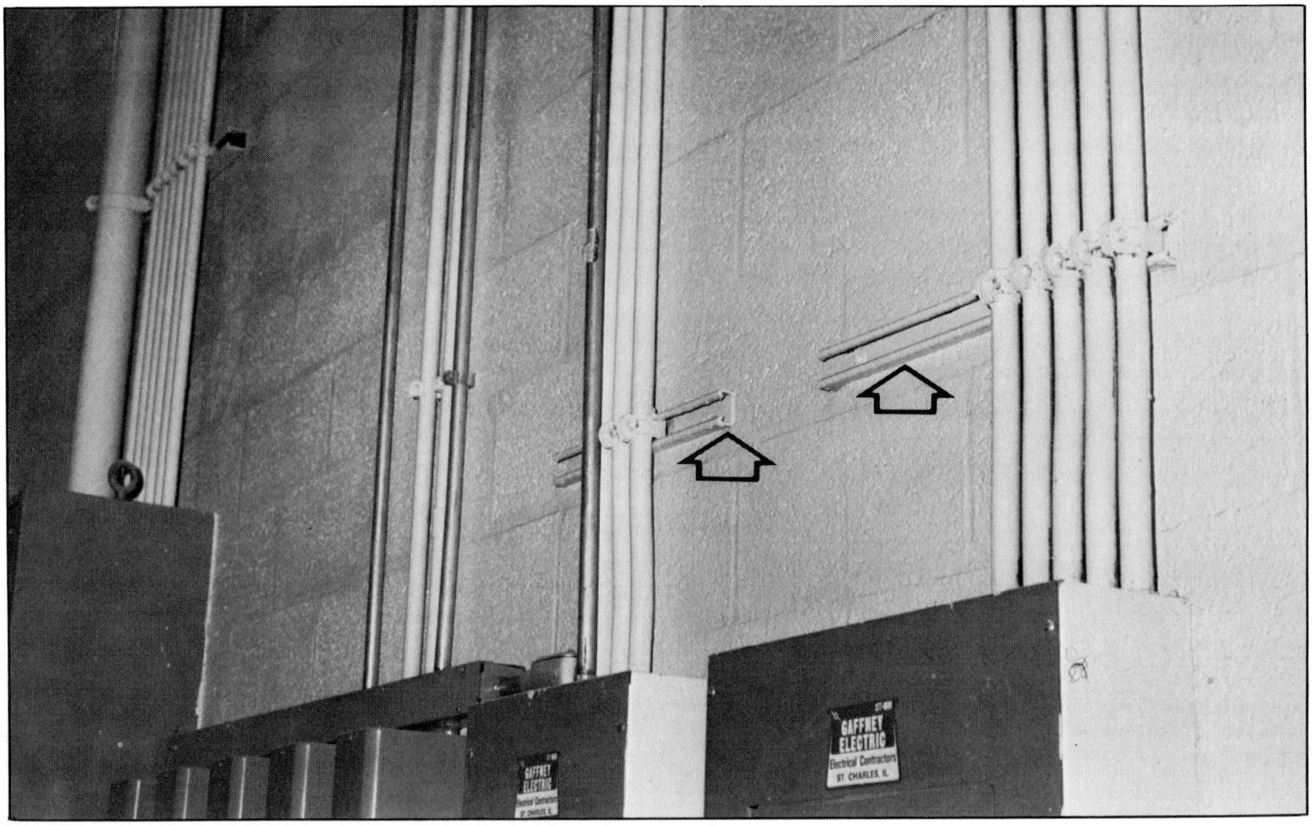

FIG. 5-8 When formed channel conduit supports with J-like cups are used in a horizontal position, sanitation problems are created.

is formed in which dirt can accumulate. Horizontal conduit runs should be held off flat surfaces with stand-off clips. Do not use single hole straps and back straps for stand-off clips. These straps are generally cast-iron and their design leaves a void area between the strap and the flat mounting surface. *(Figure 5-9)* The Minerlac stand-off clip, also pictured, is more sanitary. Small void areas created by flat surface mounting of the single hole or back strap often support insect infestation, the source of which can remain undetected for a long time.

The single hole strap is sometimes used alone to support both vertical and horizontal runs of conduit against walls or other flat surfaces. If you look closely, however, when the strap is used this way a small void area is created. Furthermore, when used to support horizontal runs, the conduit fits tightly against the flat surface, creating a longitudinal crevice, like those mentioned earlier.

As in the case of conduits, conduit supports should be fabricated and installed according to the recommendations for utility piping found in Chapter Three.

FIG. 5-9 Plated cast iron single-hole straps and back straps pictured on the left leave insect harboring void areas when mounted against flat surfaces. The Minerlac stand-off clips pictured on the right are a better choice.

CHAPTER 6 CONSTRUCTION

CHAPTER CONTENTS

BUILDING MATERIAL AND EQUIPMENT INSPECTIONS

CONSTRUCTION PROGRESS INSPECTIONS

FIELD FABRICATION

Every construction project, regardless of size, requires a construction or resident engineer to supervise the project from the time it begins through the start-up phase. The construction engineer must have a thorough acquaintance with the sanitary design requirements advocated by the food industry, and with federal, state, and local regulations that govern the production of human foods.

The construction engineer has one of the most important jobs in the entire project process. Particularly in the sanitation area, he can make the difference between a job well done and one that is not. The construction engineer constantly judges the quality of the job and decides on many engineering details. As he directs construction activities, he determines the location of some equipment, and of the controls; he also determines piping routes, and he resolves layout conflicts. The construction engineer is also instrumental in the design and construction of field fabricated equipment. Because he has the final word in so many decisions, it is important to train him in food sanitation design requirements.

THE CONTRACTOR

The familiarity of contractors with food sanitation requirements varies widely. Many contractors specialize in food plant construction and have a basic understanding of sanitation requirements while others have no experience. Regardless of their experience level, all contractors need instruction on the importance of food sanitation. Construction job superintendents, general foremen, and foremen should also receive sanitary design training. These supervisors will be responsible for directing construction activities. Their familiarity with sanitary needs will result in the job being done as it was intended.

Construction Personnel. Most construction personnel have little or no exposure to the sanitation requirements of the food industry. Indoctrination of key construction personnel will be very beneficial to the job at hand. Familiarize them with the problems, the consequences, and the solutions. In essence, put them on your team, make use of their many abilities, and take advantage of their knowledge. Above all, get them to work with you.

Job Organization. A clean, well-organized construction project is a mark of excellence and generally indicates that the end results will also be satisfying. Equally important to personnel indoctrination is job organization. Construction offices, small storage rooms, and locker or change rooms should be provided and kept in a neat, orderly condition. Provisions of this nature impress the construction employee who will in turn become neater in his working habits. Set aside designated smoking areas and equip them with ash cans and fire extinguishers. Lunch rooms and toilet facilities with the appropriate equipment and disposal containers are also a necessity. All of these facilities should be kept clean and in good repair.

Remove crating, excelsior, and other forms of debris from the building as soon as possible. Accumulations of this kind cause infestations and other sanitation problems as well as fire hazards. Keep floors and workbench areas as clean as possible. Be sure to provide a sufficient quantity of metal trash containers throughout the building; they should be emptied daily or at least every other day.

Consider floor protection in areas where pipe machines or similar types of equipment are used to contain oil spills and metal fragments. Remember, a clean, neat, and well-organized job is a well-run job.

Employee Hygiene. Employee awareness of good personal hygiene practices in and around a food plant is a necessity. A thorough orientation program, supplemented by publications on hygiene can instill this awareness in construction personnel. Clear rules should be established concerning such practices as spitting, garbage disposal, smoking, disposal of cigar and cigarette butts, and use of rest rooms. In addition, you should stress the effects of bad hygiene on the ultimate sanitary quality of the plant.

BUILDING MATERIAL AND EQUIPMENT INSPECTIONS

BUILDING MATERIAL INSPECTIONS

As building materials arrive on the job, they should be checked for damage as well as quality. These materials should be checked against the engineering specifications and those considered to be substandard should be rejected and replaced with quality materials.

EQUIPMENT INSPECTIONS

As equipment arrives on the job, it should be inspected for damage and to determine if it has been built to specifications. The inspector should be on the lookout for sanitary design discrepancies including the use of glass, lead-based paints, and lap seams. Following is a partial list of features to look for:

Construction to Facilitate Cleaning. Give special attention to the ease with which the equipment can be cleaned. Quick fasteners and couplings, and other easy disassembly features are particularly important. Implement any minor revisions which might facilitate cleaning.

Product Zone Surfaces. Inspect product zone surfaces to verify that they are of nonabsorbent, nontoxic, and odorless materials. Be sure the surface is smooth and free from visible scale, pits, checks, crevices, and other imperfections.

Seals and Gaskets. Seals or gaskets used in areas in or adjoining the product zone must be nontoxic, odorless, nonabsorbent, nonexuding, and of food grade quality.

Materials. Give particular attention to construction materials such as glass or cadmium which can contaminate the product. Remove glass parts and substitute with parts made of plastic. If the use of glass is unavoidable, be sure to provide adequate protection to prevent breaking and to contain the glass should breaking occur.

Nonproduct Zone Surfaces. Be sure that all surfaces are accessible or effectively sealed and reasonably smooth. Seal any joints and edges where two members are permanently jointed or lapped. Note loose parts which might fall into product zone areas and take necessary corrective measures to secure them.

Supporting Framework. Check the supporting framework for difficult-to-clean crevices and hollow areas. See that all open ends are properly sealed. Check the clearance below horizontal members and lateral braces. Watch in particular for hollow-base equipment, and make a notation to properly seal the unit to the floor or to fill the void area.

Rolled-under Edges. Watch for metal edges which are rolled under. Food particles can lodge in such areas and promote bacterial growth or infestation. Take the action necessary to eliminate them.

Junction Points and Dead Ends. Make sure that junctions between two members, a member and a tank, two pieces of sheet metal, or any two surfaces are smooth and flush with no seams or cracks which might hold food. Fill and seal pockets, recesses, or other void areas where a product might become trapped.

Equipment Covers. Check to see that covers can be removed easily.

Bearings. Note the location of bearings. Observe those which are located in the product zone area and make sure they are approved for product zone application. Bearings which do not meet this condition should be replaced, relocated, or properly protected to prevent contamination of the product with lubricants or bearing fragments. Many bearing housings are cast with recesses. These recesses form difficult-to-clean areas when bolted in place. Fill such areas with approved caulking material.

Motor Bases and Rails. Because many sanitation requirements are overlooked in motor mounting techniques, thorough inspections are needed. Motor rails which are flush-mounted on solid, flat equipment surfaces create inaccessible harbors. Mount these rails above the flat surfaces to eliminate the harbors. Formed-steel motor bases that resemble inverted cake pans should be elevated a minimum of six inches or sealed to the mounting surface.

Drives. Replace or modify sheaves, sprockets, or pulleys with deep wells.

Drive Guards. When checking the drive guard, remember that the primary function of the guard is to ensure the employee's safety. Many federal, state, and local codes require that all sides of a drive be enclosed. Also check their sanitation features; the accepted standard construction consists of solid steel top, front, back, and ends, with a semi-open bottom.

Electrical Controls. Inspect electrical controls for dust-tight features or other features indicated in the job specifications. Make sure that switches, panels, pilot lights, or other electrical components are adequately sealed to the machine, or that a distance of two inches is left between the control and the adjacent surface.

Transitions. Eliminate ledges and seal seams properly.

Name Plates. Seal name plates to the machines.

Paint. Repair nonproduct zone surfaces containing scratches, blemishes, or other stains. Product zone surfaces which have received coatings of epoxy paints require checking for chipping, for peeling, and for bonding characteristics.

Along with the appropriate drawings and specifications, the photographs, illustrations, and design recommendations on equipment found in Chapter Four should be used as a handy reference source for checking equipment as it arrives.

CONSTRUCTION PROGRESS INSPECTIONS

Regular inspections of the construction progress should be made by a knowledgeable food sanitarian. Because progress is made rapidly on construction projects, these inspections should be made at least weekly. The sanitarian should prepare an ongoing list of observations and discuss this list with the construction engineer. When necessary, they should plan possible solutions to prevent recurring sanitation violations. The value of this inspection cannot be stressed enough; its success depends directly on the harmonious working relationship between the sanitarian and the construction engineer. Both the food sanitarian and the construction engineer have contributions to make in heading off potential problems.

CONSTRUCTION QUALITY

Construction quality can be measured by taking frequent walks through the entire area and, in this way, many problems can be identified in their early stages and immediate corrective action can be taken. Make certain in your daily assessment that the work is being completed in accordance with the working drawings and specifications. Many small details such as sealing, fastening, and welding are not always spelled out on the drawings and need to be checked. Inspect the quality of welding, cutting, drilling, cement finishing, and block laying frequently. Attention to these details makes a real difference in the final results of the project.

Make progress checks to be sure that equipment is being installed in accordance with the drawings. Once again, give particular attention to installation details that do not commonly appear on working drawings. The following installation details are important ones that do not often appear on working drawings.

Clearances. Allow sufficient clearance for the use of access and inspection openings. The clearance should allow the removal or opening of all cover plates, inspection doors, and screw conveyor drop bottoms. Provide sufficient space for cleaning all adjacent areas and equipment. Do not install equipment less than 36 inches from a wall or from a piece of equipment. Keep in mind that some equipment requires extra clearance to facilitate removal of various components such as extruder augers, roll cooling rods, or mixer shafts. For equipment that passes through floors or walls, maintain an adequate clearance between it and the floor or wall to facilitate cleaning.

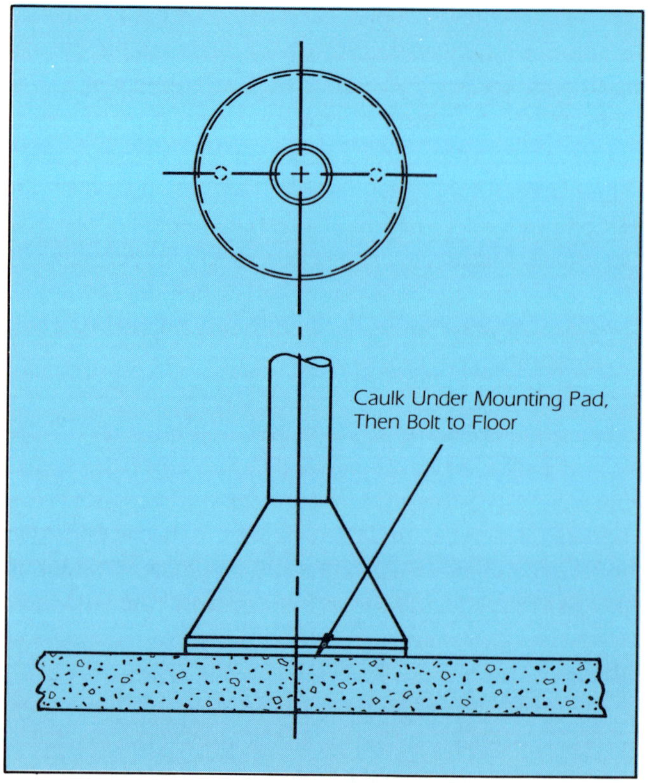

FIG. 6-1 Equipment mounting pads should be set in a caulking compound bed to obtain a good seal at the floor junction.

Floor-mounted Equipment. Secure and seal equipment to the floor or elevate it from the floor so that a distance of six inches or more is maintained between the floor and the equipment. Seal equipment mounting pads to the floor with caulking material. *(Figure 6-1)*

Wall- and Ceiling-mounted Equipment. Where equipment is wall- or ceiling-mounted, make sure all structural bracing and supports, and the equipment itself are sealed at the point of attachment.

Structural Members for Equipment Mounting. Structural members such as angles, channels, or beams, used to provide a foundation for ceiling-suspended or wall-mounted equipment, should be sealed tightly to the wall or ceiling surface. Situate structural members to eliminate crevices, dead ends, hollows, or any other void areas. Flanges of beams, channels, or angle irons should not point upward when they are installed. Flanges of angle irons or channels which are fastened to the wall should point away from the wall. If beams are to be mounted adjacent to walls, maintain a distance of four inches between the beam and the wall. If beams are to be mounted tightly to the wall, box in the backside of the beam with a piece of light-gauge sheet metal to form a sealed tubular shape.

Aisles and Traffic Ways. Make sure you maintain all established aisles and traffic ways; avoid extending equipment into designated aisles.

Building Obstructions. Before mounting equipment to the floor, observe the surrounding area for possible obstructions from doors, windows, pilasters, or ducts which might cause crowding or congestion.

Concrete Equipment Foundations. Concrete equipment foundations should be properly bonded to the existing floor by first roughing the floor in the area of the foundation and then applying a proper bonding agent. Make sure the concrete foundation is at least four inches high. Horizontal and vertical corners should be properly protected by casting a 2'' x 2'' x 3/16'' angle iron into the foundation. *(Figure 6-2)* Foundation and floor junction points should be coved with a minimum radius of a half inch.

Equipment Pits or Trenches. Avoid the use of equipment pits and trenches, except where absolutely necessary. Pits and trenches that are backfilled and closed with a knockout concrete slab can be used for groupings of pipes or conduits that otherwise would have to be installed in the overheads where regularly scheduled cleaning would be required.

Catwalks and Access Platforms. Catwalks and access platforms should be constructed in accordance with the recommendations found in Chapter Two. Sealing at the point of attachment is required where access ways are fastened to the ceiling, wall, or floor.

FIG. 6-2 Concrete equipment base

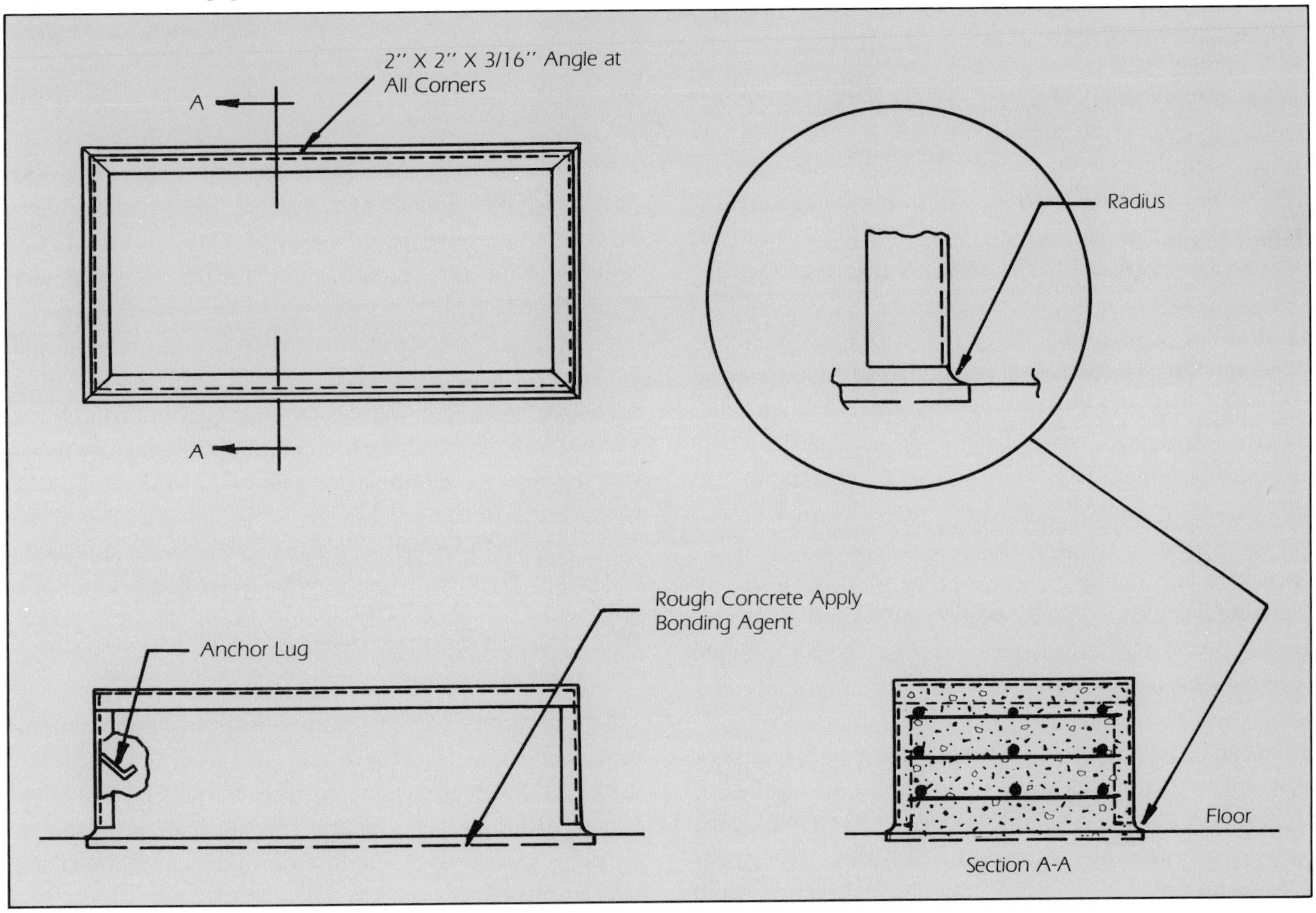

FIG. 6-3 Seamless spouting

Quick Couplings. Quick or quick release couplings commonly used for pneumatic conveying lines or liquid lines should be installed so that crevices or other areas which are difficult to clean are eliminated. Generally this requires the flanges to be installed in a vertical position or with the flanges pointing down when installed in a horizontal plane.

Ventilation Ductwork. Place ductwork to facilitate cleaning. Locate inspection openings in accessible areas and place openings no more than 20 feet apart. Do not place ductwork within 18 inches of a large adjacent surface such as a wall or ceiling unless it is to be sealed directly to that surface. Make sure all joints are tightly fastened.

Dust Pipe and Hoods. Place the dust pipe and hoods to facilitate cleaning. Locate inspection openings in accessible areas no more than 20 feet apart. All joints should be butted together and continuously welded.

Spouting. Make sure all spouting is made of seamless or machine-welded stainless steel, aluminum, or mild steel. *(Figure 6-3)* The most desirable construction methods for elbows are either cast-aluminum, cast-iron, or segment construction with all joints continuously welded and ground smooth. Locate hand holes in accessible areas on each floor. Bond hand hole gaskets to the hand hole door by vulcanizing or cementing with a water-repellent cement. Choose gaskets which are made of closed cell, food-grade rubber, or synthetic rubber. Make sure this material is nontoxic, nonabsorbent, nonexuding, and not affected by ordinary cleaning methods.

All spout-to-spout or spout-to-fitting joints should be tightly butted together and welded or fastened with a take-apart coupling. Where spouting passes through the wall or floor, tightly grout the spouting with a cement grout mixture. Secure spouting that passes through steel decks with steel floor flanges. Locate these flanges on top of the deck only and properly seal them to the spout and deck.

FIELD FABRICATION

Make sure that field fabrication done on the job site, or otherwise contracted for at the field level, conforms to drawings and specification requirements. If no drawings or specifications are available, give special attention to creating specific fabrication instructions that meet the sanitation needs.

CHAPTER 7
MAINTENANCE

CHAPTER CONTENTS

PREVENTIVE MAINTENANCE

BUILDING MAINTENANCE

EQUIPMENT MAINTENANCE

TEMPORARY REPAIRS

EQUIPMENT CLEANOUT

MAINTENANCE SHOP AND STORAGE AREAS

MAINTENANCE DURING OPERATIONS

MISCELLANEOUS MATERIALS

NEW INSTALLATIONS

APPROVED LUBRICANTS

A giant step has been taken when the building and equipment have met sanitation design needs and when the sanitation program reflects this effort in lower cleaning and pest control costs. Once these goals are accomplished, we must now meet the challenge of maintaining the structure and equipment in a sanitary condition. Because the maintenance department personnel are the key to this challenge, their technical skills and knowledge must include good *food safety and sanitation* practices.

Government regulations such as the Umbrella Good Manufacturing Practice (excerpt following) specify how to maintain a food manufacturing plant in top condition. *The design, construction and use of equipment and utensils shall preclude the adulterations of food with lubricants, fuels, metal fragments, contaminated water, or any other contaminants. All equipment should be so installed and maintained as to facilitate the cleaning of the equipment and of all adjacent spaces.*

This statement strongly emphasizes the importance of maintenance to the continuing success of the product safety effort. Maintenance affects both the cleaning aspects of the sanitation program and the safe quality of the finished food product. There have been many serious and costly product recalls due to direct product contamination from an equipment malfunction.

PREVENTIVE MAINTENANCE

Preventive maintenance is commonly regarded as a necessary element of high productivity. While its contribution to productivity is undisputed, it is also paramount for product safety. Preventive maintenance avoids costly and painstaking product recalls or product defects. Its value cannot be overemphasized.

INSPECTIONS AND HISTORICAL RECORDS

Preventive maintenance requires regular, planned equipment and building inspections which identify those components that are likely to fail and which include them in an inspection and repair plan. The equipment manufacturer's repair manual can be helpful in identifying these components. Computerized work history or repair records provide another excellent method for predicting equipment failures and for planning a preventive maintenance program. *(Figure 7-1)* These records also establish good cost histories for equipment.

LUBRICATION

The lubrication program is included as a primary part of the preventive maintenance program. The proper use of good lubricants does much to prevent direct product contamination from equipment failure. Nonetheless, the lubricant itself can be a contaminant. It is therefore imperative that only approved lubricants are used in the proper places and in the proper amounts. Far too many sanitation inspection reports cite incidents of overlubrication as possible sources of product or product-zone contamination.

BUILDING MAINTENANCE

Building maintenance involves timely repair of building materials as the need occurs. Deferring building maintenance can compound sanitation problems. Vermin gain access through and harbor in holes or cracks in the structure. Structure damage readily supports dust and dirt, and creates areas which are difficult to clean. Areas in need of repair frequently increase the risk of direct product contamination from such things as peeling paint, torn insulation, and broken light fixtures. Building repairs should be given a high priority, and they should be made as close to the date of damage as practical.

FLOORS

Repair floors immediately. Floor cracks and holes commonly harbor insects and other vermin such as rodents. They quickly become filled with dust, dirt, and sometimes moisture, all of which provide an excellent environment for bacteria, in particular. To prevent the repair material itself from becoming a source of contamination, make all repairs carefully.

Recoat or upgrade surfaced or coated floors as the need develops. It is more than likely that these floors were surfaced in the first place to make them easier to clean. Proper maintenance will retain this cleaning advantage.

Replace damaged brick and tile floor materials. Scrape, clean, and regrout failing joints. *(Figure 7-2)* Joints which have failed or broken down readily support numerous types of insects, including cockroaches and booklice.

Patch holes in asphalt surfaces such as track wells or truck approaches as soon as they appear. Dunnage from boxcars and truck trailers easily becomes lodged in these damaged areas.

WALLS

Wall cracks and holes can also harbor insects and other vermin, and, therefore, should be repaired promptly. Cracks and holes in concrete block walls, glazed tile walls, and mortar joints provide vermin with an avenue of access to the hollow core areas of the block. *(Figure 7-3)* Once they gain access to these areas, the vermin can move freely from one end of the wall to the other.

Recaulk precast wall panels as needed. Make certain that only good, pliable caulking materials are used.

In addition, recaulk metal sandwich panel wall joints as required. This wall surface should be kept free of any support bracket mounting holes. Quickly seal any holes which might have been made in this wall surface.

FIG. 7-1 Work history records play a role in preventive maintenance, an important component of the product safety program.

A B C FOOD CO.

EQUIPMENT HISTORY ANALYSIS

PAGE 1

EQUIP #: 022-10	DEPT : Proc	PURCHASED : 12/71
MFG : Doe Corp.	MODEL: 21V2	ORIG. COST: 98,000
DWG NO : 246-12	CLASS: II	REPL COST : 130,000

CLOSED	JOB DESCRIPTION	W/O #	HOURS	MAT'L	CONTR	TOTAL	REP
11/1	Repl Heater	MWO-24612	17.8	374	-	819	
12/1	Seal Auto Valve	MWO-7024	4.0	10	-	110	X
05/16	Repl Solenoid	MWO-7150	7.3	256	-	439	
07/17	Seal Water Line	MWO-7200	24.8	1,507	-	2,127	

* * * * * * * * * * * * * * * * SUMMARY * * * * * * * * * * * * * * * * *

YEAR	# W/O	LABOR HOURS	LABOR COST	MATERIAL COST	CONTRACT COST	TOTAL COST	% MAINT VS REPL
1981	440	54	1,350	2,143	-	3.493	2%
1980	882	142	3,550	1,456	-	5,006	3%
1979	136	281	7,025	2,732	1,575	11,332	6%
1978	211	354	8,850	6,952	-	15,802	8%
1977	101	98	2,450	1,165	255	3,870	2%
PRIOR	115	1,858	46,450	28,896	3,660	79,006	42%
TOTAL ...		2,787	69,675	43,344	5,490	118,509	63%

FIG. 7-2 Broken or damaged floor tiles or joints should be repaired promptly.

FIG. 7-3 Broken wall tiles should be replaced.

Resurface painted walls as needed. Remove peeling paint immediately. Make sure that paints such as epoxy or latex-base are used for repainting. Do not use any paints that contain heavy metals, such as lead.

CEILINGS

Ceiling holes and cracks require prompt repair. Remove such things as unused anchor bolts or support brackets from the ceiling and fill all holes. Make overhead ceiling repairs carefully; falling repair materials increase the risk of product zone contamination.

Peeling paint from ceiling surfaces also represents an increased chance for product zone contamination. *(Figure 7-4)* The immediate removal of peeling paint from overheads in manufacturing areas is imperative. If peeling paint is a prevailing problem, consider removing all paint in the immediate area of concern and leaving the surface bare.

Concrete ceilings may be stained rather than painted. Stains do not peel because they penetrate the surface rather than form a film over it. Painted structural steel surfaces which are peeling should be scraped, sand-blasted and recoated with epoxy enamels. Where feasible, the structural member should be replaced with a new corrosion-resistant steel member, such as a mild steel member that has been zinc-plated.

ROOFS

Roof leaks are critical contamination concerns that require immediate attention. Water seeping through the roof has ample opportunity to become badly contaminated, and, as it drops from the ceiling, it can easily find its way into any number of product zone areas below.

DOORS, WINDOWS, AND SCREENS

Damage to doors, windows, and screens generally provides an opening for insects and rodents to enter the plant; consequently they must be taken care of without delay. *(Figure 7-5)* Remember that rodents only need an opening as small as 3/8 inch to get into the plant.

LIGHT FIXTURES

Damage to light globes or protective shields requires immediate attention, since glass is a very dangerous food contaminant. Damage to the light fixture itself can provide access for insects. Many light fixtures become natural insect incubators because of the temperatures they produce.

FIG. 7-4 Peeling paint is an ongoing threat to all food products.

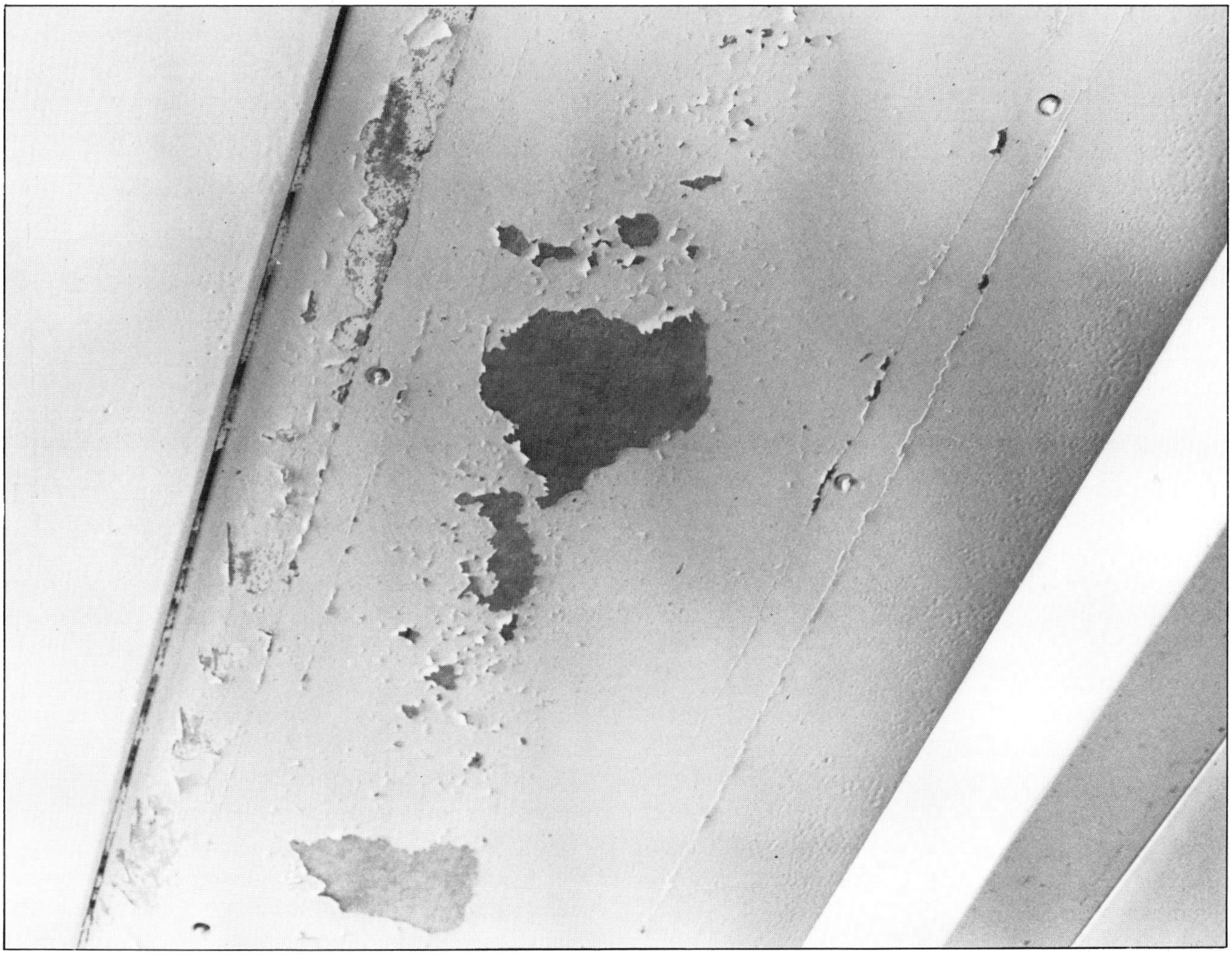

FIG. 7-5 Broken door screen

FLOOR DRAINS AND LIQUID WASTE LINES

Keep floor drains open and maintain liquid waste lines so that they freely carry the water from the drain area, according to the original system design. Backed-up liquid waste represents both a sanitation problem and a food safety problem.

PLUMBING FIXTURES AND SANITARY SYSTEMS

Maintain water closets, urinals, and lavatories in the best possible working order. When making any modifications or additions to the plumbing system, keep in mind that the plumbing system must remain independent of the process liquid waste handling system, as recommended in the plumbing section of Chapter Three. Also make sure that plumbing changes are made in accordance with national and local codes.

VENTILATION SYSTEMS

In addition to the need for a good preventive maintenance program on all of the heating and ventilating air handling equipment, the incoming air filters should be appropriately monitored and cleaned or replaced as needed. *(Figure 7-6)* Screened air intake openings should be programmed for cleaning and repair. Dirty screens and filters not only represent a contamination threat, but they frequently starve the area of badly needed air, causing increased moisture and odor problems as well as mold formations. In addition, manufacturing systems which are under a negative atmosphere are harder to keep dust-tight.

INSULATION

Torn pipe and duct insulation located in the overheads of manufacturing and storage areas represents a contamination threat. Make insulation repairs quickly and carefully. Where it appears necessary, include protective shields to prevent damage from recurring.

EQUIPMENT MAINTENANCE

Food manufacturing equipment is in itself a product safety concern. Malfunctioning equipment can cause direct product contamination in many ways. The regulatory statement quoted previously leaves little doubt that the manufacturing equipment is carefully and cautiously observed throughout the course of all sanitation inspections. Equipment repairs must be made immediately, even if the system must be shut down to do so. Failure to make prompt repairs can result in critical health hazards or in contamination of the product in process with injurious materials.

The use of proper repair materials is as important as promptness in making equipment repairs. Use only food-grade materials which have been approved for that specific use. If the original components are not available, make sure that the materials substituted are of a satisfactory food-grade quality.

It is also important to thoroughly test-run and to clean repaired equipment before putting it back into service. Make certain that the equipment is running properly and that nothing else is wrong with it. In situations where the work load is excessive, the equipment should be tested under load conditions and dry-run tested.

FIG. 7-6 Air filters should be on a regular cleaning or replacement program.

CONTAMINATION COMPONENTS

The following equipment components are sometimes involved in contamination incidents:

1. Bearings of all types, including ball, roller, and bushed. Metal fragments from the bearings and bearing grease can be serious contamination threats. It is of particular importance to avoid placing bearings in product zones. A good bearing inspection and replacement program is important to the product safety effort.

2. Scalper and sifter screens. The wear factor on scalper and sifter screens depends on the screen material. Because cloth screens tend to wear faster and are more sensitive to damage from foreign objects, they need frequent inspections. On the other hand, metal wire screens are not as sensitive to damage; consequently they require less inspection. Nonetheless, needle-like wires from metal screens can cause critical contamination. If at all possible the screen wire should have a magnetic property. *(Figure 7-7)* Chapter Four discusses this point further.

3. Motor seals. When motor seals give way, particularly those in gear motors, the gear box oil leak becomes a serious contamination threat. Repair leaking seals promptly.

4. Peeling paint. Contamination caused by peeling paint from equipment and from overhead supports can be prevented by scraping off any loose paint. While peeling paint of any type is a potential contaminant, those paints containing toxic substances are especially dangerous. Non-lead base paints can reduce the seriousness of this type of contamination.

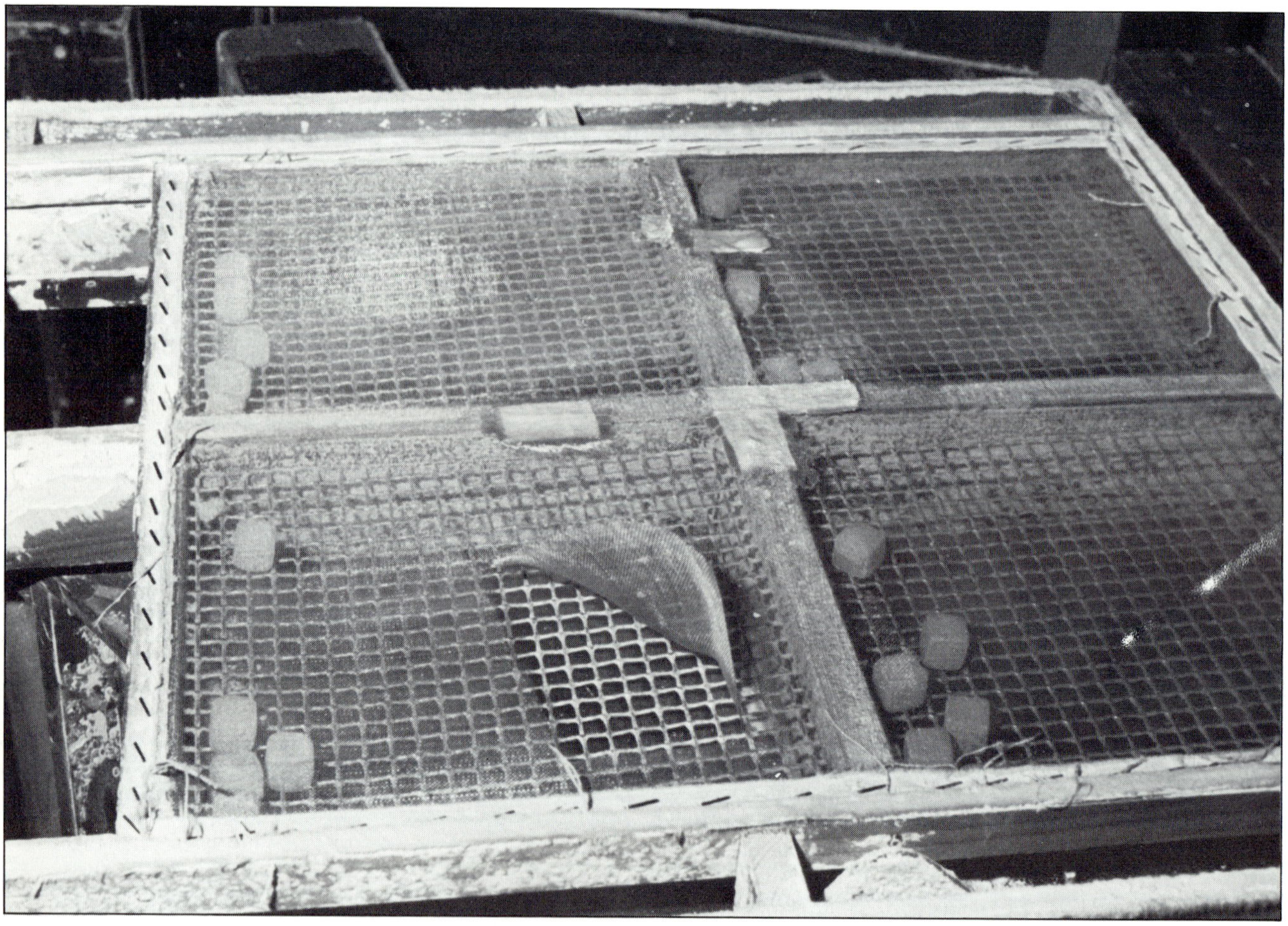

FIG. 7-7 Scalper and sifter screens should be routinely checked for damage.

5. Cutters and slicers. Implement a frequent preventive maintenance inspection schedule for all cutting and slicing equipment. Carefully observe cutting knives for nicks, gouges, and other signs of damage. Replace damaged blades immediately before they cause additional problems.

6. Grinders/hammermills/pinmills. Closely examine this equipment for hammer wear, missing pins, and screen damage.

7. Dust collectors. Dust collectors with cloth filters require filter sock changes. These changes should be programmed to the pressure drop across the filter. When the filter media becomes heavily laden with dust encrusted or imbedded in the cloth pores, the pressure drop increases significantly. The dust collecting ability of the system is greatly affected as a result.

8. Controls. Check temperature and pressure controls frequently. Do all necessary calibration. Temperature controls such as those found on retorts or pasteurization equipment require frequent maintenance checks, as does temperature equipment used on refrigerators and freezers.

9. Weighing and metering equipment. Check the accuracy of all equipment used to automatically weigh and meter food ingredients. Failure of this equipment to perform according to formula requirements might not necessarily result in a product safety problem. However, such failure almost always results in a discrepancy between the

actual and the declared quantities of ingredient contents displayed on the package. Vitamin measuring equipment in particular requires frequent accuracy checks.

10. Magnets and metal detectors. A magnet maintenance program amounts to no more and no less than a good cleaning program. Cleaning removes heavy accumulations of trapped metal that affect the overall strength of the magnet. Occasional magnet strength tests on magnets exposed to temperatures of 700° to 900°F or to various abuses are also helpful. When subjected to these conditions, some magnets lose some of their strength. Test kits are available. *(Figure 7-8)*

    Metal detectors are very sensitive pieces of equipment; check and adjust them as required to meet the plant's established reject standard.

11. Conveyors. Many mechanical conveying devices such as belt conveyors, bucket elevators, and screw conveyors require careful checking for signs of misalignment. Misalignment results in metal-to-metal contact or in belt wear-and-tear. All too frequently, sanitation inspection observations note missing pieces from a belt lacing or missing pieces of rubber conveyor belts. *(Figure 7-9)*

12. Metal-to-metal contact points. Many pieces of equipment operate with very close tolerance between moving components. The slightest misalignment or distortion of one of the components generally will result in metal contamination. Identify all equipment of this type and place it on the preventive maintenance inspection schedule.

13. In-line oil filters. Saturation often occurs on in-line oil filters used on compressed air systems. Contamination results when oil carries over with the air and makes contact with the food or food-zone surfaces. Place all filters that tend to accumulate heavy amounts of oil on the preventive maintenance schedule.

Identify cooking and cooling equipment and critical temperature controls and recording devices used in conjunction with this type of equipment. Malfunction of this equipment can lead to improper cooking or refrigeration of products, a direct cause of several types of food poisoning. This equipment needs a high maintenance priority and regular servicing. In addition, temperature controls must be checked and calibrated on a routine basis.

FIG. 7-8 Magnet test kit

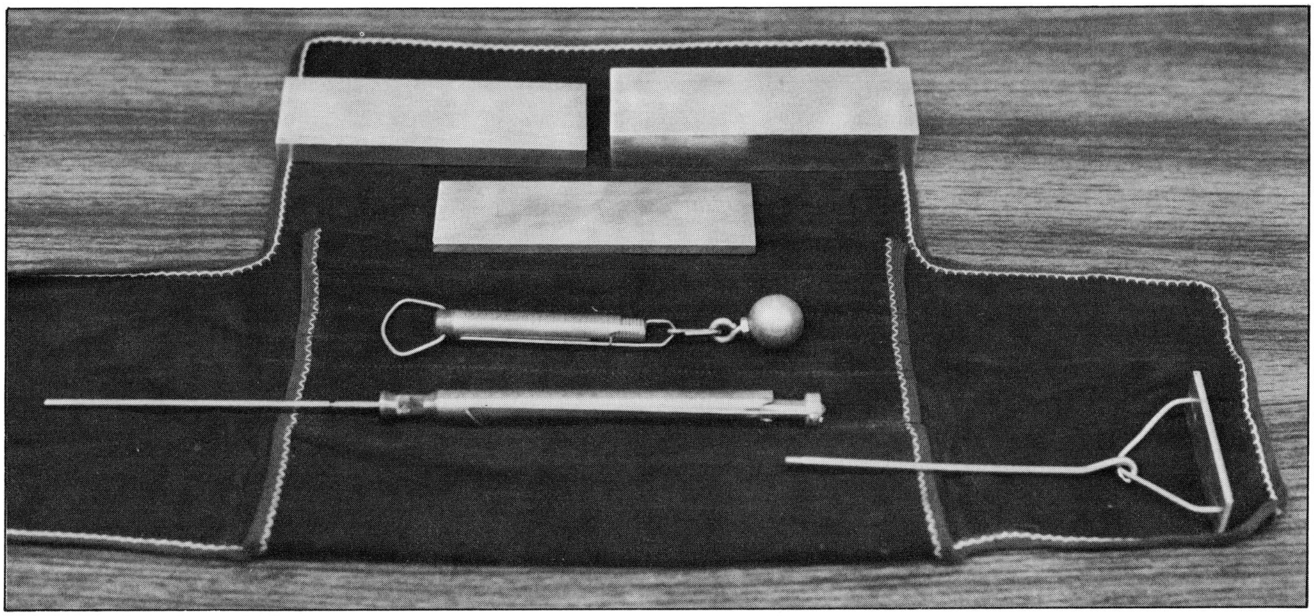

FIG. 7-9 Worn and torn conveyor belts are a product safety risk.

TEMPORARY REPAIRS

Temporary repairs to prevent leaks, dust emissions, and product spills, and to prevent metal-to-metal contact will probably always be necessary. However, temporary repairs have a way of becoming permanent. This tendency is of concern because many of the materials used to make temporary repairs, such as tape, cardboard, and wood represent a serious food contamination threat. *(Figure 7-10)* Temporary metal baffles or diverters are rarely secured adequately and, as a result, they eventually make contact with moving metal components. Caulking materials used to plug small cracks end up in the product. Baling wire is sometimes subject to vibration or constant flexing which causes it to break. String, of course, is also easily broken.

To prevent contamination, make permanent repairs as soon as possible. Dating repairs will keep them from becoming permanent.

EQUIPMENT CLEANOUT

Failure to properly clean out equipment after maintenance repair is one of the major causes of food contamination. Thorough equipment cleaning requires attention, not only to greasy and dirty surfaces, but to the following:

1. Drill shavings/metal filings. All such material must be removed.

2. Weld slag. Chip and brush welds, and remove all loose slag. Pick up welding rods, including the small ends of spent rods.

3. Repair materials. Remove nuts, bolts, screws, gaskets, and repair plates that have not been used, and return to the storeroom. Often a handful of nuts and bolts is left lying on the equipment or on a nearby ledge. Loose objects of this nature are sure to find their way into the product. *(Figure 7-11)*

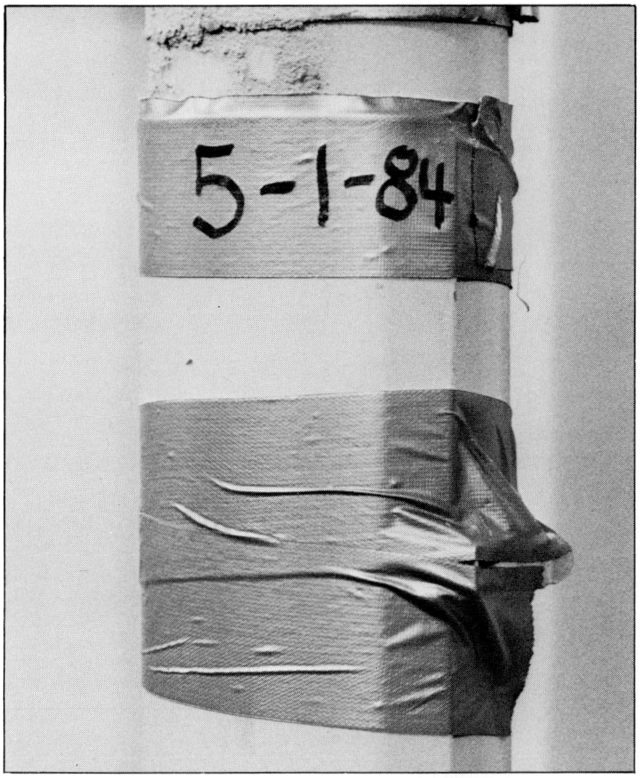

FIG. 7-10 Temporary repairs have a way of becoming permanent and for that reason should be dated.

4. Electrical components. Electrical materials such as wire and wire nuts fall into the same category as the repair materials mentioned above, and the same precautions apply.

5. Excessive lubricants. Many jobs require greasing and oiling. Make certain that all excess lubricant is wiped off and that the grease rags themselves are not left behind.

6. Tools/check out instruments. Remove all tools and instruments used in the repairs and return them to their appropriate storage areas. Provide trouble lights with plastic-coated, shatterproof bulbs, and be sure to remove them from the repaired equipment.

It is very important to make certain that the surrounding work area as well as the repaired equipment is clean. Sweep up metal filings and drill shavings, in particular. Pick up all leftover material and return to storage.

When a maintenance operation is completed, the

FIG. 7-11 Loose parts left lying around easily find their way into food products.

operating department should be informed that the equipment is ready for testing and for a thorough cleaning. Both of these steps should be part of a standing policy to minimize food contamination from equipment repairs. It is a good idea to include in this policy a final mandatory sanitation inspection. The final inspection should be accomplished only after the repair, testing, and cleaning steps have been completed, and the equipment is considered ready for production.

MAINTENANCE SHOP AND STORAGE AREAS

SHOP

The sanitation level in the maintenance shop should be comparable with the manufacturing area. Because of the day-to-day activities of equipment repair in the shop, there is a constant need to pick up both ingredients and dust particles that are removed from the equipment during the repair process. Failure to remove these materials can result in insect infestations, bacteria, or other serious sanitation problems. These same problems easily spread to the manufacturing area when the repaired equipment is returned to service. Unless they are carefully managed, many shop activities can pose serious contamination threats. Drill presses, metal lathes, grinders, and metal band saws are just a few frequently used tools that produce metal filings, shavings, and slivers. Invariably, large quantities of these metal contaminants accumulate on the floor around this type of shop equipment. The bottoms of shoes or work clothing can carry the contaminants back to manufacturing areas. Picking up filings, shavings, and slivers prevent them from developing into a sanitation problem. In addition, many lathes and drill presses can be equipped with catch pans. The use of cutting and lubricating oils must also be controlled. Excessive use can easily contaminate the product zone surface of equipment that is in the shop for repair.

Good sanitation and good housekeeping practices in the shop area result in an efficient maintenance operation.

STORAGE

Spare part storage and idle equipment rooms also call for high sanitation standards. Give particular care to the inspection of incoming parts for rodents and insects. A critical plant rodent infiltration can begin with a nest of rodents shipped into the plant along with some spare parts. Packing materials such as excelsior make excellent rodent nests. Also, make sure that oils or other fluids in spare parts are not toxic.

Sanitary maintenance storage practices require the thorough cleaning and inspecting of all equipment before it is put in storage. Used items to be placed in idle equipment storage should receive particular attention. If various food ingredients or products are not removed from stored items, all types of problems from serious odors to critical vermin infestations inevitably develop. These problems can then spread to other plant areas.

In one respect, maintenance people can be typical scavengers: they hesitate to throw something out for fear there will be a need for it tomorrow. To some extent this practice is encouraged by equipment depreciation schedules based on the economic life of the equipment. Because of a reluctance to retire equipment that is not fully depreciated, many obsolete, unused pieces of equipment are kept. The money spent trying to keep it clean, however, far outweighs any undepreciated value it may have. Equipment that is no longer of use should be disposed of.

Segregate metal, and identify any reusable materials properly and return them to their original storage area. Dispose of remaining materials. Accumulations of unmanaged scrap quickly gather dirt and filth, and frequently become a living area for insects and rodents.

MAINTENANCE DURING OPERATIONS

Sometimes it becomes necessary to perform maintenance repairs as a system or a nearby system or individual piece of equipment continues to operate. In such situations, extreme care must be taken to prevent contamination of the food being processed in the nearby system, and of open product zones. Such things as metal shavings, filings, and flying particles of hot melting metals from a welding operation must be contained. Many paints produce odors. Do not apply them during operations if there is a danger of the product in process absorbing the paint odors.

MISCELLANEOUS MATERIALS

Materials used in the course of normal maintenance, like those used in temporary repair operations, must be suitable for use in a food plant. Materials coming into contact with food or food contact surfaces should be food-grade. Some of these miscellaneous materials include, but are not limited to the following:

1. Cleaning solvents and paints. There are numerous cleaning solvents, all with varying chemical compositions. Select and use them with food safety in mind. Remove solvents from manufacturing areas and store them in controlled areas. Make sure paints are nonlead. Use and store them so they do not contaminate the food in process.

2. Caulking materials. It is a good idea to control the use of caulking materials, use them only when absolutely necessary, and never use them in or near product zones if a possibility of contamination exists. Only food-grade caulking materials should be considered. Food-grade caulking materials contain no known toxic substance.

3. Silicone sprays. Silicone sprays or other materials are frequently used to free frozen parts or squelch squealing belts. Select these sprays carefully and use them discriminately. Not only are silicones contaminants, in some cases they can affect the quality of the product. For example, silicones can affect the leavening action of cakes.

4. Boiler feedwater additives. There are many boiler feedwater additives that contain harmful or toxic chemicals. Boiler feedwater chemicals used in a food plant should be food-grade and approved for use.

Not enough can be said about the importance of *maintenance for sanitation* to the product safety program. This section has by no means covered all possible maintenance issues. Each plant has special problems which require careful attention and prompt maintenance action. These must be identified and incorporated into the overall maintenance program.

NEW INSTALLATIONS

On numerous occasions, the maintenance department is called upon to install new systems or even to make building additions. It goes without saying that these new installations or additions must conform to good *food safety and sanitation* practices.

APPROVED LUBRICANTS

An approved lubricant program is very important to every food manufacturing facility, and every plant should consider one. Many petroleum base lubricants contain toxic substances such as lead and arsenic. In large enough quantities, any oil is harmful. In any quantity, oil is objectionable and considered a food contaminant by all food regulatory agencies.

The United States Department of Agriculture (U. S. D. A.) has established a program for grading or classifying lubricants. Many of the major oil companies manufacture a line of products that meet the requirements of the U. S. D. A. program.

An approved lubricant program should include at least three distinct lubricant classifications:

- Food grade (edible)
- Nonfood grade (nontoxic)
- Nonfood grade (toxic)

Each of these classifications should be clearly defined. Products should be listed under these classifications and used in the prescribed manner.

FOOD GRADE (EDIBLE)

Food-grade edible lubricants may be used in food contact zones. Food-grade lubricants are not approved by the F. D. A., but the components of these lubricants are considered safe for use around food products. Lubricants acceptable for incidental food contact appear on the U. S. D. A. *List of Chemical Compounds*; they are classified as H-1. Mineral and vegetable oils generally qualify for this classification. There is an F. D. A. tolerance of 10 parts per million for mineral oil in food products.

NONFOOD GRADE (NONTOXIC)

Nonfood-grade nontoxic lubricants are considered contaminants and carry a 0 tolerance in food products. Their use should be restricted to nonproduct zones. In situations where these lubricants must be used in equipment components located in product zones, take extra precautions to prevent accidental contamination of the food product. If accidental contamination does occur, hold the product and take immediate corrective action. Nonfood-grade nontoxic lubricants should have an oral LD_{50} of 5g/kg or greater. This classification applies to the most widely used lubricants.

NONFOOD GRADE (TOXIC)

Do not use nonfood-grade toxic lubricants anywhere within the food manufacturing facility. If for any reason, a plant cannot comply with this requirement, establish a special approvals program. A special approvals program should clearly establish the conditions and limits of use; it should also include an inspection program, which will absolutely guarantee that no contamination could occur.

The best remedy for lubrication contamination problems is to avoid locating bearings, motor, and other equipment requiring lubrication in or over product zones. Where there are recognized contamination possibilities, establish extra precautions, including a good maintenance program. These safety steps, coupled with an approved lubricant program which lists all the approved products, including hydraulic fluids, should minimize the contamination concerns.

CHAPTER 8
CLEANING

CHAPTER CONTENTS

WET CLEANING

CLEAN-OUT-OF-PLACE (C. O. P.) SYSTEMS

CLEAN-IN-PLACE (C. I. P.) SYSTEMS

CENTRAL CLEANING SYSTEMS

INDIVIDUAL HOT WATER HOSE STATION

DRY CLEANING

Just as a great many economic benefits result from the sanitary design of buildings and equipment, economic benefits also come from the careful planning of a cleaning system. A good design can greatly reduce cleaning costs and at the same time vastly improve the quality of cleaning. To accomplish this, it is necessary to define the cleaning job and then select the right cleaning system to get it done. Technological advances in cleaning equipment and cleaning materials will make this process easier.

Because it sometimes seems difficult to justify the capital investment required for various cleaning systems, it is important not to overlook possible productivity improvements. Properly applied, cleaning systems get the cleaning job done faster, putting the manufacturing system back on stream more quickly. This results in optimum case outputs and reduces the need for overtime wages. Cleaning functions can be productive!

In Chapter One we discussed the importance of linking the sanitary design requirements to the type of risk a specific food product represents. Likewise, the cleaning requirements must also be linked to this risk. Some food manufacturing systems are very sensitive to microbial problems and must be thoroughly wet cleaned and sanitized on a frequent schedule. Some systems are basically dry processing systems; in the interest of minimizing microbial risk, they should be dry cleaned. Then there are those that will require dry cleaning and wet cleaning because of the nature of the process or the stickiness of the ingredients.

WET CLEANING

Wet cleaning is a science which includes the chemistry of cleaning and the proper application of cleaning equipment. The chemistry of cleaning involves the proper selection of chemicals and the quality of water. In this chapter we will not discuss cleaning chemicals. This subject is covered in many food sanitation texts. However, you should be aware that many different types of dirt and surface film collect on many different food manufacturing equipment surfaces. The proper selection of cleaning chemicals can save many cleaning hours and improve the quality of the cleaning job dramatically.

The proper application of cleaning equipment begins with becoming familiar with some of the various cleaning systems and retrofitting them to the cleaning task as required. Wet cleaning systems experts have more or less adopted a language of their own, and specific terminology is used to define the system. Some of the more common systems are:

- Clean-out-of-place systems—C. O. P.
- Clean-in-place systems—C. I. P.
- Central cleaning systems—C. C. S.
- Hose stations

Each of these systems has a number of advantages. In most cases they generally offer a high quality of cleaning and cleaning economics. Remember that it is important to understand the basic cleaning principles involved so you can put them to work—and so you can retrofit them! These principles are discussed in more detail in the sections that follow.

WATER QUALITY

It goes without saying that cleaning water must be of potable quality. In addition, of course, the water must be soft. Hard water greatly reduces the cleaning action of many detergents. Hard water in itself can create a number of surface scale or surface film problems such as rust formations or calcium deposits. An effective cleaning program must have an adequate supply of soft, hot water. Conditioned water reduces the corrosive effects of hard water and extends the economic life of plumbing systems as well as that of many other pieces of equipment using water. Many efforts have been made to classify water hardness. The U. S. Geological Survey Water Supply Paper number 658 classifies hardness as follows:

- Soft. 0-3.5 grains/gallon
- Moderately Hard. 3.5-7.0 grains/gallon
- Hard. 7.0-10.5 grains/gallon
- Very Hard. Over 10.5 grains/gallon

CONDITIONING EQUIPMENT

The plant water treatment system should include zeolite and dealkalizing equipment where water quality requires it. Sodium zeolite softening is the most common conditioning system. The primary purpose of conditioning is to remove scale-forming calcium and magnesium ions, which are constituents of hard water. Dealkalizing equipment reduces the alkalinity of the treated water and the subsequent formation of carbon dioxide in the steam. This reduces steam and condensate line corrosion as well as excessive boiler tube scaling. Corrosion is a compounding steam impurity problem when steam is used in direct contact with the food product.

WATER TEMPERATURE

There are many opinions about the correct temperature for cleaning water. Whereas there may be an application or two for exceptionally hot water (165° to 200°F), generally speaking, water at 110° to 145°F is more than adequate. In fact, many detergents are most effective in the presence of warm water. Extremely hot water can set or fix soil to the surface which is to be cleaned.

CLEAN-OUT-OF-PLACE (C. O. P.) SYSTEMS

Clean-out-of-place systems do exactly what the term suggests, cleaning out of place. Essentially, this means moving equipment from its normal location (and/or disassembling it and removing it from its normal location) for cleaning. The removed equipment or components are generally taken to a cleaning room that has been specially designed and equipped for cleaning out of place. This room is sometimes used to clean utensils, portable equipment, trays, and pans as well.

CLEANING ROOM

The equipment cleaning room should be constructed of moisture-impervious materials that are easy to clean. *(Figure 8-1)* Use design criteria that meet requirements for a wet cleaning area. Make sure the room is large enough to accommodate the largest piece of equipment likely to be cleaned in it.

Building Materials for Cleaning Room. Construct walls, ceiling, and floor of materials that can be easily wet cleaned. Use ceramic or glazed tile for the walls and ceilings. Floors should be quarry tile or acid-proof brick, including epoxy grouting mixtures.

Provide adequate lighting approximately 80-100 foot-candles and use lighting fixtures suitable for wet areas. Install at least two 110-volt NEMA 4 receptacles for electricity. The floor should have a floor drain (complete with grate, screen, and trap) and should slope to the drain at the rate of 3/16 inch per foot.

FIG. 8-1 This equipment cleaning room features building materials which are impervious to moisture and the proper cleaning equipment to get the *clean-out-of-place* accomplished as easily as possible.

In the wash area, use double or single doors, made of metal or rubber. (Do not consider using wood.) The doors must be wide enough to allow clearance for the largest piece of equipment to be cleaned.

The room will need an exhaust fan to expel steam vapors. Drip legs with shut-off valves should be installed in the fan and exhaust ducts to drain off accumulated condensate.

Equipment for Cleaning Room. The cleaning room should have a large, deep-well utensil cleaning sink constructed of stainless steel. The sink should be equipped with a water mixing valve supplied with an adequate quantity of hot and cold water. *(Figure 8-2)* Mount a separate hose cleaning station on the wall in a convenient location. (This station may be part of, or similar to, the stations described later in this chapter under Central Cleaning Systems.) Depending on your preference, it may be either a high-pressure, low-volume system or a low-pressure, high-volume system. Since the room is specifically designed for wet cleaning you might find a low-pressure, high-volume system more useful. The room should also have a separate wall-mounted sanitizing

FIG. 8-2 Large deep-well stainless steel sinks are used for utensil cleaning.

station with a water supply and a self-contained spray sanitizing unit. *(Figure 8-3)* These units are designed to draw predetermined amounts of concentrated sanitizing solutions into the water stream. Equip the station with an adequate length of hose complete with an automatic shut-off type nozzle. Backflow prevention equipment should be provided in potable water supply lines. A portable wet vacuum unit would be useful for this area as well.

Shelves, wall hangers, and drying racks provide space for cleaned utensils. Detergents, wiping towels, and other supplies can be kept in storage cabinets. Waste receptacles are also necessary.

Food manufacturing operations requiring a lot of disassembly of smaller equipment parts for out-of-place cleaning would benefit from equipping the cleaning room with a specially designed recirculating tank-type washer. *(Figure 8-4)* Recirculating tank washers have a series of turbo-action jet nozzles and a high-efficiency stainless steel pump to circulate the water. Detergents are added and cleaning takes place as the water flows turbulently through the tank. Parts to be cleaned are set in the tank and left until they are clean. Constructed of heavy-gauge stainless steel, the tank must be of sufficient size to accommodate the largest components intended to be cleaned. Specially designed stainless steel baskets permit cleaning of smaller parts. Important tank construction details are rounded corner junctions, automatic temperature regulating equipment, direct-reading thermometers, strainers on the water inlet, and valves to drain the tank. Optional equipment may include a chemical feed system and a heat exchanger for heating the water.

Recirculating tank washers can be purchased from manufacturers specializing in the design of this equipment or they can be designed by most food plant engineering departments.

CLEAN-IN-PLACE (C. I. P.) SYSTEMS

Clean-in-place systems can be used to clean process piping systems, bins, tanks, mixing equipment, or larger pieces of equipment in place (without disassembly) where interior product zones are fully exposed and soil can be readily washed away by the flow of the cleaning solution. *(Figure 8-5)* These systems use high volumes of water to literally wash away the soil. The success of the system is keyed to proper water flow, temperature, and selection of the right cleaning chemicals. If the equipment being cleaned has been correctly designed in the beginning, this method can do a better and quicker job of cleaning than standard methods, and a great deal of time and effort can be saved. Designs for C. I. P. systems range from very basic manual or semi-automatic recirculating spray systems to highly sophisticated and fully automated systems for cleaning bacteria-sensitive process piping systems, depending on your needs.

RECIRCULATING SPRAY CLEANING

When bins, tanks, mixers, or similar equipment are to be C. I. P. cleaned, spray cleaning is involved. Successful spray cleaning applications require a properly designed spray assembly with a constant water flow at the required pressure and temperature. Spraying assemblies that consist of spray balls or nozzles can be stationary, water driven, electric, or air propelled. These units may be permanently installed in the equipment or they may be portable and put in place at the time of cleaning. The spray assemblies must be designed and located to assure that the cleaning solution contacts the entire surface to be cleaned. *(Figure 8-5)* The minimum C. I. P. solution circulation rate should be 10 gallons per hour (GPH) for each square foot of surface to be cleaned. Large-size storage tanks having a capacity of 3,000 gallons require a minimum cleaning solution circulation rate of 4,000 GPH.

The use of recirculated spray cleaning techniques really begins with the design of the process equipment. Bins, tanks, mixers, or other equipment which is to be spray cleaned should be designed with spray cleaning in mind. Equipment components such as access doors, sight glasses, temperature wells, mixer blades, inlet and outlet connections should be designed and located so they, too, can be cleaned by the sprayed solution.

The primary element to the C. I. P. system is the recirculating equipment. This consists of a solution tank, centrifugal pump, and a series of valves which are normally air operated. *(Figure 8-6)* Cleaning chemicals can be injected into the water stream or added directly at the recirculating tank. Steam is introduced into the water to heat and maintain predetermined water temperatures through temperature sensing probes and control valves located in the system. A typical C. I. P. spray cleaning cycle would consist of:

1. Pre-rinse cycle

2. Wash cycle

3. Post-rinse cycle

4. Sanitize cycle

FIG. 8-3 Wall-mounted, spray-type sanitizing units should be included in every equipment cleaning room.

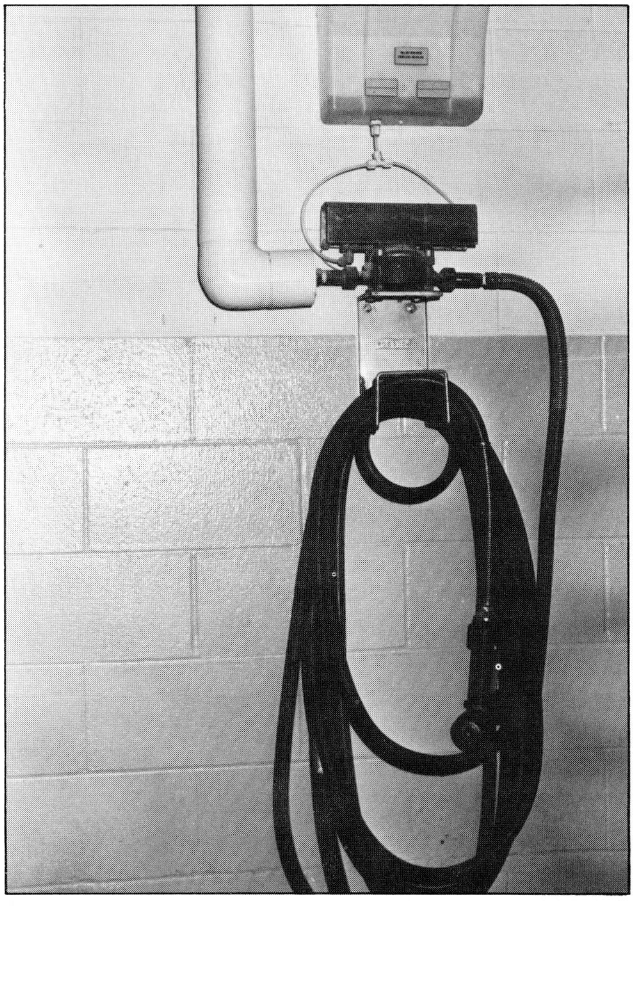

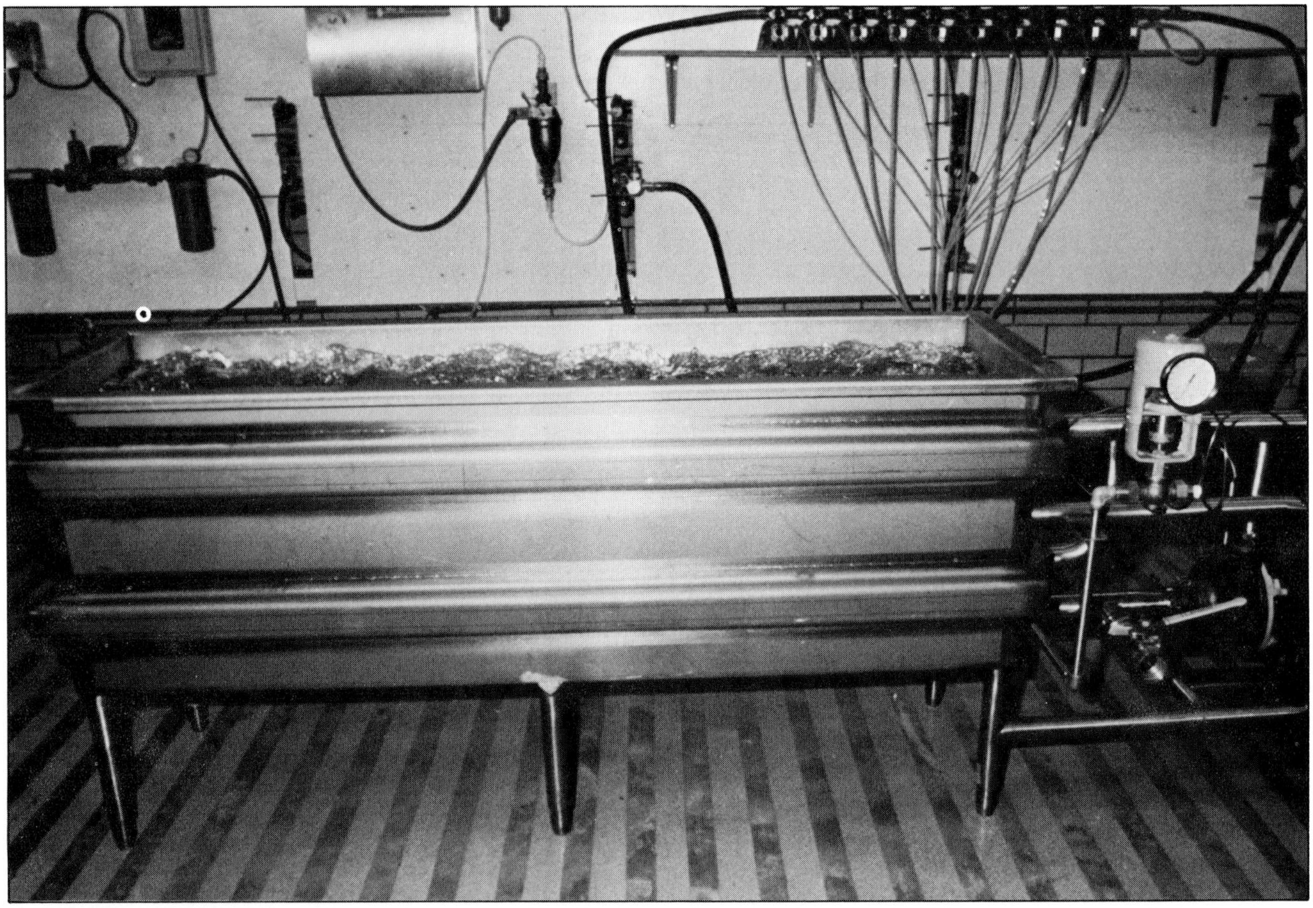

FIG. 8-4 The turbulent cleaning action of the water in *clean-out-of- place* tanks can clean small parts with little effort.

All cleaning operations begin and end at the primary recirculating unit. A centrifugal return pump moves the solution back to the recirculating unit where it is either placed back in the tank or directed into the drain. To minimize pump cavitation, the pump should be equipped with an air vent. The line to the suction side of the return pump should be adequately sized so that it does not restrict the return solution flow, as this can also cause cavitation.

The recirculating package can be permanently installed or mounted on wheels so that it can be moved. The system can be fully automated to perform rinse, wash, and sanitizing operations by adding a control panel to the unit. Supply-return piping can also be permanently installed if there are a number of cleaning applications in a given area.

Tanks or equipment to be spray cleaned must be properly vented to prevent them from collapsing from a vacuum condition that can occur when a cold water rinse follows a hot cleaning cycle. It is very important to provide vents that are the right size. The 3-A Dairy Standards for Storage Tank Design can be helpful in determining vent size, which in some instances might require a man-sized or larger access opening. Vents must also be designed to open before the cleaning cycle begins. It is important to remember that whereas properly designed and vented round storage tanks will withstand most vacuum conditions, square or rectangular tanks might not. This is because it is somewhat difficult to design square or rectangular tanks that will resist collapsing in vacuum situations. Recirculating spray systems are effective cleaning systems that make good use of water and chemicals through the recirculating feature. There are many applications for this cleaning approach in almost every food plant.

AUTOMATIC CLEAN-IN-PLACE FOR PROCESS PIPE SYSTEMS

Widely used in the dairy industry for many years, this system automatically controls all the variables of a mechanical

FIG. 8-5 A typical clean-in-place (C. I. P.) cleaning system used for liquid line or tank cleaning is shown. C. I. P. cleaning reduces cleaning costs, and improves cleaning quality.

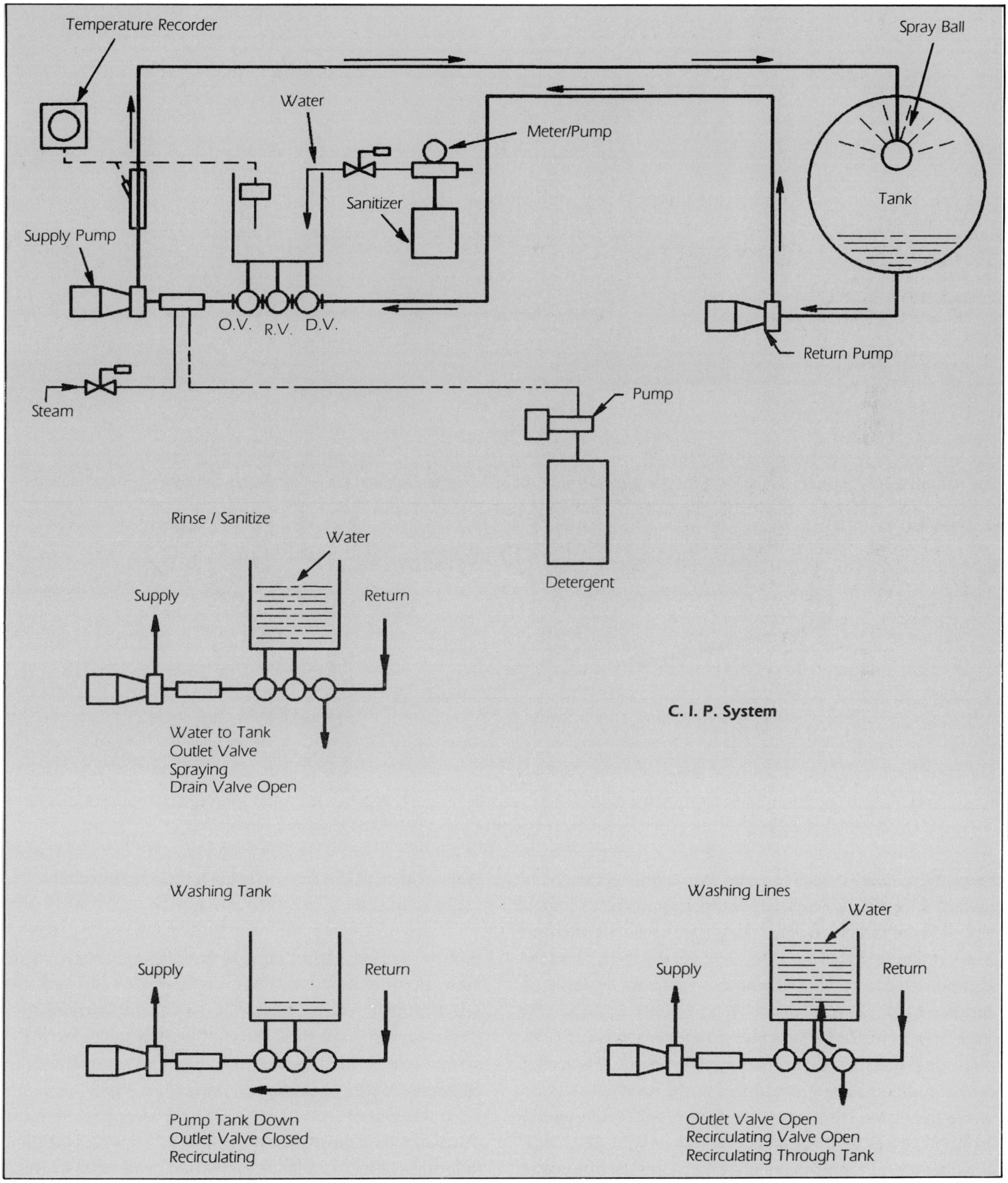

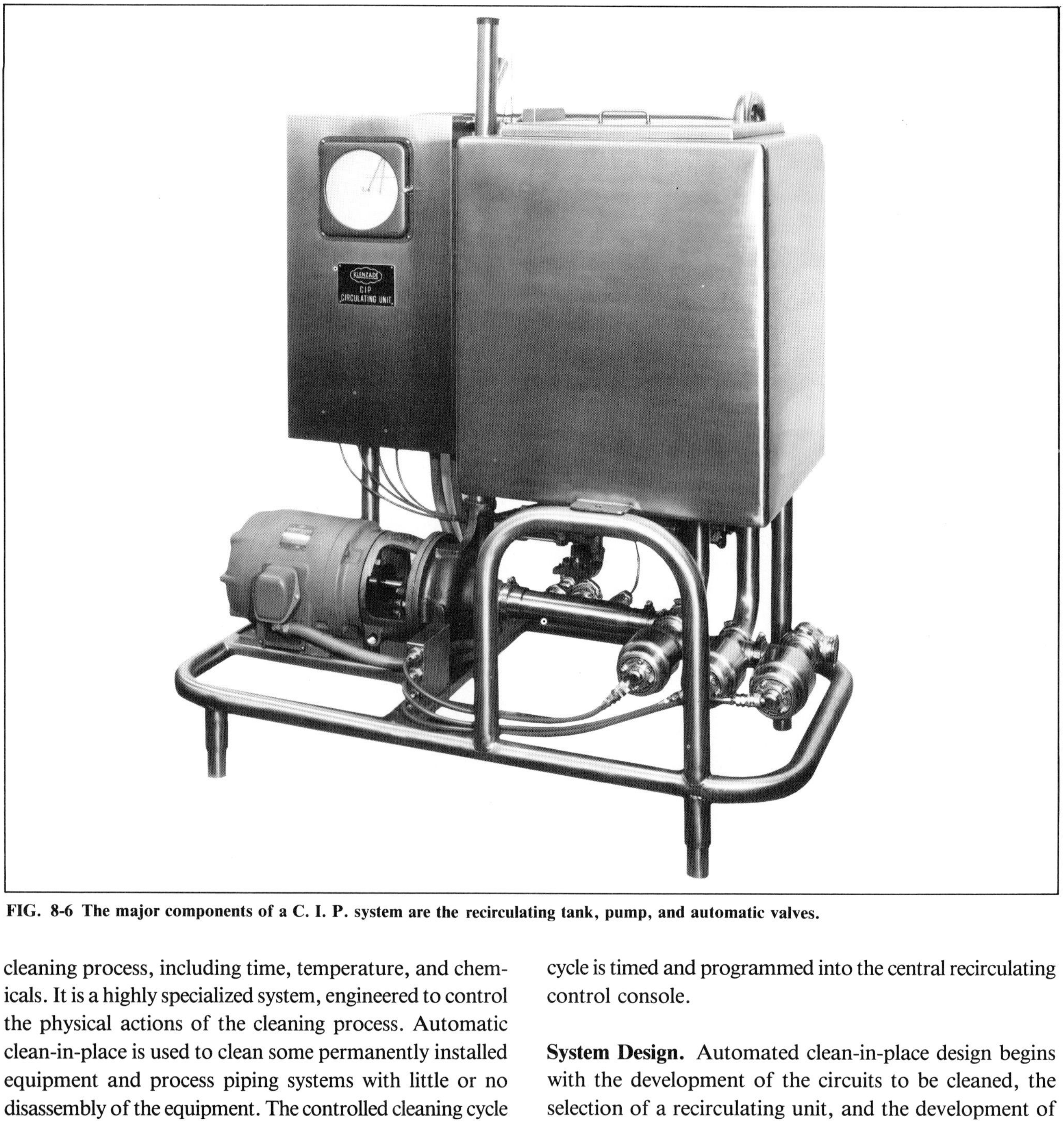

FIG. 8-6 The major components of a C. I. P. system are the recirculating tank, pump, and automatic valves.

cleaning process, including time, temperature, and chemicals. It is a highly specialized system, engineered to control the physical actions of the cleaning process. Automatic clean-in-place is used to clean some permanently installed equipment and process piping systems with little or no disassembly of the equipment. The controlled cleaning cycle generally consists of a pre-rinse, a detergent wash, a post-rinse, and a sanitizing step. The cleaning action is controlled by the pumping rates and velocities of the various solutions, as well as by the selection of chemicals. A fully automatic recirculating system permanently located in a designated area is used as the primary unit, and each clean-in-place cycle is timed and programmed into the central recirculating control console.

System Design. Automated clean-in-place design begins with the development of the circuits to be cleaned, the selection of a recirculating unit, and the development of a cleaning program that will get the equipment clean. The system must be designed so that all surfaces are effectively contacted by the cleaning solution.

Layout of the C. I. P. lines for processing functions involves using both the process lines and the cleaning solution lines effectively. *(Figure 8-7)* The object is to switch from the processing mode to the cleaning mode with the least number of crossover connections and without bypasses which would make manual cleaning necessary for some parts of the system. It is also important that there be no dead ends, and the pipe be properly pitched toward the return lines to assure complete drainage. A pitch of one inch in ten feet should be considered, and for the most part should not be less than one inch in twenty feet. All piping should be well supported. It is best to provide supports on approximately ten to fifteen-foot centers so that there are no sags for solutions to lie dormant, and to place pipe supports adjacent to air-operated valves. Include expansion joints, as many C. I. P. systems will use hot water at 165°F. For a temperature rise of 100°F, the expansion could be approximately one inch in 1,000 lineal feet of pipe.

Process and return piping systems should be designed with sanitary pipe and fittings that meet the requirements of the 3-A Dairy Standards and are C. I. P. cleanable. Processing and return systems may also be designed with welded pipe. Modern welding techniques make it possible to achieve very smooth, crevice-free interior welded joints which are easily cleanable. Welded piping systems should have occasional sanitary union connections so that lines can be taken apart and inspected.

Recirculating Equipment. Recirculating equipment will be similar, if not identical, to that used for recirculating spray cleaning. It will consist of solution tanks, pump, and a series of automatic valves. Recirculating solution tanks vary in size from 160 to 350 gallons. The size and number of tanks required are based on the size of the total C. I. P. system. The solution tank must be large enough to provide the pump with a constant supply of hot water, and it must serve as a place for introducing detergents and sanitizers into the system. The unit will be fully programmed to complete all phases of the cleaning operation—the pre-rinse, wash, post-rinse, and sanitizing steps—in a timed sequence.

The physical action of the cleaning operation is greatly influenced by the rate of flow and the velocity of the solution. A velocity of five feet per second is considered to be ideal for cleaning process piping. C. I. P. pumps should be stainless steel centrifugal units carefully chosen to meet design conditions. Pump performance is important if the system is to operate satisfactorily. Fully automatic C. I. P. systems can fit the cleaning needs of the dairy process as well as those of many other food processors. C. I. P. systems should be given serious consideration because they will do a thorough cleaning job in the least amount of time. One most important C. I. P. design consideration, however, is to make certain that the basic design includes fail-safe features that will prevent cleaning solutions from contaminating food products in process or storage when adjacent equipment is being cleaned. For example, if tank 'A' is empty and adjacent to tank 'B', which is full of a food liquid, an error made while spray cleaning tank 'A' could cause the cleaning solution to enter tank 'B' and contaminate the entire food contents.

CENTRAL CLEANING SYSTEMS

In central cleaning systems, a supply of hot water is piped to a number of cleaning stations placed strategically throughout the manufacturing area. Cleaning chemicals can also be pumped from a central make-up system to these stations, or if preferred, fed directly into the hot water supply at each cleaning station. Central systems can be designed as low-pressure, high-volume systems or high-pressure, low-volume systems; if needed, they can also be high-pressure, high-volume systems. Each design offers some advantages for different types of cleaning problems. It is fairly well accepted that the cleaning quality is directly proportionate to the quantity of water used. Many cleaning jobs involve bacteria-sensitive products and somewhat difficult-to-remove soils. These jobs are best done with low pressure and high volumes of water, in which case the soiled surfaces are cleaned by a larger amount of water flowing over them. Extremely stubborn soils which cling or stick to surfaces and resist removal may be more easily removed by a high-pressure, high-volume system. The high-volume feature of a high-pressure system adds impingement or striking force due to the mass and weight of the additional water. Nine to twelve gallons per minute (GPM) or higher is considered high volume.

Most cleaning problems, however, fall into the category of *difficult-to-remove soils with a tendency to stick to almost any surface.* These soils can generally be removed with a high-pressure, low-volume system. High-pressure, low-volume systems operating in the 600 to 800 psi ranges and delivering three to five GPM at each station have been accepted by many segments of the food industry. These

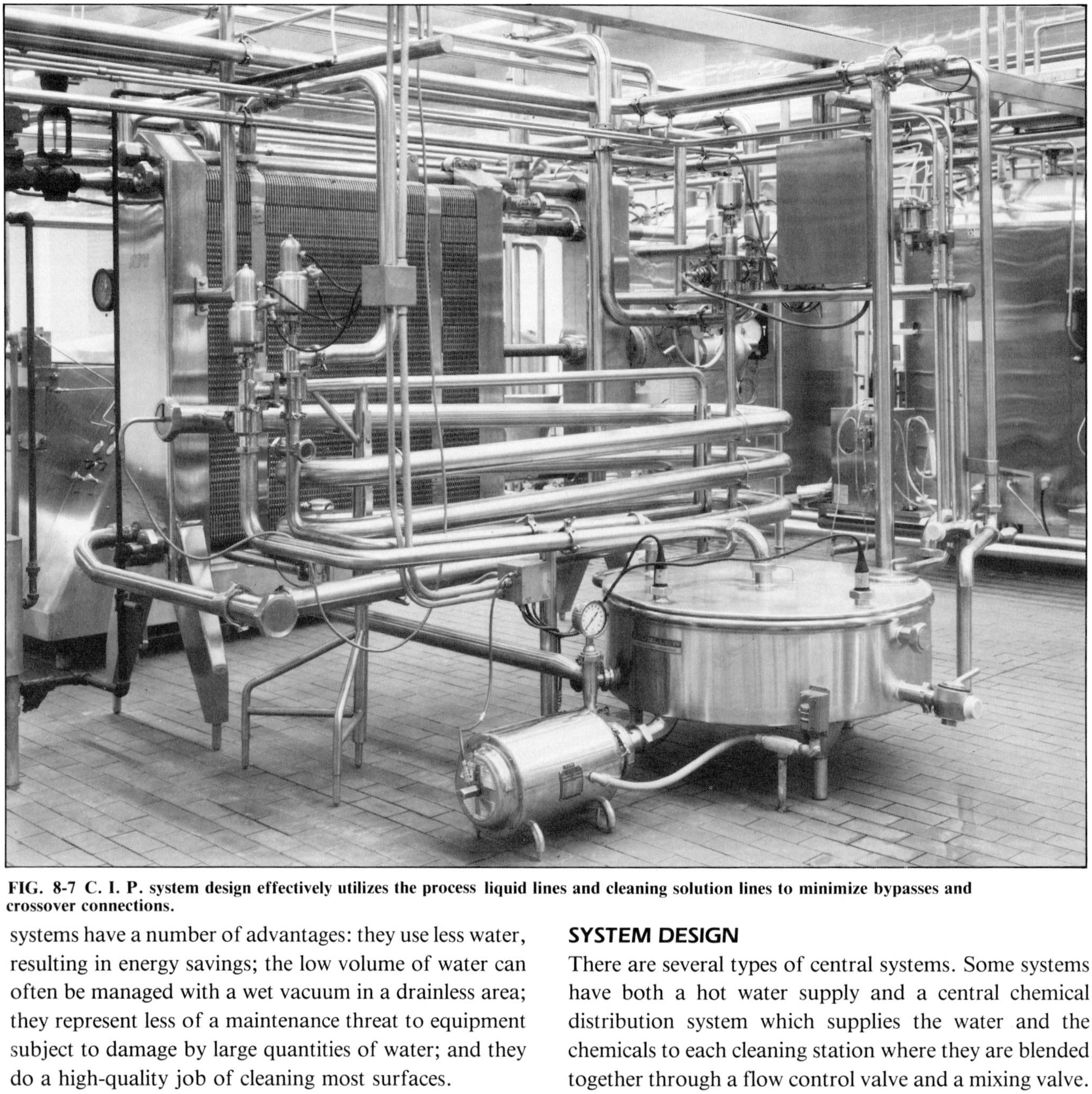

FIG. 8-7 C. I. P. system design effectively utilizes the process liquid lines and cleaning solution lines to minimize bypasses and crossover connections.

systems have a number of advantages: they use less water, resulting in energy savings; the low volume of water can often be managed with a wet vacuum in a drainless area; they represent less of a maintenance threat to equipment subject to damage by large quantities of water; and they do a high-quality job of cleaning most surfaces.

Central cleaning systems provide a unique and quick way to assure an adequate supply of hot water for all areas requiring it. Central systems are used to clean equipment, walls, floors, and ceilings. There is very little that these systems cannot clean—they can remove even the toughest soil. Combined with the proper cleaning and sanitizing chemicals, central cleaning systems are both efficient and effective.

SYSTEM DESIGN

There are several types of central systems. Some systems have both a hot water supply and a central chemical distribution system which supplies the water and the chemicals to each cleaning station where they are blended together through a flow control valve and a mixing valve. *(Figure 8-8)* Another system offers hot and cold water in addition to a central chemical supply. Still another system provides hot water with chemical additions from an individual injector system at the cleaning station. *(Figure 8-9)* Each of these systems has something to offer. Features should be selected on an *as needed* basis. However, remember that most cleaning tasks will require a pre-rinse, a wash, a post-rinse, and a sanitizing step. Also, it is hard to do without the convenience of being able to add or omit chemicals as desired at the cleaning station.

SIZE AND LAYOUT

The size of the system will depend on the total number of stations and—most importantly—the total number of stations that will be operating at any one time. The total number of hoses to be used at one time coupled with the flow rate at each station determines the size of the pump and/or the number of pumps needed. The location of individual stations is obviously determined by need, but each station should be convenient to the equipment and areas it will clean. A station should be three to five feet off the floor, and there should be ample room in the immediate vicinity to operate and service the station.

Water and chemical lines should be installed in such a manner that they themselves do not create a sanitation problem. Insulate the hot water line, and if a central chemical distribution system is used, identify the chemical lines. Make sure the pipe lines are the proper size for carrying the needed quantities of water and chemicals at the required pressure.

PUMPS

The heart of a central system is the pump or pumps. One kind of pump is the centrifugal pump, which is normally used on low-pressure systems. Sometimes a centrifugal pump functions as a distribution pump for a high-pressure system, and the high-pressure pump is used at the cleaning station as a booster pump. A centrifugal pump should be sized to accommodate the required water needs. In cases where the pump moves water containing cleaning or sanitizing chemicals, the pump should be constructed of stainless steel.

In small-to-medium sized high-pressure, low-volume cleaning systems, a positive displacement piston pump is

FIG. 8-8 Cleaning system with central chemical distribution system

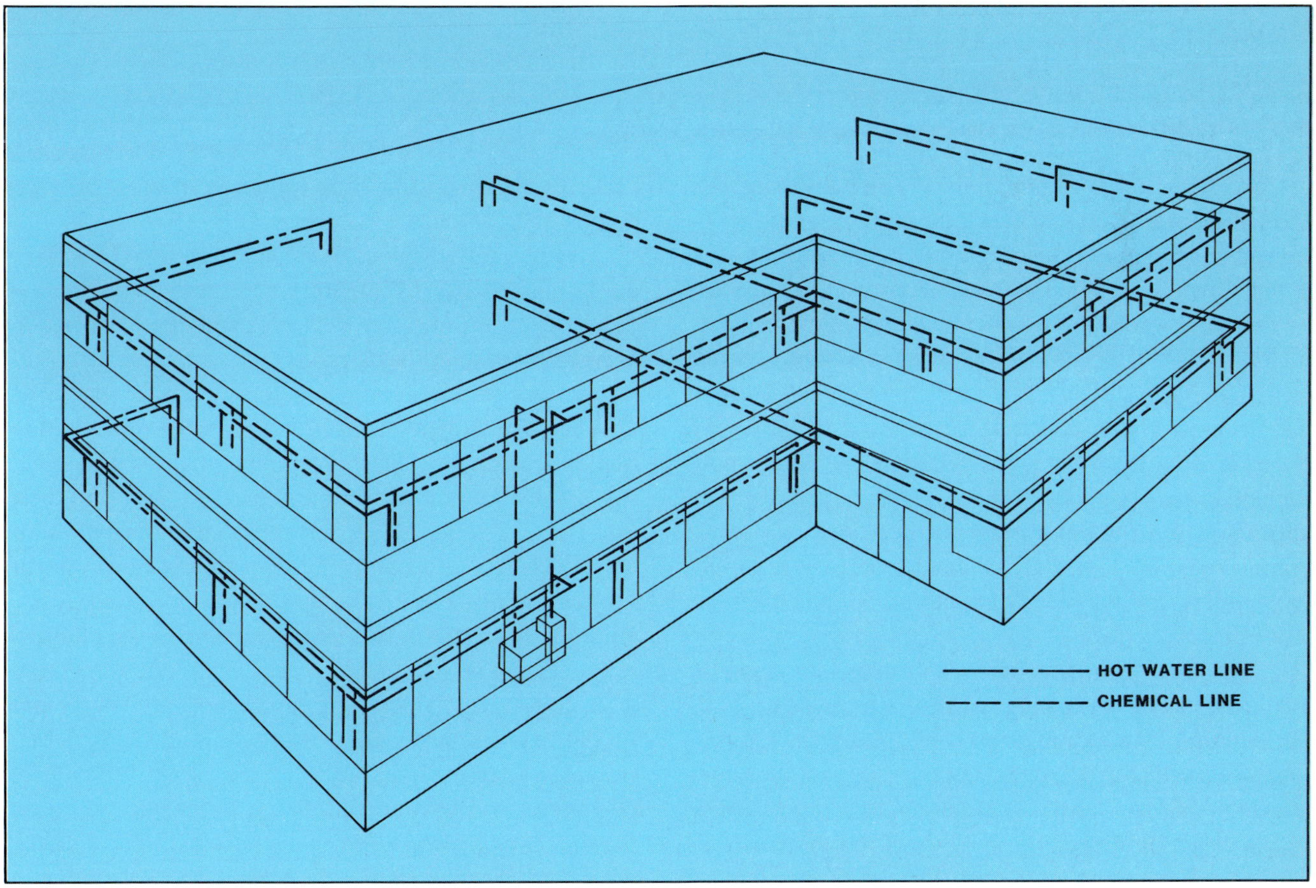

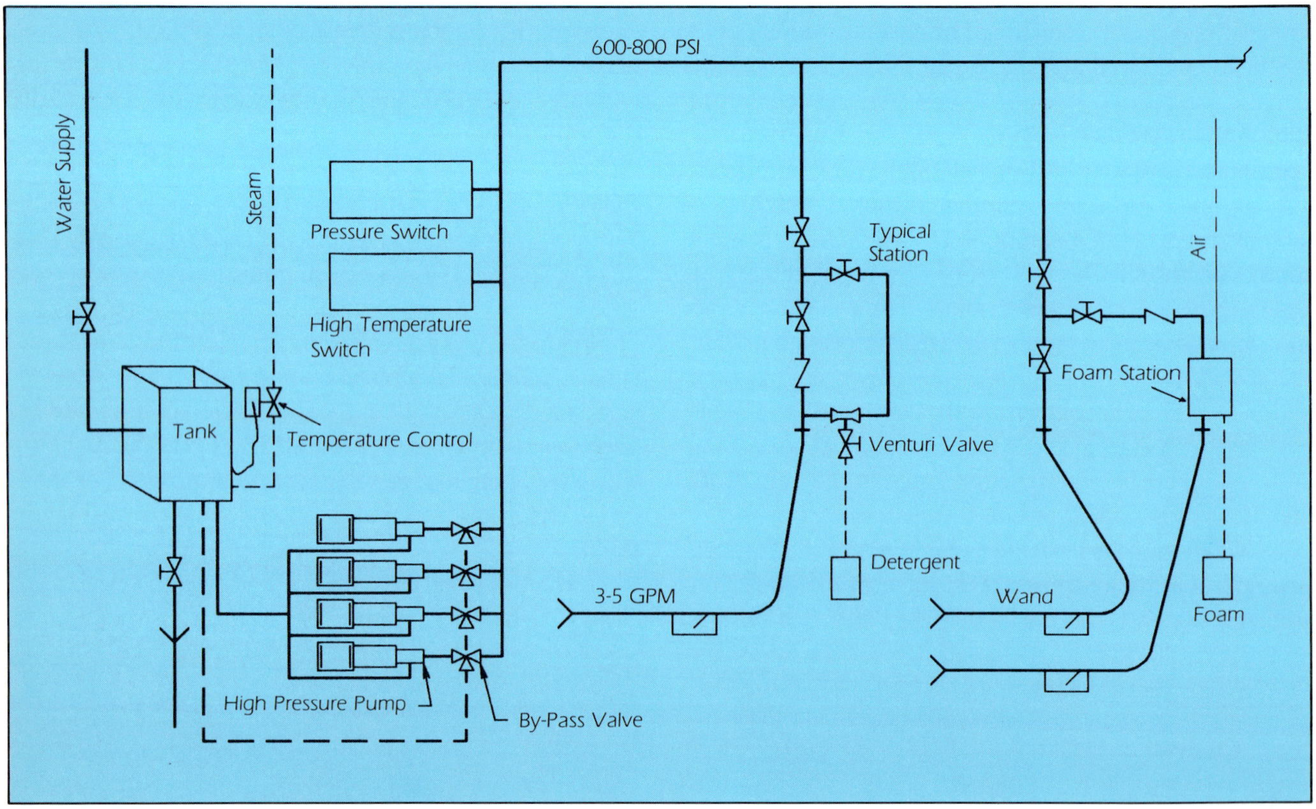

FIG. 8-9 Central cleaning system with individual use point chemical system

often the main distribution pump. Piston pumps are also used as booster pumps at individual cleaning stations. In a medium-sized system that is designed to run intermittently or that runs at something less than full capacity of the system on occasion, a group of two to four piston pumps may be staged together as a main distribution pump system to better satisfy the capacity variables of the system. Positive displacement piston pumps displace given quantities of water. If all of the water is not being used at the cleaning stations, it must be bypassed back to the supply source. The advantage of a multiple pump system is that only those pumps required to fulfill the immediate water supply requirement have to run. Other pumps are phased in on a demand basis.

Multi-stage centrifugals or jet pumps are used on medium-to-large size central systems as the primary distribution pump. These pumps not only satisfy the requirements of the higher pressures, but they can provide larger quantities of water from a single pump. Multi-stage centrifugals or jet pumps are normally used on systems that need to run continuously rather than on an intermittent basis; frequent starting and stopping can damage large pumps and motors.

High-pressure pumps should be considered for heavy-duty applications. All high-pressure pump installations should feature protective devices, including bypasses or pressure cut-out switches to protect the pump from damage. At one time or another any system is likely to operate at less than the designed capacity, and pump protection is needed accordingly. Filters or strainers on the water supply will protect the pumps from any particles that find their way into the water. Also, install accumulator-pulsation dampeners on piston pump systems or any other system subject to fluid shock. Place the dampeners as close to the pump as possible to avert shock damage to the pump.

Neoprene or rubber pump seals are sometimes affected by high temperatures (approximately 160°F) and should be protected by high temperature cut-out switches. Because detergents will also affect seals, add them downstream of the pump discharge. Some pumps require the water supply pressure to be controlled to protect the seals; and, in such cases, place pressure regulators and pressure cut-out switches on the suction side of the pump.

The pump electrical circuit should be designed so that the pump protective devices cannot be bypassed, such as with a hand/off/automatic switch. If a hand running position is desired for maintenance purposes, consider a specially keyed switch. For multi-state piston pump systems, include a sequence switch which will permit a change in the starting sequence lead pump. Without this, the number one or two pump will have many more running hours which could increase maintenance costs and result in more system failures.

CLEANING STATIONS

The desirable features at a typical cleaning station will vary with the basic design of the pumping system. For systems using a central centrifugal distribution pump and a booster pump at the cleaning hose station, the station should include a start/stop switch for the booster pump and pilot lights to indicate if the pump is running. All stations need clearly identified valves for controlling water and chemicals. *(Figure 8-10)*

Equip each cleaning station with 50 to 75 feet of high-pressure water hose and hose racks or automatic hose reels. These hoses must be able to handle high water pressure as well as withstand high temperatures and the caustic or acid effects of chemicals used in the system. Hoses must also be resistant to lubricants and oils found in many areas waiting to be cleaned.

Although not imperative, it is a good idea to equip the cleaning station with a direct-reading temperature gauge.

APPLICATOR GUNS/NOZZLES

Choose applicator guns and spray nozzles to accommodate the pressure conditions of the system. *(Figure 8-11)* These guns are available with wands of different lengths. Many come with automatic (fail-safe) shut-off features which are highly recommended. Applicator guns constructed of corrosive-resistant materials for use with highly corrosive cleaning chemicals are the best choice. Nozzles are available in a variety of spray patterns. A spray pattern of 15° will satisfy most cleaning jobs. Nozzles ranging from 0° to 40° and wider may be required for special jobs. The nozzle orifice is important in obtaining the desired flow rate at the design pressure. Many high-pressure systems are designed to produce the required pressure six inches from a nozzle with a 15° spray pattern. At a distance of twelve inches, the pressure drops to approximately one half; and at eighteen inches, to approximately one quarter of the design pressure.

FIG. 8-10 Hose stations; note the clearly identified valves.

High-pressure low-volume systems need adjustable nozzles that can be easily adjusted to suit the cleaning requirement. A multiple spray pattern selection nozzle that offers several angles of spray (such as 0°, 15°, and 40°) and a selection of different orifices may be worthwhile if the cleaning tasks are somewhat different. Orifice size selection features allow chemicals to be applied at low pressures, which is desirable for eliminating splash back and chemical waste.

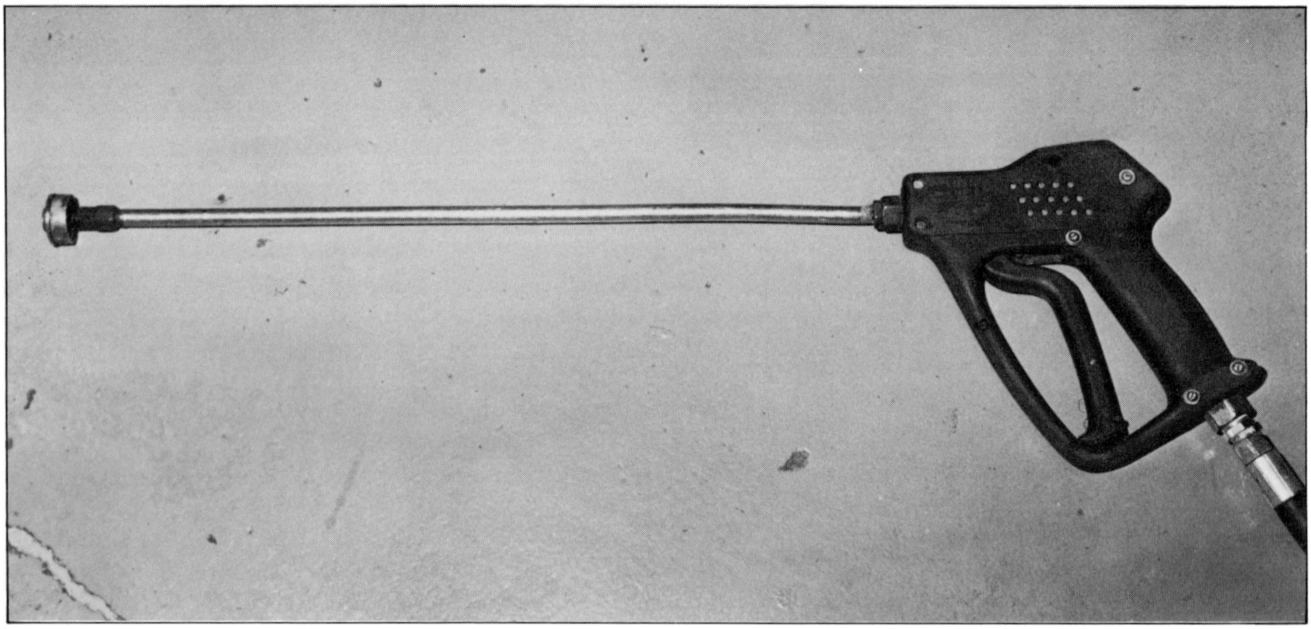

FIG. 8-11 A cleaning gun assembly with a fail-safe water valve automatically closes when the trigger is released.

CHEMICAL FEEDERS

As previously mentioned, chemicals can be added in several places: at a central mixing point, from which they are distributed to the use points, or specifically at the use points. Chemical feed pumps can be used to dispense chemicals into the system, or a simple Venturi valve can be used to siphon the chemicals into the system. Chemical feed pumps should be considered when accuracy is important. *(Figure 8-12)* Venturi valves require a flow control orifice or flow control valve to regulate the chemical quantity. Orifices can be easily changed as indicated by the chemical requirements.

Remember that whenever chemicals are added to a potable water system, backflow prevention equipment will be needed to prevent contamination of the water system.

HOT WATER SUPPLY UNITS

Hot water supply units are the backbone of the central system. *(Figure 8-13)* It is essential to have a system that can dependably supply an adequate amount of hot water at the proper pressure. Hot water units can be direct gas fired, electric, or steam heated. Small gas-fired hot water boilers or large water heaters should be considered if cleaning activities are planned for production downtimes. This eliminates running larger boilers just to provide hot water for cleaning.

The size of the hot water heating units will depend on the number of cleaning stations to be operated at any one time. Give careful consideration to this in order to ensure satisfactory performance. Operating more stations than specified will result in a hot water supply problem.

The hot water supply units should be installed in enclosed areas with adequate space for servicing. Boiler rooms or areas near boiler rooms are desirable.

FOAM AND JELL

Foam and jell applicators can be added to central cleaning systems. These applicators require special foaming and jelling detergents and compressed air at around 60 to 90 psi, which is used as the expelling agent. The foam and jell materials virtually encapsulate the cleaning detergent, trapping or holding it against the surface to be cleaned and allowing it to do its cleaning work through soil breakdown. *(Figure 8-14)* The length of time this takes depends largely on the toughness of the soil and the strength of the cleaner. Once the surface is considered clean, the foam or jell material is rinsed off with clear water.

INDIVIDUAL HOT WATER HOSE STATION

Although different from a central cleaning system in size, for all practical purposes, the individual hot water hose station is very similar to the central system. The individual hot water hose station can be high pressure or low pressure. It can consist of the same components found in a central system or of specially designed water or water/steam mixing equipment. In this section we will discuss primarily the specially designed equipment.

The individual low-pressure, high-volume hot water hose station utilizing steam/water mixing valves offers one of the more economical ways to provide hot cleaning water at a given location in the plant. This station is particularly

FIG. 8-12 A chemical feed pump can assure chemical accuracy.

suited for remote areas of the plant and has been used by virtually all segments of the food industry for many years. These units provide an instantaneous supply of hot water at the desired temperature. *(Figure 8-15)* The mixing valve provides for positive and complete mixing of steam and water and can also be used to blend hot and cold water. Stations operate at city water pressures and do not normally use distribution or booster pumps.

EQUIPMENT

Mixing valve assemblies will need to be equipped with direct reading thermometers which display the temperature of the water in the blending chamber. Select steam/water mixing valves with safety features that make it impossible to draw live steam only, thus eliminating the hazard of steam burns. Assemblies should also include backflow devices to prevent cross-contamination of the water supply system. For each hose station, provide 25 to 50 feet of high-pressure hose, hose racks, and adjustable fail-safe nozzles to regulate the water from a solid stream to a fine mist. The fail-safe nozzle should cause the water flow to stop when the lever is released. Select hoses that can withstand the water temperature and will not be affected by normal cleaning solutions. In addition, make certain that hoses are oil resistant, lightweight, extremely flexible, and resistant to kinking.

Individual hot water hose stations of this type work well for supplying cleaning water to a wet operation phase of what is otherwise considered a dry process. They are also ideal for satisfying the wet cleaning needs of certain smaller wet manufacturing processes.

FIG. 8-13 Separate hot water boilers, properly sized, can provide the needed quantity of water to all central cleaning hose stations.

FIG. 8-14 Foam and jell agents hold the cleaning chemicals to the surface, allowing the cleaning action of the chemical to take place. A foam application is shown.

CHEMICAL SYSTEMS

Individual hot water hose stations can be made more effective by adding a chemical system so that detergents or sanitizers can be injected. A chemical system includes an atmospheric tank, a booster pump, a chemical tank, and an injector mechanism with a flow control valve. *(Figure 8-16)* The atmospheric tank receives the water from the steam/water mixing valve. Its primary purpose, however, is to isolate the potable water, since chemicals are added to the water and contamination of the water supply is possible. The booster pump restores the system pressure lost at the atmospheric tank. The chemical tank can be supplied with liquid detergent concentrate or a liquid sanitizer. When this tank is pressurized with compressed air, the chemical is introduced into the water stream just behind the hose nozzle. The chemical goes through an injector mechanism which has a flow control valve to regulate the quantity. The addition of the chemical system improves the effectiveness of individual hose washing stations. Without this system, detergents must be applied from a bucket with a sponge and some

FIG. 8-15 Individual hose stations with steam/water mixing valves provide an instant supply of hot water.

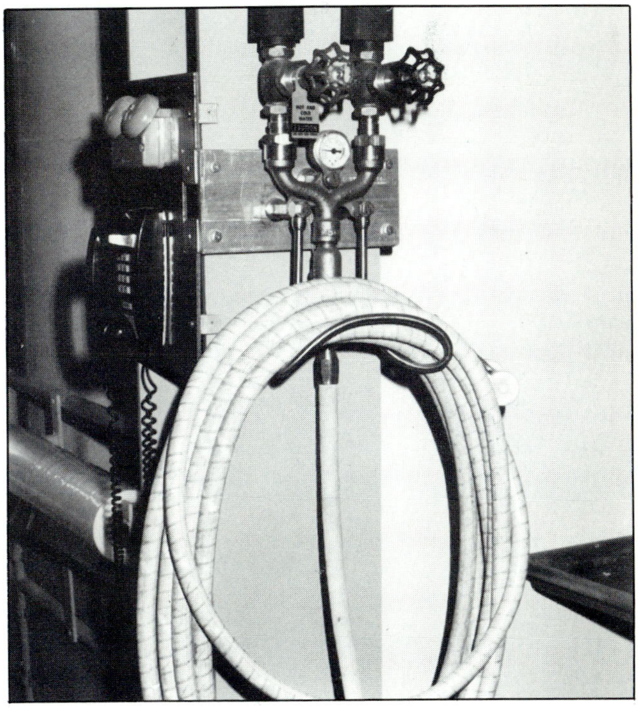

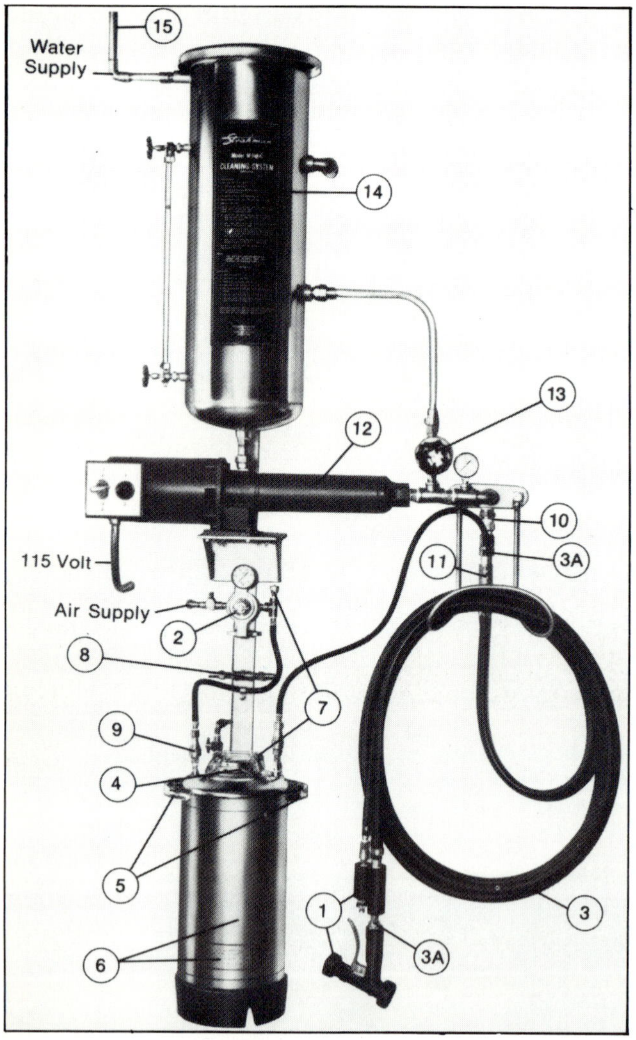

FIG. 8-16 Chemical feed systems and booster pumps increase the effectiveness of this type of hose station.

1. Chemical Injector
2. Air Regulator
3. Twin Hose (3A Swivel)
4. Safety Lid
5. Tank Disconnect
6. Pressure Tank
7. Relief Valve
8. Chemical Supply Line
9. Check Valve
10. Backflow Preventer
11. Check Valve, Water
12. Pump
13. Recirculating Valve
14. Atmospheric Break Tank
15. Water Line

hand scrubbing action. Sanitizing would also have to be accomplished separately.

DRY CLEANING

A great many of the cleaning jobs around a food plant are dry cleaning jobs which require sweeping, brushing, wiping, blowing, and vacuuming. Vacuuming offers a few advantages over the other methods: it does not spread dust and it picks up as it goes along. It also removes deeply imbedded dust and dirt. There are two types of industrial vacuum systems available: portable and central. Each has a place in the average food plant and each has desirable features. A central vacuum system probably will offer a number of important advantages over many of the portable systems. *(Figure 8-17)* Central systems have more power, and several cleaning people can use the system at the same time. In addition, the central system can be used to clean up product spills, to assist in emergency dust control, or to safely clean equipment that under other circumstances might have to shut down for cleaning. The efficiency of a sanitation program in an average size food plant will increase noticeably with the use of a central vacuum cleaning system.

SYSTEM DESIGN REQUIREMENTS

The design of an industrial system must satisfy the needs; therefore, each need must be clearly listed and defined. These systems can be used for wall, floor, and ceiling cleaning; they can be used to clean equipment and pick up spills; they can be used at inspection tables to remove objectionable materials; and they can be designed for wet pickups as well. There are any number of tasks that can be performed by this equipment, but it is important to plan the intended use into the design of the system. It is important that each specific job describe the handling characteristics and densities of the material to be picked up by the system.

How the system will be used must also be known. Will it be used to clean overheads? Will it be used to clean running equipment or energized electrical equipment? How many operators will use the equipment at the same time? Will the system be used in a hazardous atmosphere? How the system will be used will determine its size and location as well as the selection of various components.

Size and Layout. Once you determine what the system is to do and how it is to do it, then it must be properly

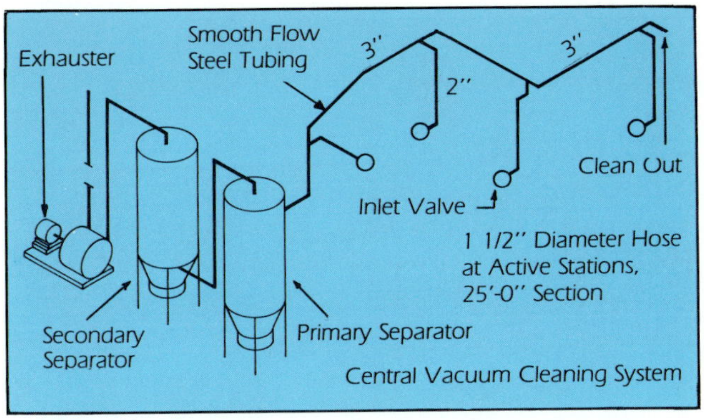

FIG. 8-17 Central vacuum cleaning systems offer a number of cleaning advantages to most food plants.

sized and conveniently located. The number and location of inlet valves needs to be decided. Valves used for floor cleaning should be about three to five feet from the floor. For inspection tables or other industrial uses, inlet valves should be adjacent to the point of operations, and for high ceilings or roof cleaning the valve should be located near the use point.

Stations must be located so that no more than 50 feet of hose is required to clean an entire area from one valve station. Hoses should be purchased and used in 25-foot lengths whenever possible. Pressure in flexible hoses can drop significantly; this can impair the overall efficiency of the system.

In congested areas, or where heavy floor deposits or spills are anticipated, or where heavy spot cleaning is required, the inlet valves should be placed closer together so that it is only necessary to use 25-foot hoses.

Hoses. One and a half inch inside diameter (I. D.) rubber hose is satisfactory for most cleaning requirements. It is very flexible, relatively lightweight, and easy to handle. This size hose has ample capacity for cleaning large areas quickly and efficiently. The larger sizes (2 inch and 2-1/2 inch) have correspondingly greater capacities and are normally used to pick up large spills, clean large tanks, or clean railroad cars and ship holds.

Tubing/Fittings. Smooth flow tubing and fittings are preferred for central vacuum lines over schedule 40 pipe. Tubing surfaces are smoother and easier to clean. Mild steel tubing is generally satisfactory for most installations. Consider stainless steel, aluminum, and zinc-coated tubing where corrosive conditions justify their use. In addition to being adequately sized to handle anticipated cleaning requirements, the tubing network must be installed in keeping with the good sanitary design practices described in previous chapters.

Separators. The usual type of central vacuum cleaner applications in the food plant will benefit from a primary separator and a secondary separator to separate out the collected material from the air. *(Figure 8-18)* The function of the primary unit is to separate the majority of the solid bulk material from the stream, particularly the larger particles and heavier dust. Air proceeds to the secondary unit where final filtering occurs as it passes through cloth bags. This two-step separation greatly improves separating efficiency and reduces system resistance, which optimizes the overall performance level of the total system. Single-stage separation should be used only for extremely light dust loads.

Wet pickup systems require separators specially designed for wet material separation. Wet separators must be placed where they will not be affected by freezing in winter.

Separators are generally sized to handle an anticipated capacity of collected material. They can be equipped with different types of discharge or emptying devices, from manual dump valves to fully automatic units which continually discharge separated material. A manual dumping system is adequate for most applications, but make certain that the primary separator has sufficient capacity to hold the collected material between dumps.

Position separators where it will be easy to dispose of the collected material. Be sure to leave sufficient room under the discharge valve for the collection containers. These containers will need covers and flexible connections to contain the puffs of dust which result when the material is discharged into the container. Not only should the separators be placed where the collection containers can be easily moved in and out, but it is also desirable to have them near the solid waste disposal area. This reduces the time it takes to dispose of the collected material.

Exhausters. The exhauster, or vacuum cleaner pump, should be located adjacent to the separators. The exhauster and drive should be mounted on a four-inch high concrete pad with sufficient space for cleaning and maintenance. Installation should include vibration and noise control equipment as required.

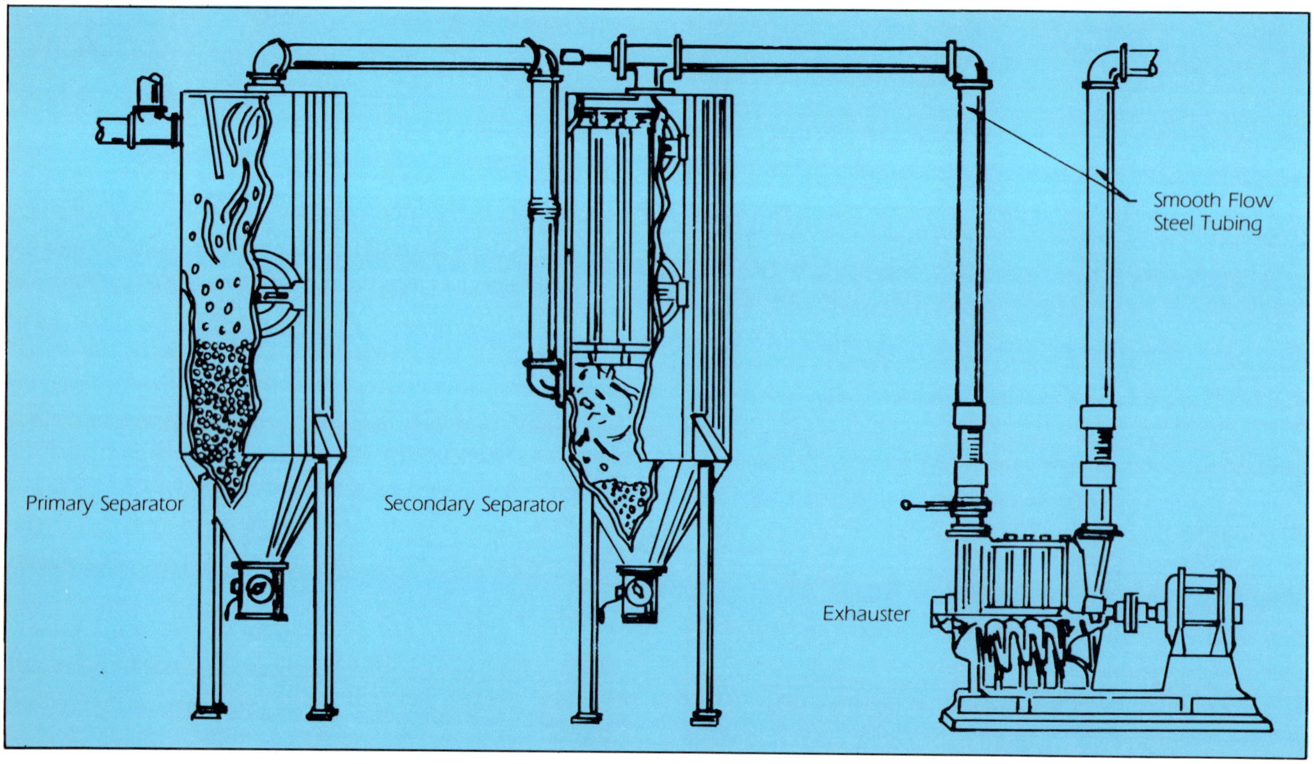

FIG. 8-18 Primary/secondary separator used on many central vacuum systems

Cleaning Tools. Floor cleaning systems must have sufficient capacity so that only one pass is needed to clean a given area. An operator should not be required to run the cleaning tool over the same area time and time again. The first and only pass should be at a speed that allows the tool to do an effective job. This speed will depend on the depth of the dirt deposit, the operator's skill with the tool, and the design of the tool. The effectiveness of cleaning tools varies with the cube of the intake velocity. For example, a tool entrance volume of 80 cfm is over 2-1/3 times more effective than a system having a tool entrance volume of only 60 cfm per cleaning tool. It only makes good sense to use tools with high intake velocities to increase the scouring, picking up, and conveying capacity of the system. It is therefore important that all applications of industrial vacuum systems be selected to provide adequate inlet nozzle velocities; otherwise an unsatisfactory system will result.

It must be understood that the velocity of the air moving over the surface of the floor is in fact the cleaning agent. It does the scouring and picking up. Therefore, the operator must not bear down or hold the tool in a manner that seals off the air. Light, deliberate movement will allow the air to do its work.

Certain jobs require special tools. It is advisable to have a complete set of tools including those constructed of materials which make them somewhat safer for use around electrical components. Place cabinets for hose and tool storage in strategic locations throughout the plant.

OPERATION

Once the total number, type, size, and location of inlet valves is determined, the user must decide how many inlet valves will be used at any one time. The importance of this concurrent use cannot be overemphasized. If the number of inlets used at one time is greater than the system was designed for, the system will be ineffective. If the original design assumes greater concurrent use than is really necessary, the system will be oversized, resulting in higher installation and operating costs.

Central vacuum systems should not be used as pneumatic conveying systems. Doing so will significantly reduce the effectiveness of what otherwise can be an important and time-saving sanitation tool. Use central vacuum systems only for their intended purpose.

CHAPTER 9
SOLID WASTE HANDLING SYSTEMS

CHAPTER CONTENTS

LOCATION

BUILDINGS AND EQUIPMENT

OUTSIDE LOCATIONS

Beyond any doubt, the handling of solid waste is a major challenge for plant management. Only human waste presents a greater hazard to food manufacturing facilities. Solid waste and the solid waste environment attract and harbor microogranisms, insects and rodents. Solid waste areas frequently are housekeeping eyesores that create overbearing odor problems. Nonetheless, solid waste handling need not present overwhelming difficulties, if the potential problems and health implications are recognized from the start. A sanitation program beginning with the sanitary design of the solid waste handling area can then be implemented. A good program will control the problem and a good design will make control much easier and more economical.

LOCATION

Solid waste disposal areas should be located out of view, preferably at the rear of the plant. Even if they are well designed and maintained, such areas hardly enhance the food plant image. At the same time, these areas must be conveniently located and of sufficient size to handle current as well as future wastes. The waste handling facilities may be in a separate building or in a specially designed area on the grounds. They may also be located in a remote, enclosed, controlled location of the plant itself. The chief aim of any layout plan is to isolate waste handling systems from the manufacturing operations, including ingredient and packaging material storage areas. A possible solid waste layout is illustrated. *(Figure 9-1)*

TYPE OF WASTE

Solid waste, as referred to here, covers both recoverable and nonrecoverable waste. Neither of these types of waste are byproducts of the manufacturing process that are destined for other uses, such as for animal feed or fertilizer. Recoverable waste comprises paper, metal cans, or glass which can be salvaged for recycling purposes. Nonrecoverable waste is that which is destined for sanitary landfills or for incinerators.

AREA SIZE

The area designated should be of sufficient size to allow for the staging of several containers of refuse removed from various plant areas. If the solid waste operation is to include a paper salvaging unit, adequate room must be allowed for

FIG. 9-1 Solid waste handling areas require careful sanitary design as they represent a major sanitation problem.

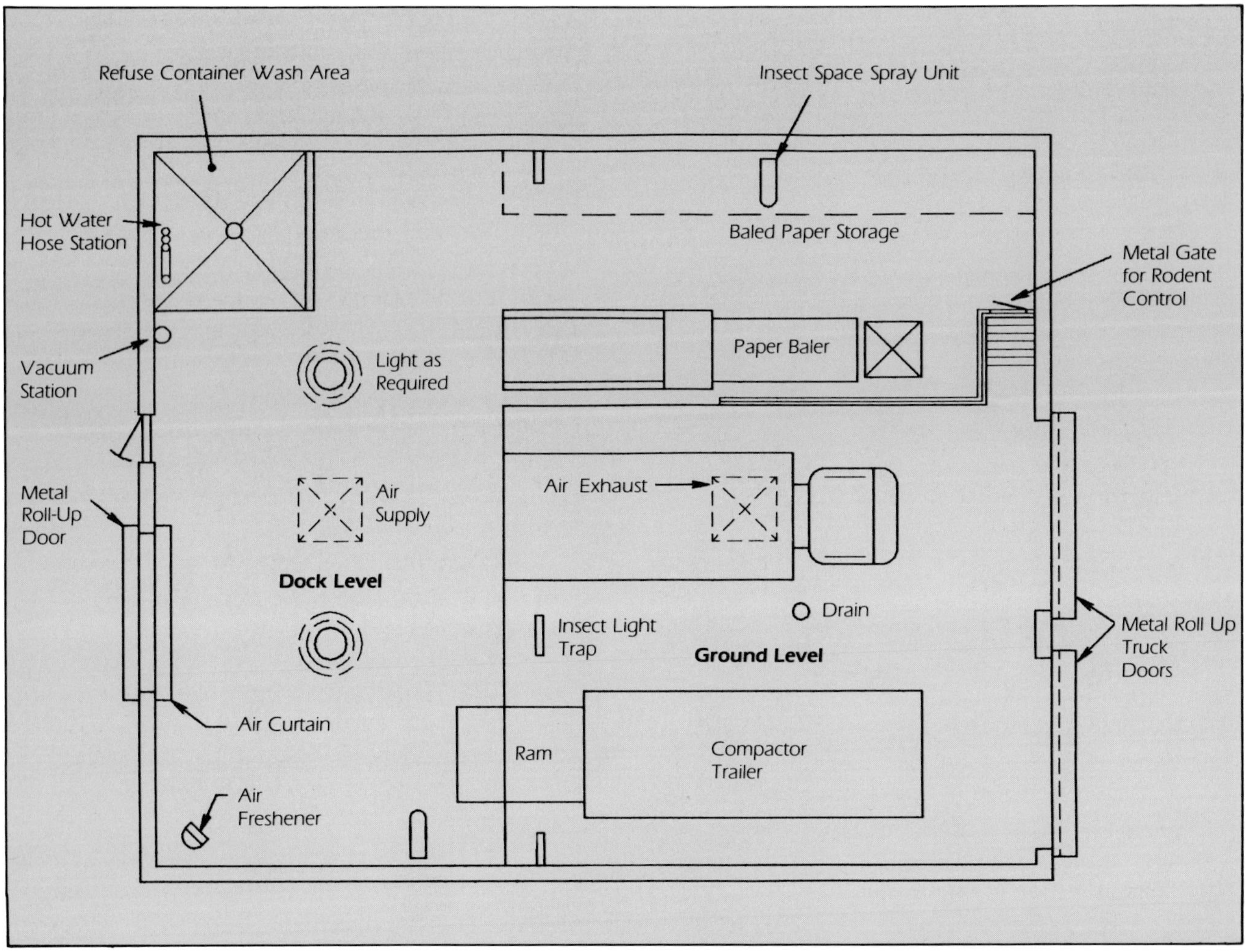

FIG. 9-2 Most plants handle a large quantity of solid waste and benefit from large enclosed waste containers.

the paper baling equipment as well as the baled paper.

Many plant solid waste operations will be of sufficient size to warrant a contract with a local refuse contractor who will provide large compactor refuse containers on the site. The area plans must include an allowance for this large equipment.

BUILDINGS AND EQUIPMENT

BUILDING MATERIALS

The designated area must be separated from adjacent areas by a wall or partition. The building materials selected must be impervious to moisture, have a smooth surface, and be readily cleaned. Floors should be constructed of concrete, have a smooth, steel trowel finish, and be treated with a concrete sealer. Floor drains should be provided.

Preferred wall materials are concrete or concrete block. Concrete block walls should be of a less porous medium to heavy weight block that is coated with a latex block filler and painted with a finish epoxy enamel. A concrete ceiling is also desirable.

Interior and exterior doors should be made of steel. Large interior equipment doorways should be equipped with air curtains or strip doors to control odors and flying insects. Exterior doors, truck docks, and pedestrian access-ways must be rodent-proof.

Adequate lighting of approximately 60 to 80 foot-candles should be furnished for this area. High pressure sodium lights should be considered, since they are less of a flying insect attractor. Two or more 110-volt electrical receptacles should be included.

The area ventilation system should be designed to provide a minimum of three air exchanges per hour. More exchanges should be provided if a particularly offensive odor problem prevails. In fact, in these extreme cases, odor-masking dispensing units should be provided. This equipment releases an air sweetener on a programmed basis. Odor absorbing equipment can also be helpful.

EQUIPMENT

With large quantities of solid waste being handled daily, refuse contractors often suggest that large size compactor truck trailers be used. *(Figure 9-2)* These self-closing compactor trailers are filled by an automatic feed unit which includes a hydraulic ram to compact the waste in the storage container. If the hydraulic ram is installed in a well or pit under the dock area, it is important to include access openings to this area. Access openings permit the removal of spilled refuse that accumulates in this pit. The compactor and ram equipment should be elevated approximately six inches. Self-closing compactor trailer storage systems greatly reduce the potential sanitation problems associated with this area.

A hose washing station complete with a fail-safe steam/water mixing valve and detergent feed system should be provided. One of the most important tools needed to maintain a solid waste area at an acceptable sanitation level is wet cleaning. For solid waste areas to be maintained at this level, they must be cleaned often and thoroughly with hot water and soap.

A small area should be provided to clean and sanitize the portable solid waste containers that are used to transport the waste from the plant. This area should be curbed and equipped with a floor drain. It should also be within range of the hot water hose station. Portable solid waste containers are one of the major sources of bacterial contamination. Since these containers are used all over the plant, they can easily spread bacteria across the plant. Waste containers should be cleaned and sanitized frequently. Other equipment needed for this area includes electric flying insect light traps and automatic insecticide fogging devices. *(Figure 9-3)*

OUTSIDE LOCATIONS

In many sections of the country, year-round climate conditions might permit the outside location of the solid waste area. There are advantages and disadvantages to this kind of an arrangement. On one hand, more frequent air exchanges make odor control easier. On the other hand,

FIG. 9-3 Insect light traps and automatic insecticide dispensing equipment are critical in solid waste handling areas.

FIG. 9-4 The salvage value of paper waste can justify the cost of purchasing a baler.

vermin control might present more of problem, and there is an even greater need for closed refuse containers and good housekeeping.

Some design features remain the same whether a waste facility is located inside or outside the plant. Both types require an accessible location that, at the same time, does not detract from the appearance or cleanliness of the plant. Both types must be designed with cleaning ease in mind.

While indoor waste handling facilities are closed off with walls and a ceiling, a metal canopy-style roof and wall partitions suffice for an outdoor facility. Use smooth, moisture-impervious materials for the partitions, floor, and canopy. Make sure that the canopy is free of perches where birds can roost.

The equipment requirements for an outdoor waste area are the same as those for an indoor facility. These requirements include a hose station for wet cleaning with hot water and soap, an automatic handling and closed storage system for solid wastes and a high pressure sodium lighting system to reduce the flying insect population. Electric insect light traps located in or adjacent to this area provide additional protection.

SALVAGEABLE WASTE EQUIPMENT

Salvageable waste equipment, particularly paper balers, are a good addition to the solid waste handling program. A large portion of the solid waste consists of corrugated paper containers, flat cartons, and miscellaneous wrapping paper. All of this material has a recycle value and can, in most cases, provide a return on the investment in salvageable waste equipment. *(Figure 9-4)*

CHAPTER 10
PEST CONTROL SYSTEMS

CHAPTER CONTENTS

COMMON INSECTS

CONTROL SYSTEMS

RODENT CONTROL WITH ULTRA-SONICS

Controlling pests around a food manufacturing plant is a full-time job. All segments of the industry are bothered by insects to some degree, and for certain segments, the problems can be serious. The entire industry is troubled by rodents as well.

The control of pests is very complex and requires the use of many tools. The process starts with good sanitation practices and a high level of knowledge of insect and pest animal behavior. In this chapter we will begin by briefly discussing the behavioral habits and characteristics of common pests. For the most part, the chapter will concentrate on pest control systems, including the design of pesticide storage areas.

There are several proven pest control systems which can be very helpful in controlling various insect problems found around most food plants. Most of these systems are designed to kill pests instead of repelling them.

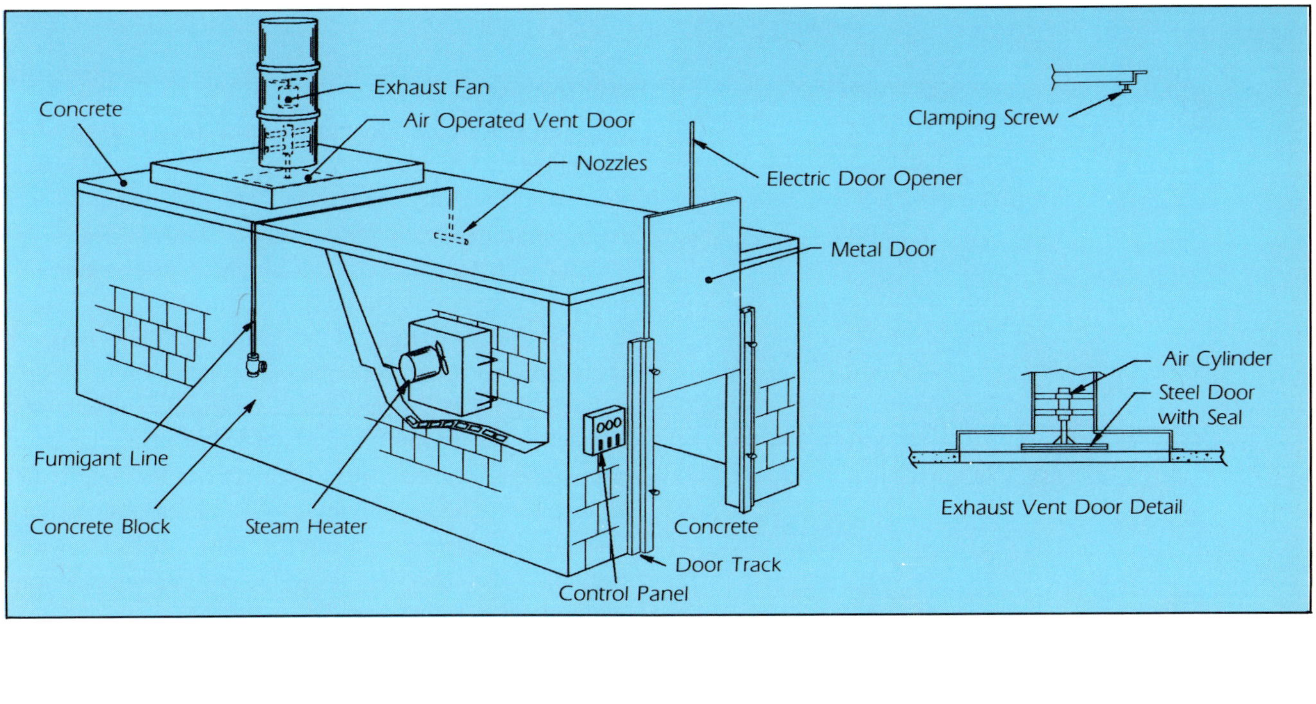

COMMON INSECTS

FLIES

The common housefly (musca domestica) and related fly species are by far the most annoying and concerning flying insects that plague the entire food industry. *(Figure 10-1)* Houseflies are filthy pests that transport many diseases. Their life often begins in animal waste or human garbage. They move from the egg to larva and into the adult stage in a very short period of time. As adults they are capable of carrying tremendous amounts of bacteria from garbage and animal excreta to human foods. Because houseflies have sponging mouth parts, they have the appalling habit of vomiting on their food to help dissolve it. Houseflies are warm weather insects; they prefer temperatures of around 82°F, and for this reason the fly population is heaviest in the summer season.

Flies are controlled by eliminating attractors, by fly-proofing buildings and by using mechanical control devices and chemicals. A flying insect control program does not depend on any one of these controls in particular, but on all of them used simultaneously.

Fig. 10-1 Most insects undergo a complete metamorphosis through four life stages: egg, larva, pupa, and adult.

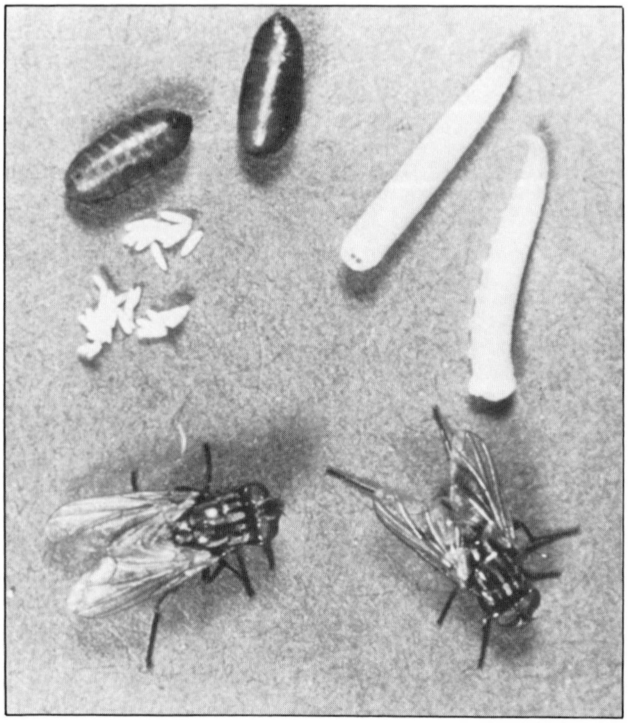

DERMESTIDAE

In addition to flies, much of the industry is bothered by a large group of insects known as dermestidae, whose feeding habits fall into three groups:

- Species feeding on animal matter only
- Species feeding on either animal or vegetable matter
- Species feeding on vegetable matter only

Members of the dermestid group include the warehouse beetle and the black carpet beetle. The black carpet beetle, in particular, is a well-known dermestid in this country. *(Figure 10-2)* Along with the majority of the dermistidae, it falls into the second of the above categories.

STORED-PRODUCT INSECTS

Certain segments of the industry are bothered by a select group of insects known as the stored-product insects. They include the confused and red flour beetles *(Figure 10-3)*, the sawtooth grain beetle *(Figure 10-4)*, and the granary weevil *(Figure 10-5)*. As was noted in Chapter One, the stored-product insects are extremely troublesome to the grain processing, baking, cereal and dry mix, and pasta industries. Insect control in these segments is an important part of the sanitation program.

COCKROACHES

Cockroaches are also troublesome to a large segment of the industry. Of the several species, the more common roaches are the American *(Figure 10-6)*, the German *(Figure 10-7)*, and the Oriental *(Figure 10-8)*. The American is the largest at 1-1/2 inches long. It is frequently found in sewers and gains access to the plant through drains. The German and the Oriental roaches are smaller. They measure approximately one half inch and one inch in length, respectively. The German roach is found in cracks and crevices in bathrooms and in kitchens or in similar areas where food is handled. The Oriental roach commonly settles in basements, under buildings or in similar damp, dark places. Cockroaches are also very filthy insects known to carry many bacteria, and they always represent a serious contamination threat to food.

FIG. 10-2 Black carpet beetle (actual size 1/8 inch)

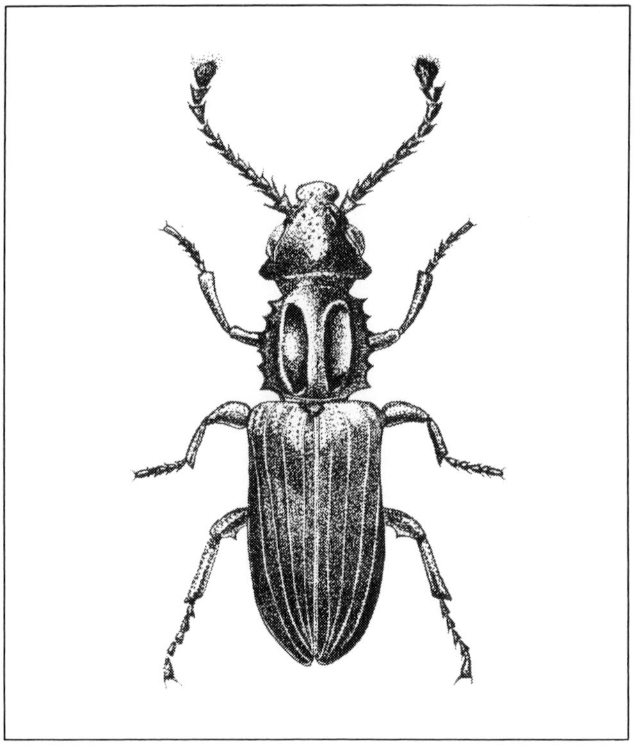

FIG. 10-4 Sawtooth grain beetle (actual size 1/10 inch)

FIG. 10-3 Red flour beetle (actual size 1/8 inch)

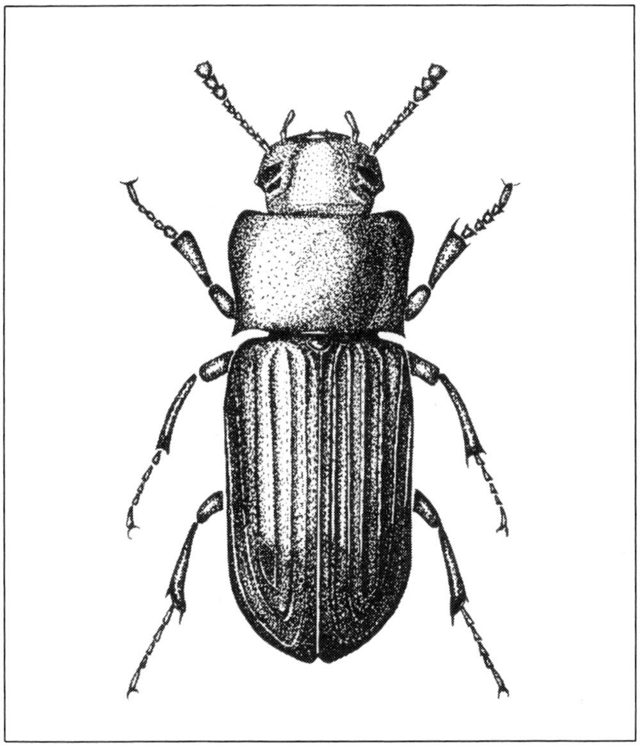

FIG. 10-5 Granary weevil (actual size 3/16 inch)

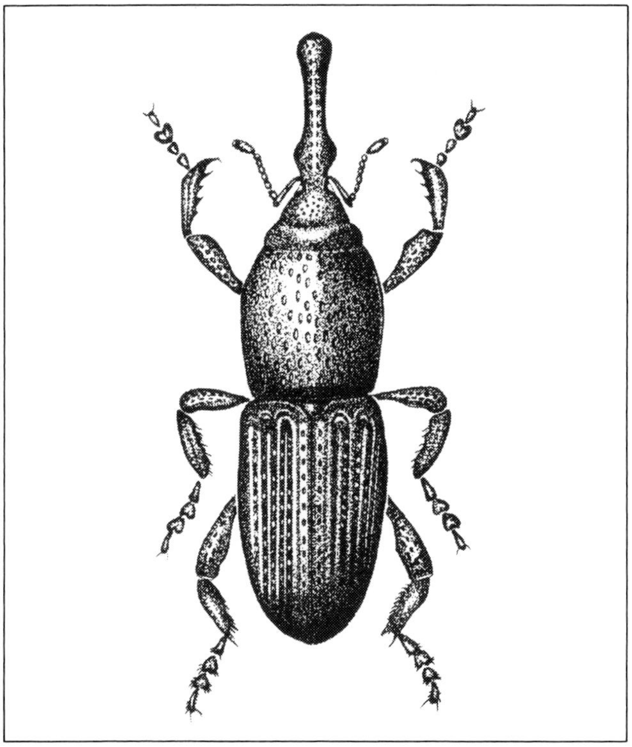

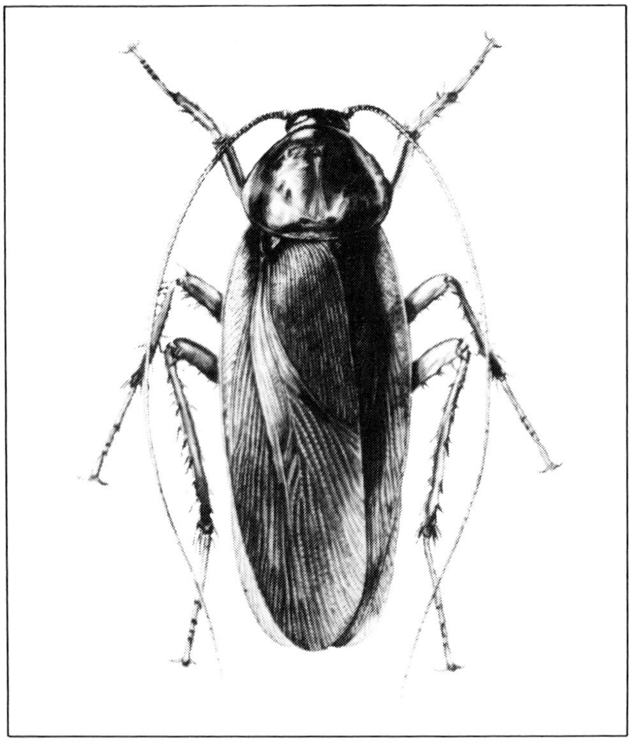

FIG. 10-6 American cockroach (actual size 1-1/2 inches)

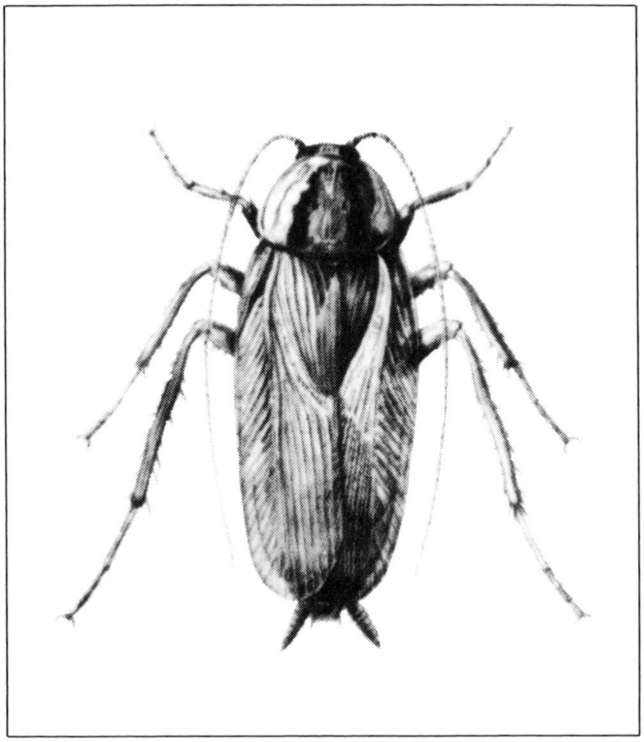

FIG. 10-7 German cockroach (actual size 1/2 inch)

FIG. 10-8 Oriental cockroach (actual size 1 inch)

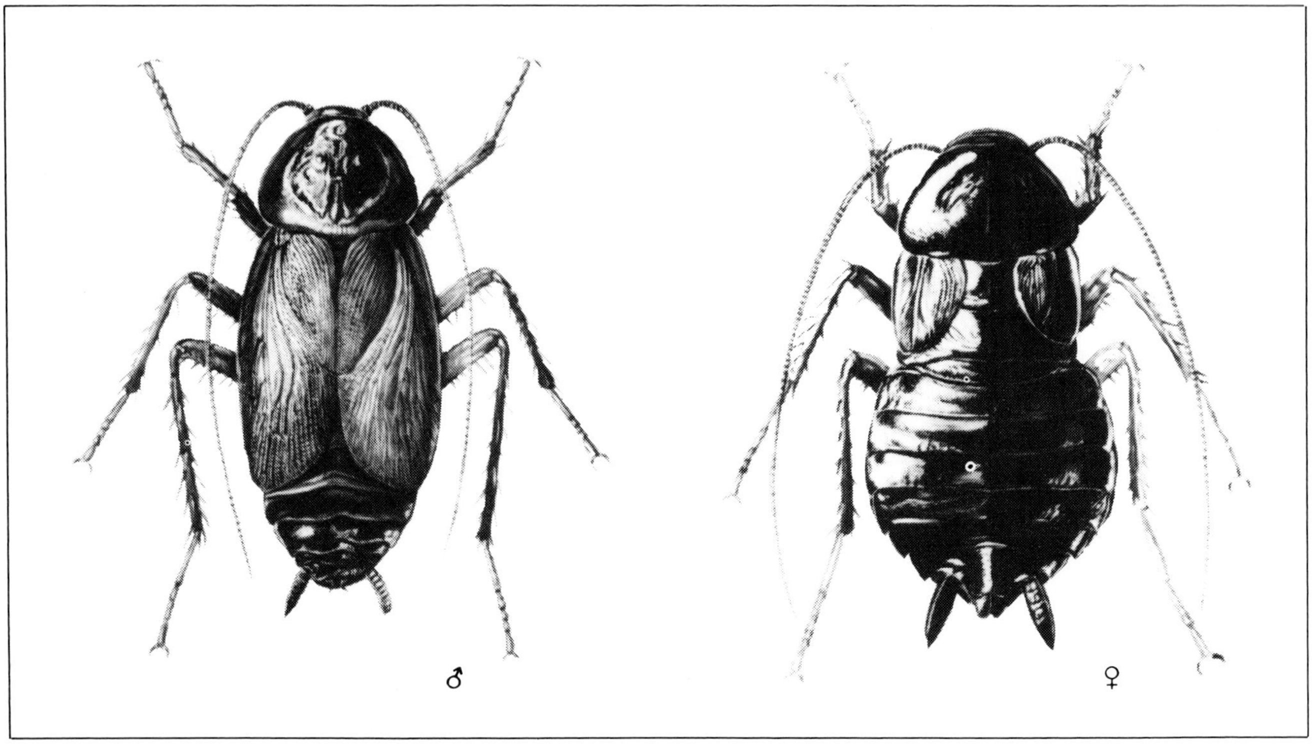

CONTROL SYSTEMS

While there are no exact figures, it is estimated that the number of species of insects in the entire world totals well into the millions. For this reason, insect control is critical to the production of safe and sanitary food.

There are several proven pest control systems which can be very helpful in controlling insects. As you go about designing these systems, it is important that you learn as much as you can about the habits and characteristics of the various species you are trying to control. The professional food sanitarian and the pest control operator can be very helpful in providing you with this information. Manufacturers of various insect control equipment can also be helpful.

Fig. 10-9 This electric insect light trap features a catch pan latched to the bottom of the unit.

ELECTRIC LIGHT TRAPS

Insect light traps have proven highly effective in the control of the common housefly and many other flying insects. Some flying insects are attracted to the ultra-violet emissions produced by fluorescent and mercury vapor lights. Flying wildly at this attractor, the fly ultimately makes contact with a low-voltage electric grid device, which electrocutes it. The fly then drops into a catch tray suspended from the base of the light trap. *(Figure 10-9)* The response range depends on the visual capability of the particular flying insect, but it is usually less than 25 feet from the light attractor.

Selection and Location. The selection of light traps requires that two questions be answered: what species will be trapped, and where will the traps be placed? For example, night flying insects entering the plant through doors are attracted to the light and warmth of ceiling areas. Large traps with catch pans mounted eight feet or more from the floor do a good job of controlling night fliers. *(Figure 10-10)*

Day fliers such as the housefly tend to look for the perfect temperature of 82°F when they enter the plant. Along with ideal temperatures, they look for food and flymates, which are likely to be found at the floor level. Light traps to catch day fliers should be located around five feet from the floor, preferably no more than eight to sixteen inches from the floor and extending upward 60 to 70 inches. *(Figure 10-11)*

The fly is reasonably comfortable outside so long as the temperature remains at 82°F. At this temperature, the fly has no interest in going inside. Nevertheless, the temperature seldom remains at this ideal level for a prolonged period of time. When it changes, the fly seeks a more comfortable climate and heads for the nearest building entrance. Flies do not adjust well to varying light intensity, and become very responsive to an ultra-violet light source produced by a nearby trap. This means that the best time to trap flies is as soon after they enter the plant as possible. Light traps should be located 12 to 15 feet inside entrances and in a position where the light cannot be seen from the outside, whereby it becomes an attractor. In theory a significant number of the flies should be trapped by units located just inside the doors, provided they are properly located. The remaining number should be trapped by subsequent units strategically placed throughout the plant.

Flies do not respond in the same manner to ultra-violet light, as evidenced by an interesting test designed to measure this response. One hundred flies were turned loose in a 20-foot square room where food and water were placed on the floor and a light trap was positioned near the center. In the first 90 seconds, 12 to 13 flies were trapped; 60 more were trapped over the next two to three hours, and all were trapped in six to seven hours. These response times remain fairly constant whether 100 or 1000 flies are involved.

The lesson here is that flies are not all alike in their time-related responses to ultra-violet light; in fact, they have responsive and nonresponsive periods. No one knows why, but sooner or later, as the test indicates, they all become responsive. The implication is that light traps, properly positioned, will eventually trap virtually all flies.

To have a better understanding of trap location, we need only witness what happens when two traps are placed in a room: one three feet from the floor, and one more than eight feet from the floor. Over any period of time, the trap near the floor will trap ten times as many flies as the unit nearer to the ceiling. As previously stated, the reason for this is that the active and responsive flies buzz around at levels less than five feet from the floor. High-flying insects are on their way to a resting spot or they are scared away.

One important reminder pertaining to trap location should be made. Light traps are not to be used in Class II, Division I (explosion-proof) electrical areas.

This information will not make you an expert on light trap location, but it will give you a better understanding of the problems of trap selection and location. Light traps are an important part of flying insect control and are most efficient when they have been selected and located by

Fig. 10-10 Electric insect traps, properly located, are very effective against flying insects gaining access through nearby doors.

Fig. 10-11 Active and responsive flies generally move around at low levels, and for that reason, insect light traps located closer to the floor are extremely effective.

knowledgeable light trap engineers. Trap layouts should have the approval of these experts.

SPACE SPRAYS OR MISTING SYSTEMS

Space spraying or misting with liquid insecticides during flying insect peak periods in the summer months is also a fairly effective means of control. Several automatic dispensing systems are available from several pest control equipment suppliers. These systems can be programmed to automatically release a given quantity of insecticide into a given area at a predetermined time. This equipment can treat both large and small areas. Most liquid space sprays or mists, as they are sometimes called, depend upon making contact with the insect to obtain a kill. Since these insecticides can be harmful to humans, the treated area must be free of people when the sprays are released. Food products must be removed from the area or covered. Product zones must be covered or washed prior to start-up.

There are three principal automatic space spray systems available: mechanical ULD, pneumatic, and aerosol or pressurized cylinders. In addition there are conventional portable mechanical units and thermo fogging units. We will look at the three principal systems.

Mechanical Ultra-Low Dosage. ULD stands for ultra-low dosage. Ultra-low dosage involves the reduction of liquid insecticide droplets to a fine particle size. This reduction is made possible by the mechanics of the ULD equipment. ULD equipment can reduce liquid insecticide droplets to a point where the majority are in the 15 to 20 micron range where it is most effective. The significance of micron particle size reduction is that many more droplets of the right size are airborne, thereby increasing the chance of a droplet coming into contact with an insect. To better understand the theory of ULD space spraying, visualize a quart size jar filled with 24 ping-pong balls. If the balls are thrown into the air in a room containing 24 people, the odds are slim that anyone would be hit. Take the same quart jar and

Fig. 10-12 The wall-mounted ULD (ultra low dosage) insecticide dispensing unit is an important insect control device.

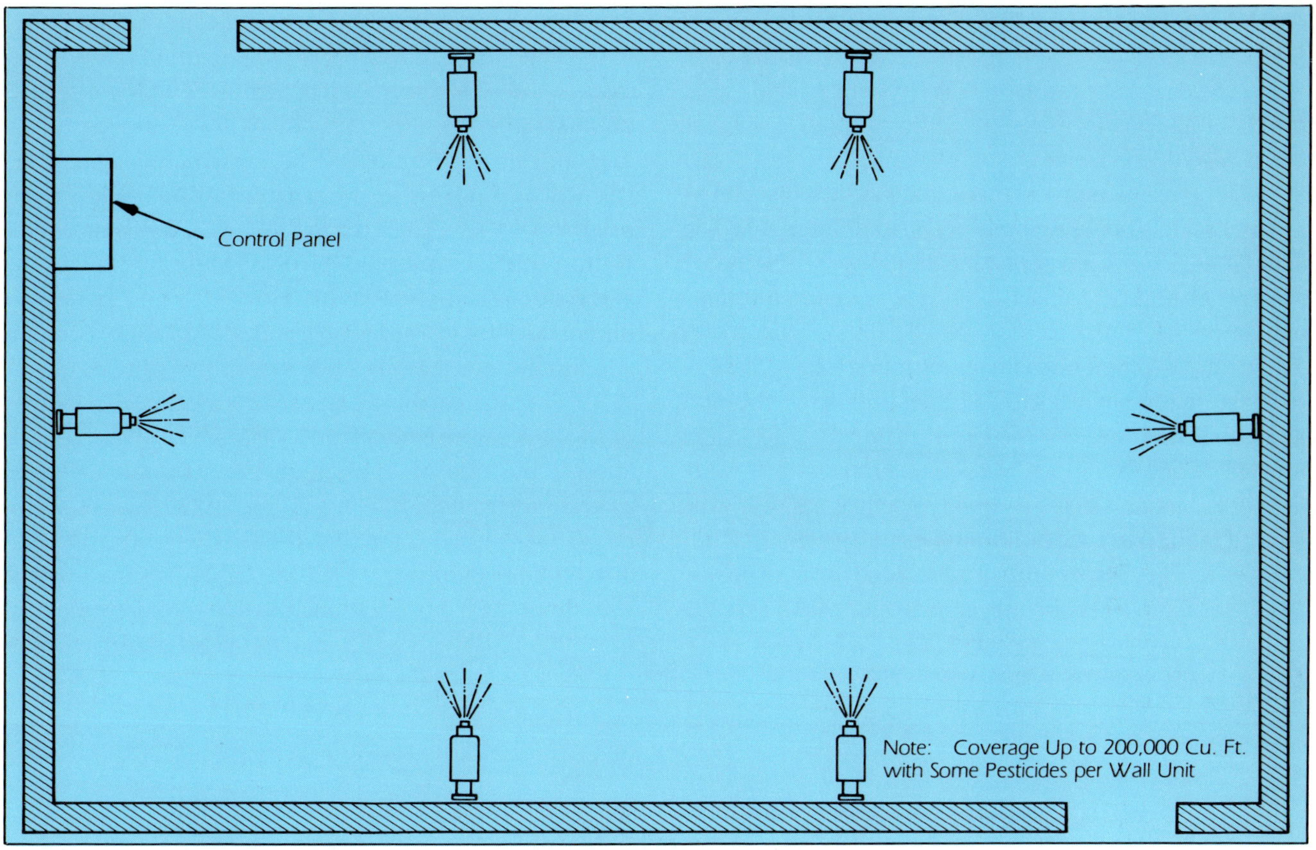

Fig. 10-13 Wall-mounted units must be properly located so that the insecticide is uniformly distributed.

fill it with thousands of bird shot. Throw it into the air in the same room containing 24 people and chances are that everyone will be hit.

ULD space spraying equipment uses a smaller quantity of insecticides having a higher percentage of active or toxic material; as a result, each micron-size droplet is more lethal. Therefore, it could be said that this equipment uses insecticides more effectively. Additionally, the smaller droplet size does remain airborne longer, which increases the possibility of contact with flying insects.

Automatic ULD dispensing equipment is available in modular units. Each unit has specified coverage capacity, which is usually specified in cubic feet. *(Figure 10-12)* Some modular units can cover up to 200,000 cubic feet. Control units furnished separately can operate up to 15 dispensing units, which are normally wall or ceiling mounted. It is only necessary to calculate the total volume of the area involved and to determine the number required on the basis of the capacity of each unit. Units should be equally spaced throughout the total area to be treated. *(Figure 10-13)* The electrical requirements are 110 volts.

Control units include timing relays that permit the equipment to disperse a set amount of liquid spray in a given time. The dispersion period is controlled with a timer. Before beginning, fill the dispensing unit bottles with insecticide, set the timers, and prepare and clear the area. When it is safe to reenter the area, usually several hours after treatment, the excess pesticides should be removed from the supply bottles.

The preparation of an area to be treated and the application rates are the keys to the safe application of any insecticide. The direction of a trained sanitarian is essential to this operation. The same proviso applies to the next two systems.

Pneumatic System. The working principle of the pneumatic space spray system differs from that of the ULD system. The insecticide is dispersed through a series of air atomizing nozzles. The system uses a 20 to 40 psi air supply to disperse the insecticide. The micron measurements of the majority of insecticide droplets are larger. The system generally uses standard insecticide formulas which contain a smaller percentage of the active or toxic chemical.

A typical pneumatic space spray system consists of an insecticide storage container, a one half-inch schedule 40 stainless steel welded insecticide pipeline, a series of multiple siphon-type air activated spray nozzles, one half-inch air piping, and an air pressure regulator. *(Figure 10-14)* The multiple siphon nozzle requires four outlets. The size of the storage container should be determined by the total. number of spray heads in the system. The complete system is installed one to five feet below the ceiling. By following aisleways, column lines, or walls, pipelines can often be located in the overheads without passing over equipment, open product zones, or open product zone surfaces below. Problems from leaks are thus minimized. Make sure the liquid pipeline is free of sags and adequately pitched back toward the storage container. Provide a drain valve on the storage tank. This arrangement permits complete drainage of the system. It is also important to install nozzles so that they are a little higher than the storage tank full-liquid level. Use welded stainless steel piping to further minimize the possibility of pesticide leaks. Stainless steel is recommended because pesticides are quite corrosive. Make sure the system is pressure-tested prior to use. However slight or remote the potential for leaks, it simply makes good sense to follow good safety precautions when working with pesticides.

To calculate the capacity of the system, the volume of the area to be treated is first determined. When this is known, the rate of application per 1000 cubic feet is ascertained from the insecticide label. At 30 psi each spray nozzle can propel the liquid droplets eight feet in four different directions for a total coverage of approximately 250 square feet. To determine the total number of spray

Fig. 10-14 Pneumatic space spray systems utilizing air atomizing spray nozzles are also effective in dispensing insecticides into large areas.

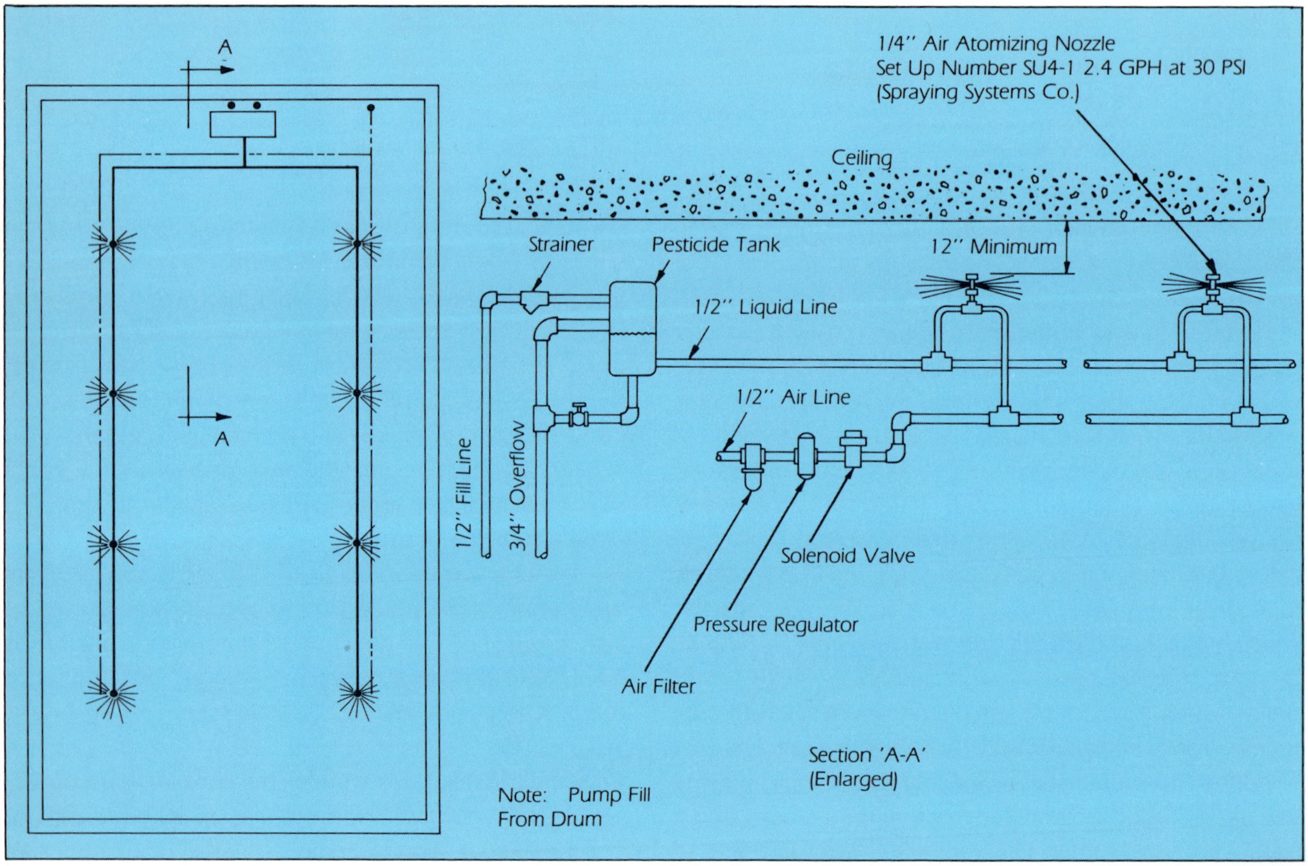

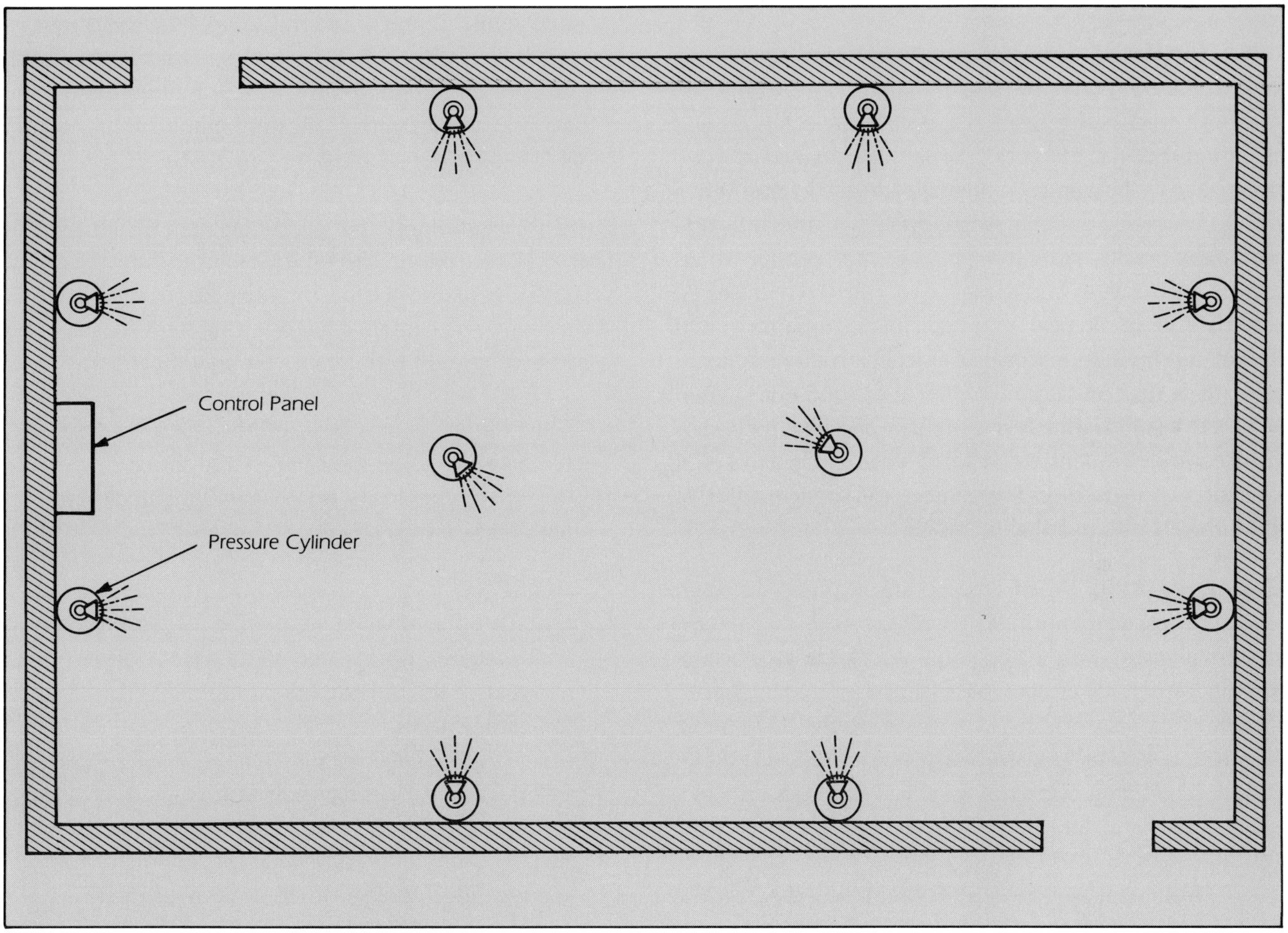

Fig. 10-15 Cylinder systems can be used in large areas, too.

nozzles needed, determine the total square footage of the area to be treated and divide by 250 square feet (the coverage of one nozzle).

The system can be totally automated by using normally closed electric solenoid valves activated from a timing mechanism. Both the duration of the dispersion of insecticide and the activation time can be programmed. Upon completion of the application and at a time safe to do so, drain the surplus pesticide from the supply container back into the supply drum.

Aerosol/Pressurized Cylinder. The aerosol or pressurized bottle systems have the added convenience of never needing manual refills. The cylinder is simply replaced with a full one when it becomes empty. Some of these systems employ discharge nozzles that, with assistance from high internal pressures, reduce droplet size to that of the mechanical ULD units (15 to 20 microns). These systems also can be fully automated with the addition of electrically operated release valves and programming equipment.

Like the ULD system, the size of this system is determined by the volume of the area and the capacity or release rate of the cylinder nozzles. *(Figure 10-15)* The required number of cylinders are mounted on walls and ceilings so that insecticide is distributed evenly throughout the space. A 110-volt electrical service is then provided to activate the electrically operated valves.

HEAT STERILIZATION

Not only is heat sterilization the most effective nonchemical control used to control dermestidae, stored-product, and other insects common to the grain processing industry, it could very well be one of the better insect control tools ever used. There are very few places that heat will not penetrate given adequate time to do so.

Heat sterilization for insect control is not new. It was used in the early days of flour milling to control insects. Although an effective means of control, this technique was quickly abandoned because of the damage it caused to wood floors, and to the wood equipment in use at the time. When wood is exposed to high temperatures, it dries out and eventually cracks. Additionally, line shaft belting would stretch and grease would melt and run from unsealed bearings. Technological improvements in equipment and the virtual elimination of wood as a construction material have made heat an attractive control method once again.

Heat sterilization is an extremely efficient method of insect control. Most insects are killed when exposed to 120°F for short periods of time. Some insects can survive at 120°F, but virtually all are killed at 140°F.

How Heat Kills. When insects are exposed to high temperatures, death normally is caused by the coagulation of soluble proteins in the body tissues. Death also occurs at temperatures lower than those which cause coagulation. In such cases death is caused by injury to the body tissue enzymes, which takes place when body fluids are dehydrated. The lethal effects of heat will take place in five minutes at 131°F, and in less than an hour at 120°F. Heat kills all stages of insect life: egg, larva, pupa, and adult.

Although heat sterilization obtains kills quickly, the procedure itself requires considerable time. It is necessary to heat an entire food plant or area of the food plant to 140°F and to hold this temperature for a minimum of 20 to 24 hours. The reason for this is that most building materials are not good conductors of heat, and it takes a long time to raise the hollow cores of walls to the 120° to 140°F level. Furthermore, thermo-shock can be avoided with a slow rise.

The feasibility of heat sterilization depends on the type of facility and on the type of manufacturing process. Many manufacturing processes utilize a great deal of drying, toasting, or baking equipment, which can also be used to heat the building for insect control. In such situations, heat sterilization can be accomplished with very little additional heat from supplemental heating units. Old facilities with wood floors and wood framing are not good choices for heat sterilization.

Energy Requirements. When considering heat sterilization for insect control, one of the most important questions is how much energy is required. This question can be answered by a detailed design study. However, for a rough estimate of the energy required, the following simplified procedure for determining the amount may be useful.

Energy Calculations. To determine the BTUs (British Thermal Units) required to raise the temperature of a room, it is necessary to calculate the total heat loss from all surfaces of the area to be heated. The following steps may be used to arrive at a good estimate of the total heat loss:

1. Take the difference between the inside heat-up temperature to be maintained and the average outside temperature, for this will be the design temperature difference.

2. Calculate the areas in square feet of glass windows, wall surfaces, floors, and roofs which will be exposed to the outside temperature or other unheated spaces.

3. From Table 10-1, select a heat transfer coefficient (U value) for each exposed surface as determined in Step 2. (For floors, add 2 BTUs per square foot of ground floor area to the total number of BTUs found in this step.) Using the temperature difference ΔT from Step 1, calculate the heat transfer loss (Q) for each area (A) by using the following formula:

   $$Q = AU\Delta T$$

   Where Q = BTUs per hour, A = area in square feet, ΔT = the difference between indoor and outdoor temperatures.

   This is done for each surface. The sum of the heat losses for all the surfaces will be the total heat loss in BTUs per hour. If any of the areas are subject to prevailing winds, above 15 miles per hour, the values should be increased by 10% before being added into the total BTU value.

4. Generally, a heat sterilization job is done by

TABLE 10-1 HEAT TRANSFER COEFFICIENT VALUES

Roofs

Roofing Materials	U Value
6" Concrete slabs w/1-1/2" insulation and roofing material	.17
6" Concrete slabs w/2-1/2" insulation and roofing material	.14
6" Concrete slabs w/3-1/2" insulation and roofing material	.10
Metal ceiling w/3-1/2" insulation and roofing material	.11
2" Wood ceiling w/3-1/2" insulation and roofing material	.10

Windows

Type of Window	
Single pane glass	1.13
Double thermo pane glass (3/16" air space)	.69
4" Glass block	.60

Floors

Floor Materials

6" Concrete floor slab on grade
2 BTUs/hr. per sq. foot of floor

Walls

Wall Materials	U Value
12" Concrete block, painted	.36
12" Concrete block, 2-1/2" structural tile	.28
12" Concrete block, ceramic facing tile	.35
4" Brick/8" concrete block	.30
4" Brick/8" concrete block/ceramic facing tile	.35
8" Solid concrete	.67
8" Precast concrete, hollow core	.39
8" Precast concrete, hollow core, insulated	.23
Metal sandwich curtain wall, 2" foam insulation	.08

recirculating air with no air being exhausted. However, there will be infiltration of air, and this infiltration is normally estimated to be one or two complete air exchanges per hour. Calculate the volume of the heat-up area in cubic feet and multiply it by one or two air exchanges (one for a fairly tight building and two for a draftier building). To determine the BTUs required to raise the air from outside to the heat-up temperature of the room, use the following formula:

$$Q = \text{Cu. Ft/Hr} \times (\Delta T) \times 0.018$$

Where 0.018 is the specific heat of air (BTU/cubic foot/degree Fahrenheit).

5. To calculate this heat loss due to building steel and equipment surfaces, estimate the weight of the steel and the equipment and apply the following formula.

$$Q = .12\ M\ (\Delta T)$$

Where .12 is the specific heat of steel (BTU/pound/degree Fahrenheit), M is the weight of the steel in pounds and (ΔT) is the temperature differential.

Estimating the total weight of all the building steel and equipment can be an involved process. For a steel-framed building, a simple way to arrive at this total weight would be to assume that the steel weighs 10 pounds per square foot of the building space. A simple way to arrive at an estimated weight of the equipment is to multiply the square footage of the building by 100 pounds. Food manufacturing plants are considered to be light industry, and for light industry, floors are normally designed for a 250-pound loading. The actual loading is estimated to be around 100 pounds.

6. The total BTUs per hour to be supplied in order to maintain the desired heat-up temperature is the sum of total losses calculated in Steps 3, 4, and 5.

If there are any heat gains to the room such as processes that release large amounts of heat, the amounts should be determined as accurately as possible and subtracted from the total heat loss.

The total heat requirements in BTUs can be transposed into steam, gas, or oil requirements by dividing the total BTU requirement by 945 BTUs for 15 pound steam, 1000 BTUs for a cubic foot of natural gas, 156,000 BTUs for a gallon of number 6 oil, or 142,000 BTUs per gallon for number 2 oil.

The energy requirement obtained through these simplified calculations will serve either to obtain a general idea of how much energy is needed or to determine if your current heating system seems to have enough capacity.

However, there are many other factors that must be considered, one of which is time. Twenty-four hours is not a lot of time in which to accomplish a heat-up. If you want to accomplish a heat-up in 24 hours, additional heating capacity may be required. The final energy determination should be done by an engineer who specializes in heating and ventilation design.

Equipment Planning. If heat sterilization is to be used as a major insect control measure in a new plant, begin planning early in the design stage. Early planning can effectively account for and meet the energy needs. Boilers can be sized to produce the necessary steam, or additional quantities of natural gas can be provided. The heating and ventilating systems can be designed to produce the higher temperature which will be required. This is not to say that heat sterilization cannot be incorporated into existing facilities; it can be and it should be.

In either case, planning should include an analysis of the total heat required. Determine the portion of the total heat required that can be obtained from existing processing equipment and from existing heating systems. Finally, calculate supplemental heating needs and add the additional heating equipment accordingly.

Portable supplemental heating units have the advantage of mobility; thus, they can be located where the heat is needed. Low-pressure steam coils mounted on mobile platform trucks produce a very convenient, supplemental mobile heating unit design. *(Figure 10-16)* These units are equipped with circulating fans. Both a single coil and a double coil unit are pictured. Detailed drawings of a double coil unit are provided. *(Figure 10-17)*

Permanently valved cut-ins should be provided in the steam and condensate lines for the convenient hookup of the portable heaters. High-pressure steam hoses and pipe

Fig. 10-16 The supplemental steam heating unit for heat sterilization is a very effective insect control technique. Both double and single steam coil units can be easily built.

Fig. 10-17 Drawings of a typical double coil unit

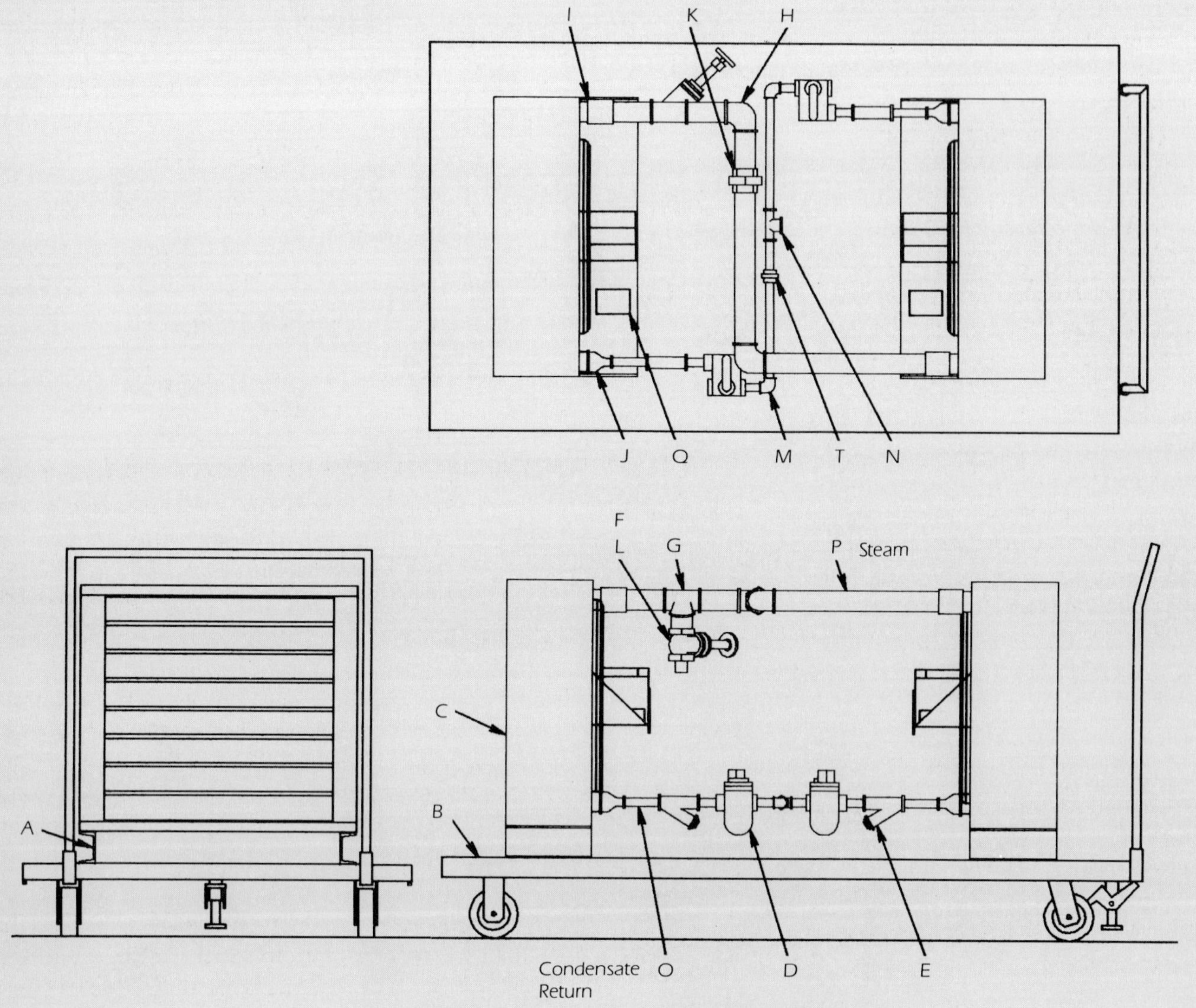

DOUBLE HEAT-UP CART

BILL OF MATERIALS

Item	Qty	Description	Item	Qty	Description
A	4	4″ at 13.8 x 10-1/2″ Lg. H. R. Channel	K	1	2″ Union
B	1	4′-0″ x 7′-0″ Cart	L	2	1-1/2″ Nipples
C	2	Young Unit Heaters	M	2	3/4″ Elbows
D	2	Steam Traps 3/4″	N	1	3/4″ Tee
E	2	Strainers 3/4″	O	1	Length 3/4″
F	1	Steam Valve 1-1/2″	P	1	Length 2″ Pipe
G	1	2″ x 2″ x 1-1/2″ Tee	Q	2	Switch
H	2	2″ Elbow	R	1	1″ I. D. x 50′-0″ Hose
I	2	2″ Couplings	S	1	1-1/2″ I. D. x 50′-0″
J	2	2″ to 3/4″ Reducers	T	1	3/4″ Union

couplings should be provided to complete the hookup. One of the more difficult areas to heat to killing temperatures is the ground floor level. Portable circulation fans can be helpful distributing the heat about the plant and, in particular, in raising floor temperatures to the killing level by pulling the warm air down from above and redirecting it at the floor. The difficulty at ground floor level results from low soil temperatures.

Process equipment such as dryers and toasters can provide a large portion of the heat needed. The doors of these units should be left open and circulating fans should be running to distribute the hot air.

Sprinkler Systems. By planning early, high-temperature sprinkler heads can be specified for areas which will be heated. High-temperature heads (286°F) should be used because hot spots do develop. Although still lower than the 165°F limit of standard sprinkler heads, the hot spot temperatures (which should never be allowed to exceed 145°F) can weaken the bi-metal fuse of the head. A weakened head can release water at temperatures less than 165°F. The high-temperature heads minimize the risk of an accidental water release by one of the sprinkler heads.

It is very important that someone completely familiar with the sprinkler system participate in the heat-up. If a head were to release, this individual can quickly identify the area involved and shut off the water without delay. The system should be repaired and returned to service quickly. Pressure in the sprinkler system might need to be adjusted as the plant is being heated. Increased temperatures at the ceiling level can cause a slight increase in water pressure.

Equipment and Material Removal. It will be necessary to remove some equipment or material from the area to be heated. Items that can be affected by higher temperatures include ingredients like cocoa, sugar and shortening; packaging materials such as waxed paper should also be removed. Either remove pressurized containers or, if possible, reduce their pressure levels. As a precaution, remove all combustible materials.

Heat sterilization procedures affect very few pieces of equipment. Depending on their temperature ratings, some electrical components should be shut off, and some circuit breakers might require resetting at start-up time. For the most part, these requirements should be interpreted as precautions, not as insurmountable problems.

Sealing. One of the advantages of heat sterilization over other major insect control efforts with deadly fumigants is that very little sealing is required. Doors must be shut, windows closed, and areas which are not to be included must be partitioned off if they cannot otherwise be isolated. Exhaust and intake stacks must be sealed by closing dampers or by covering them. Fumigations require the extensive sealing of doors and window sash joints with cloth tape, and the sealing of many openings with nylon or plastic tarps, all of which are very costly.

Thermometers. Locate thermometers throughout the area to monitor temperatures. Use portable thermometers that can be put in place for the heat-up. The thermometers should have a range of 100°F to 200°F. Take readings every hour and make any necessary adjustments to the heating equipment as needed to stay within the heating limits. For safety reasons, put two people on this job. Temperatures must be recorded by location to identify the trends and evaluate the job.

Insect Test Cages. Insect test cages containing live insects should be placed in strategic places to evaluate the overall efficiency of the completed job. These cages should be placed inside wall cavities, at the floor level, and in the overheads. They should be collected at the end of the job and the kill determined.

Cool-Down. Allow a long cooling down period to prevent problems with thermo-shock. Begin by shutting off the heating equipment, but keep the circulating fans running. Open doors and windows next. Finally, shut off the fans.

Restoring the Plant. Restoring the plant for operation will include adjustments to the sprinkler system, removal of portable heating equipment, and unsealing. It might also require that some bearings be regreased and, as mentioned earlier, some circuit breakers might have to be reset.

Heat sterilization is an exceptionally fine insect control tool. It should be done only by people who have the experience to do it. It requires good detail planning. If this procedure is being done for the first time, it is advisable to obtain the services of an experienced person to help plan and carry it out.

The information provided in this chapter does not qualify anyone to carry out a heat sterilization job. It was intended to familiarize the engineer with the procedures involved in a successful heat sterilization of a facility. Using this knowledge, the engineer can incorporate the heating requirements into the facility design in the early project stages. Heat sterilization is a safe and very efficient way to control insects in food manufacturing plants when they are confronted with major stored-product insect problems.

CONTROLLED ATMOSPHERE FUMIGATIONS

By far the greatest objection to the use of pesticides in food plants has been to those applications in which liquid fumigants were directly applied to raw grain products. These direct applications resulted in the presence of ethylene dibromide residues allegedly in excess of what were believed to be safe limits. As a result of this problem, ethylene dibromide was banned for use as a grain fumigant. Ethylene dichloride and carbon tetrachloride, two other grain fumigants commonly used in conjunction with ethylene dibromide, also appear to have an uncertain future. Furthermore, there is and probably always will be extra concern about all fumigants which contact the food products directly. The food industry has not been standing idle and has been actively pursuing new nontoxic pest control techniques such as controlled atmosphere fumigations.

Controlled atmosphere fumigations involve the use of carbon dioxide (CO_2) or nitrogen (N_2) to control insect infestations in grain and other grain-based products such as flour. CO_2 and N_2 are inert gases that reduce the atmospheric oxygen in gas-tight enclosures and kill insects by asphyxiation. Early test work with these two gases as an insect control tool seems to suggest that CO_2 is somewhat more effective than N_2 in the killing of all stages of insect life: eggs, larva, pupa, and adult. In environments where temperatures were above 70°F, a 60% concentration of CO_2 used to fumigate grain in bins consistently produced kills of over 96% of the insects exposed within four days. In contrast, the N_2 atmosphere required 10 days or more to obtain kills of all insect life forms. Temperatures are important to controlled atmosphere fumigation, as they have a direct effect on the respiration rate of the insect. Warmer temperatures increase this rate. Temperatures below 70°F generally require a longer exposure period for an efficient kill.

Applications. CO_2 fumigations can be applied to bulk food products in transit or in storage. Fumigations of bulk food in transit can be effectively done with dry ice. Dry ice has the advantage of continuously releasing CO_2 gas to replace any that may be leaking from the moving bulk hopper rail car. Applications to bulk storage bins can be done with either dry ice or vaporized liquid CO_2. Bulk storage bin applications seem to favor applications with vaporized liquid CO_2.

Bulk storage bins that typically store bulk grains or bulk grain-base flours can be found in all sizes, shapes and forms. CO_2 is heavier than air. It can be introduced into the top of the bins over a period of time and it will permeate the entire bin.

The key phrase to note is *over a period of time*. This period would undoubtedly be longer for larger bins than for smaller bins. In the case of extremely large bins, the CO_2 should be added as the bin is being filled. However, it is neither accurate nor practical to suggest that every load requires fumigation. Fumigation of many products is simply done as a precautionary measure to protect products which have been in storage for a prolonged period or to prevent a minor insect infestation of product in a bin from getting out of control. Consequently, food products stored in extra-large storage bins may require the addition of an internal CO_2 manifold to help distribute the gas throughout the entire bin, since the product is already in the bin.

Equipment. First and foremost, fumigations of bins with CO_2 require a good, tight structure. Almost all steel bins meet this condition. At most, they might require some minor sealing of bin discharges, access doors, and vents. The procedure for fumigating the bin with vaporized liquid CO_2 will require an initial large charge of the gas to completely purge the bin and then an occasional charge of CO_2 to maintain a gas concentration of 60%.

If a liquid CO_2 system is already in place, the additional equipment needed to effectively do a bin fumigation includes a vaporizer, a temperature gauge, a pressure regulator, a pressure gauge, a gas flow meter, a control valve, and a high-pressure schedule 80 steel or copper pipe line for the gas supply system. To appropriately monitor the gas, a vacuum pump, a CO_2

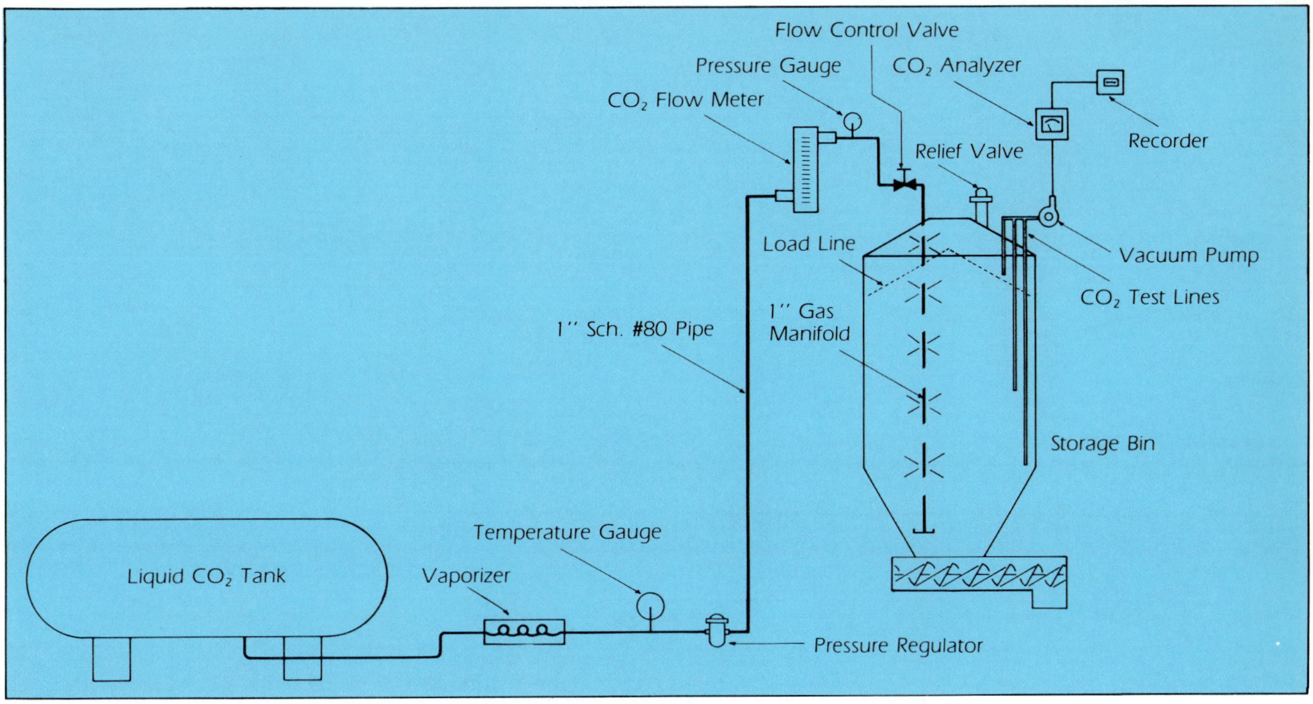

Fig. 10-18 Schematic of a CO_2 fumigation system for a product storage bin.

analyzer, and a concentration recorder should be considered. *(Figure 10-18)*

Most storage bins are not totally air-tight. In such cases, purging takes place through the tiny bin openings. On extremely tight tanks, a purging valve and a relief valve should be considered.

Liquid CO_2 is stored at 300 psi at 0°F. The vaporized CO_2 is delivered to the bin at 10 psi and around 70°F to avoid condensation problems. To assure that concentration levels are obtained throughout the bin, plastic gas concentration monitoring lines should be placed in the tank permanently to obtain readings in different areas or at different elevations. To assure safe working atmospheres in and around the bin, oxygen readings should be taken periodically, particularly in the area surrounding the bottom of the bin.

The quantity of CO_2 required will vary with a number of conditions such as the tightness of the bin and the product being treated. With wheat and a fairly tight bin, .175 to .225 pounds per bushel treated could be used to determine an approximate cost.

CO_2 could also be used as an equipment fumigant. However four days is a long time for equipment to be down and out of production. In all probability, its biggest value will be in the treatment of insect-sensitive food products.

FUMIGATION VAULTS

Some segments of the food industry still have a justifiable need for fumigation vaults. Both atmospheric and vacuum vaults can be used, but the vacuum vault has the advantage of reducing exposure time. Under a vacuum, many fumigants have a more rapid penetrating capability. On the other hand, atmospheric vaults can be made for far less money than a comparable vacuum vault.

Atmospheric Vaults. Atmospheric vaults can be constructed from many different materials, including sheet metal, concrete, concrete block, brick and wood. It is only necessary to make them gas-tight. Although atmospheric vaults need not be expensive, they should be durable enough to withstand the normal abuses that occur when products are moved in and out with conventional material handling equipment. *(Figure 10-19)*

- Construction Detail. If sheet metal or wood is used, carefully seal all joints and seams with pliable caulking materials or other

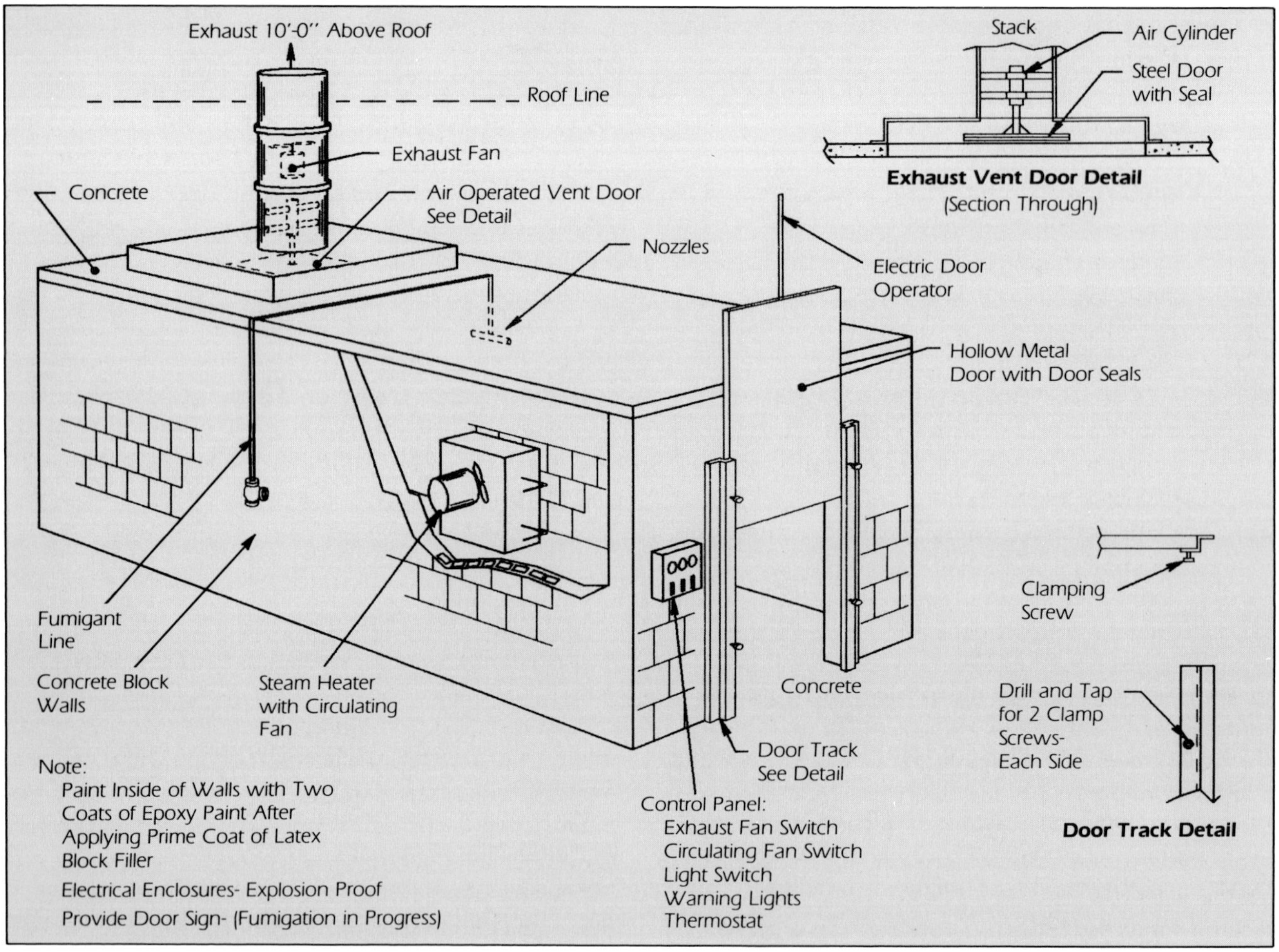

Fig. 10-19 Atmospheric fumigation vault

sealing substances. Concrete, concrete block, and brick walls can be sealed with latex sealing materials and two heavy coats of epoxy base paint. The vault door must be large enough to allow materials to move freely in and out. Take extreme care to select a gas-tight door. Many refrigerator doors will meet this condition.

A small, nonsparking, circulating fan should be mounted on one end of the vault to provide a movement of gas and air to obtain an even distribution throughout the area. An exhaust fan capable of 10 to 12 air exchanges per hour should be provided. The exhaust stack should discharge to the outside, and it should not be located near air intakes or near areas where there is danger of employee exposure. At the inlet of the exhaust system, provide a sealable vent door which can be opened and closed from the outside. Make sure the vault is well lighted and that all electrical components are suitable for explosion-proof conditions. If the vault is located where heat will be required, furnish a thermostatically controlled steam heating unit. Open flame units are unsatisfactory.

The fumigant line should be placed so that it enters the vault through the roof into a manifold. Nozzles should be located at the end of this manifold. Equip the vault with lights which warn of a fumigation in process. Locate a direct reading thermometer outside the vault and adjacent to the door, along with all of the heating, fan control, and light switches.

Mount gas masks and gas detecting equipment in cabinets outside and alongside the vault door. Be sure to select the equipment that is compatible with the gases used. Methyl bromide or ethylene oxide are two of the more common fumigants used in vault fumigations.

Select fumigants which have been approved for use with the food products being treated. They must be applied by trained personnel under the direction of a plant sanitation manager who is certified to handle fumigants and other pesticides. Again, only trained, qualified people should operate the equipment.

Vacuum Vaults. Vacuum vaults differ from atmospheric vaults in that they are designed to accommodate a vacuum of 25 to 28 inches mg. They require vacuum pumps which are capable of producing this vacuum in 10 to 15 minutes. After fumigation, the chamber is cleared of the fumigant by air washing with at least two vacuums. While atmospheric vaults require an exposure time of 24 hours, vacuum vaults can complete a fumigation in as little as one and one half hours.

Never use phosphine in a vacuum vault. This gas is prone to spontaneous combustion when used in very high concentrations or when used in a vacuum atmosphere.

- Construction Detail. Vacuum vaults are expensive to design and they should be left to the experienced vault designer. Many excellent ready-made units are available from a number of vacuum vault suppliers. These units should be purchased complete with all necessary controls and safety equipment.

PESTICIDE STORAGE ROOMS

Chemicals used to control pests can be highly toxic and must be carefully and cautiously used and stored in a food plant. Misuse or improper storage can contaminate nearby food products and food product surfaces. Pesticides can contaminate by direct contact with the food or by indirect contact, such as when a pesticide is present on food contact surfaces. Pesticides also contaminate foods which readily absorb their odors and vapors.

Many pesticides and diluents used to cut or dilute pesticides are very flammable and must be used and stored with concern for potential fire. Several federal, state, and local regulations and codes govern the storage of both toxic and flammable chemicals. Fire regulations and codes, in particular, define the allowable quantities of chemicals that can be stored, depending upon their degree of flammability or combustibility. Allowable quantities depend on occupancy levels in the building as well. All applicable codes and regulations pertaining to safety must become the prevailing design conditions for pesticide storage rooms.

Design Criteria. Pesticide storage room size requirements vary according to the type and quantities of pesticides to be used. With some food processes, pest control is limited to controlling a handful of insect pests with a small quantity of insecticides. On the other hand, insect control in the grain handling and related grain processing operations might require the use of large quantities of pesticides. When large quantities of pesticides are used, more is generally needed than just a larger storage area. Large users buy bulk quantities or 55-gallon drums, and they buy concentrated instead of ready-mixed formulas. Concentrates require the purchase and storage of oil-based diluents. Equipment is needed to handle the larger pesticide and diluent containers. Mixing rooms and storage areas for application equipment must also become part of the complete storage complex.

Large pesticide users might require more storage space and more equipment, but the design requirements for both large and small storage rooms are the same. They are simply scaled down according to the quantities of pesticide in storage.

Pesticides should never be stored in or adjacent to food manufacturing, food storage or packaging material storage areas. At the same time, the location should be reasonably accessible. Well-ventilated locations are preferable. Keep the storage room locked and post a sign that identifies it as a pesticide storage area.

Area Layout. Large pesticide storage rooms should

include a work area which is partitioned into two areas: one for the mixing of pesticides and one for minor maintenance of pesticide application equipment. These areas require additional air exchanges. Provide special rooms or cabinets for deadly fumigants and rodenticides. Some rodenticides absorb insecticide odors, which make them less appealing to the rodent. Provide a special room, free of pesticide vapors, for gas mask canisters, respirators, and other life-safety equipment. A plan view of a typical pesticide storage room is provided. *(Figure 10-20)*

Building Materials. The walls, floors and ceilings of the pesticide storage room should be constructed of smooth, impervious, and nonabsorbing materials. Concrete and concrete block are acceptable. Slope the floor toward the center, at the rate of 3/16 inch per foot, where a drain normally would be located. However, because of the concern for toxic materials in the sewage system, the drain should be omitted. Provide a four-inch high sloped concrete sill across the door openings. Install self-closing metal doors with the appropriate fire rating for a flammable storage area. Include a dead-bolt lock assembly on the door.

Lighting/Heating System. Make sure that the electrical system, including lights, is Class II, Group G for an explosion-proof atmosphere. The lighting level should be 50 to 70 footcandles. Equip the work areas of the room with an adequate number of 110-volt explosion-proof

Fig. 10-20 Toxic chemicals used to control vermin must be properly stored in specially designed pesticide storage rooms.

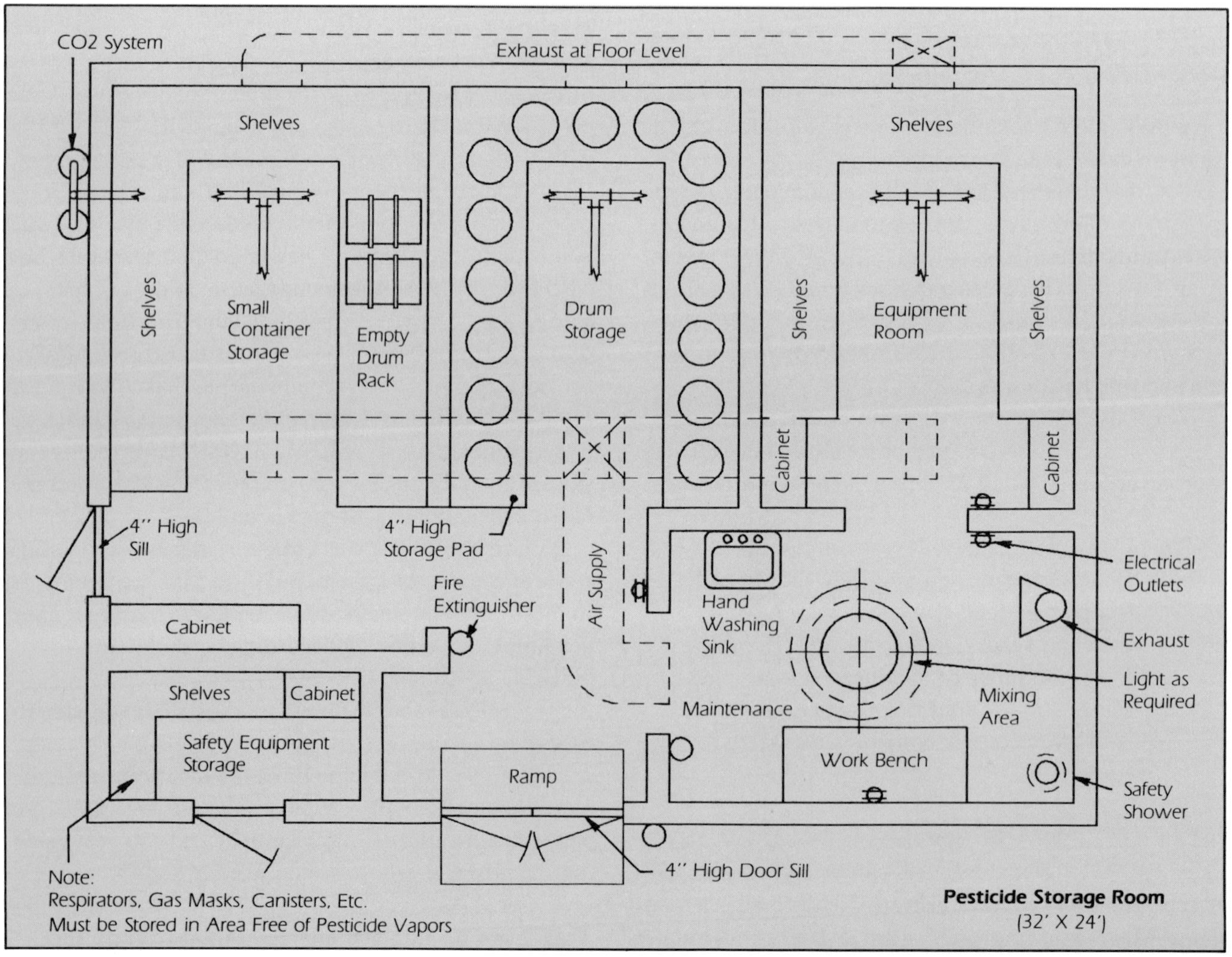

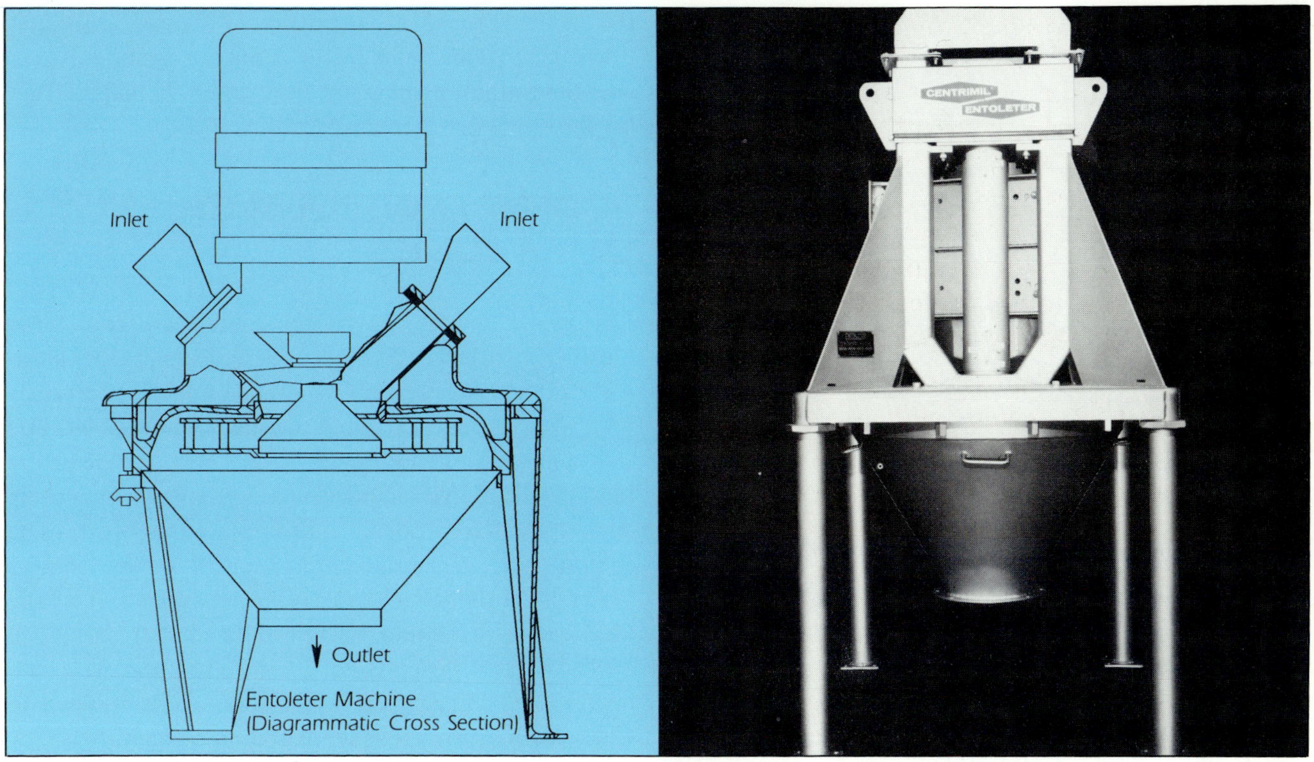

Fig. 10-21 Infestation destroyer used to kill stored product pests.

electrical receptacles. Use forced air, hot water, or steam for heating. Avoid open flame equipment. A heating system is needed to prevent stored pesticides from freezing, particularly the water-based chemicals.

Ventilation. The ventilation system should provide for 20 to 30 air exchanges per hour during area occupancy and from four to eight air exchanges per hour when unoccupied. Because most pesticide fumes are heavier than air, the air exhaust opening should be located in the wall at floor level. Locate the air supply duct in the ceiling or in the wall at ceiling level. Fan motors and electrical switches should meet explosion-proof requirements.

A specially designed wall-mounted exhaust hood should be considered in the pesticide mixing room. Mount this hood in a vertical position, and locate it at floor level directly in front of the point on the floor where the mixing is being done. The capture velocity at the face of this hood should be 200 to 300 fpm.

Plumbing. A small hand-washing lavatory should be located in or adjacent to the storage area. The lavatory is to be used for hand washing only, not for equipment cleanouts. Make sure that the lavatory is supplied with hot and cold water, and that water lines have backflow-preventing devices to prevent contamination of the potable water supply. A soap and towel dispenser is also needed.

Many liquid fumigants cause severe burns to skin and eyes and must be neutralized with water immediately. An emergency eye-wash fountain for this area is important to prevent serious eye injuries. If the frequent use of liquid fumigants is anticipated, consider the installation of emergency showers.

Fire Extinguisher. Automatic central CO_2 or halogen fire extinguishing systems should be considered when these storage areas are located within the plant. Ideally, a large pesticide storage facility would include a building of its own. Water sprinkler systems should not be used in this area, particularly if phosphine fumigants are stored in the room. In addition to automated extinguishing systems, or in the case of small facilities not equipped with automated systems, dry chemical fire extinguishers should be located on the outside wall adjacent to the door. Additional extinguishers should be provided in the mixing and

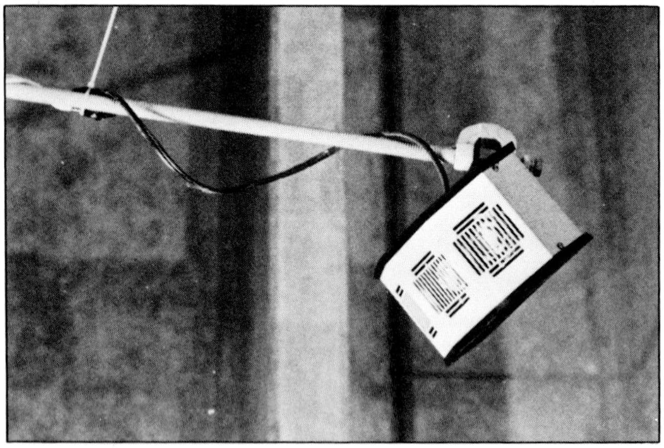

Fig. 10-22 Ultra-sonic rodent control devices transmit high-frequency sound that is annoying to many rodents.

maintenance areas of the room.

Equipment. To prevent leaking or major spills from a faulty valve assembly, elevate drums off the floor in a vertical position. Ideally, the room should be equipped with a 55-gallon drum stand or concrete pad that will allow for upright storage. Provide hand-operated drum pumps. Steel racks for the storage of small pesticide containers and a floor-mounted, lockable cabinet for the storage of special pesticides are also important.

Steel tables and work benches should be provided as required. Covered trash containers are also needed.

IMPACT INFESTATION DESTROYERS

Impact infestation destroyers are effective in the control of many stored-product pests and have been used by flour mills for this purpose for many years. They are also used by dry prepared-mix manufacturers and bakery operations which handle large quantities of bulk flour.

An impact infestation destroyer consists of a high-speed impeller revolving at around 3600 rpm. The product is fed into the unit through the inlet transition and deposited directly on the rotating impeller. The centrifugal force of the propeller throws the product and the insect against the outer housing of the machine, and the insect is killed on impact. *(Figure 10-21)* This equipment is capable of killing the insect in any stage: egg, larva, pupa, and adult. This equipment is used on flour products and on whole grains. It is particularly useful in removing insect-damaged kernels of wheat, along with any insect forms which may be inside the kernel. In this case, a specially designed rotor throws the damaged kernel against the outer housing, causing it to break into many small pieces. A sifter or aspirator is then used to separate the small pieces and insects as tailings from the healthy whole kernels of wheat.

RODENT CONTROL WITH ULTRA-SONICS

Ultra-sonic devices for rodent control are considered highly controversial. As was noted in Chapter One, ultra-sonic devices are mechanical control units that produce a very high frequency sound, usually around 30,000 H_Z. It is annoying to the rodent, yet it cannot be heard by most people. Some qualified evidence indicates that sonic devices used for rodent control do indeed work. The controversy centers around the misconception that a sound device is all that is required to keep rodents out of the food plant. It should be remembered that successful rodent control depends on a complete program which makes simultaneous use of the seven tools outlined in Chapter One. An approach based on a single device or method, whether it is baits, traps, or ultra-sonics, will not work.

EQUIPMENT SELECTION AND LOCATION

As is the case with any controversial item, the selection and application should be made with some caution and on a limited test basis. Start by testing a particular unit. For example, doorways are an excellent place to test units intended to keep out mice. Mount the device so the sound wave pattern completely covers the opening of a warehouse railroad trackwell door at the threshold. *(Figure 10-22)* Evaluate the results by noting previous catches in traps placed on both sides of the door. Compare them with catches after the sound unit has been placed in operation. The results may be surprising!

There are any number of ultra-sonic rodent control units on the market. Many can be purchased on a satisfaction guaranteed basis. Many make claims that they will work anywhere in the plant. Each unit must be carefully evaluated on its own merits. A unit's performance is affected to a great extent by the sound attenuating characteristics of materials found in many parts of the plant. Again, as with all rodent control devices, success depends on total commitment to a complete program.

CHAPTER 11 FOREIGN MATERIAL CONTROL

CHAPTER CONTENTS

MAGNET APPLICATIONS

METAL DETECTORS

SCALPING AND SIFTING

ASPIRATORS

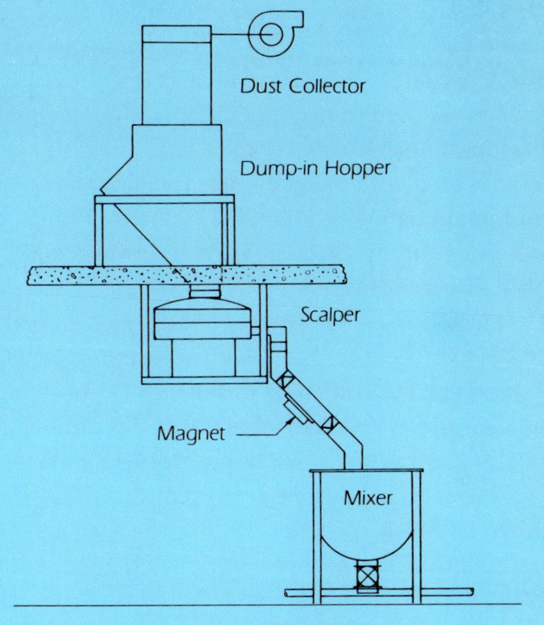

Foreign material control involves the prevention of both chemical and physical contamination in manufactured food products. Chemicals represent the most serious form of foreign material contamination, with the ever-present concern for harmful, even poisonous, chemical contamination of nationally distributed food products. Certainly the opportunity for contamination exists from the very early growth stage of the raw materials right on through processing to the finished food product. For the most part, the food industry has done a good job in controlling the use of harmful and toxic chemicals with *approved chemicals only* programs and with very detailed policies and guidelines that govern chemical use around food products and food product zones. These internal control programs, coupled with technological improvements to the food processes that eliminate the need for chemicals, make it appear that very little more can be done in the area of additional chemical control. Physical contamination, on the other hand, represents a different type of problem. For this reason, we will devote this chapter to physical contamination problems.

Magnets and metal detectors, which can effectively identify and remove metal contaminants, will be discussed in detail. The process of scalping and sifting for physical or foreign materials removal poses some very special problems. But because this equipment is exceptionally effective in detecting and removing stored-product insects, its virtues will also be reviewed. Although the stored-product insect problem is best controlled with a solid sanitation program highlighting good cleaning practice and the safe use of insecticides, scalping and sifting equipment serves as a monitoring or detecting device to evaluate the sanitation program. It further provides a safety valve for coping with any insects that escape other efforts designed to control them. We will also discuss the use of aspirators for foreign material and insect removal.

MAGNET APPLICATIONS

Misapplication and misconceptions about magnets and magnet application abound in the food industry. This probably stems in part from the fact that we do not really understand some basic magnet principles and the type of contamination we are likely to encounter. For the most part we are beyond the days when large size tramp metal was a problem. Raw materials are less likely to be contaminated with any metal, let alone horseshoes. Many food manufacturing processes begin with ingredients or raw materials that have been preprocessed. This is both good and bad. It is good in that some attempt has already been made to remove metal, and bad in that the materials have been exposed to machines that could malfunction and recontaminate them. By and large, the problems we are confronted with today involve fine ferrous or small metal contaminants—and usually in small amounts. This is not an ongoing problem, but a random or sporadic one attributable to occasional machine failure.

If the problem involves only fine or small metal and not very much of it at that, why are we concerned? We are concerned because some of the metal contaminants may be injurious. Though large quantities of unwanted metal filings may or may not be hazardous, small needle-like wires or metal fragments could cause serious injury. Frankly, all food products should be free of metal contamination. The proper selection and application of magnets can be instrumental in controlling metal contamination.

Magnets can also be helpful in protecting facilities that contain explosive food ingredients. By catching metal contaminants before they reach the explosive, dusty atmosphere found in some equipment, magnets can prevent ignition by sparking. In this chapter we will discuss location, selection, installation, inspection, cleaning, and maintenance of magnets.

LOCATION

It is suggested that magnets be used at four specific locations to insure adequate magnet protection throughout the food manufacturing system: at incoming ingredient points, for equipment protection, after equipment which can cause contamination, and at the packaging point.

Incoming Ingredients (Bulk). Install magnets on all incoming bulk ingredient handling systems (except those handling ingredients with iron enrichment). When pneumatic unloading systems are used, magnets should be placed between the unloading receiver and the distributor just prior to entering the selected bulk storage bin. *(Figure 11-1)* Alternatively, magnets can be placed after the bulk bins. *(Figure 11-2)* This requires many smaller magnets (one for each bin) and more time for checking all of them. Magnets located after the bin provide a distinct separating advantage by monitoring a smaller stream of ingredients. However, many food processors prefer having a larger magnet ahead of the bins so that one magnet can be used to monitor several ingredients. More importantly, placing magnets ahead of the bins makes it easier to link bulk lots of ingredients

FIG. 11-1 Magnets should be used to monitor all incoming bulk food ingredients as they are received.

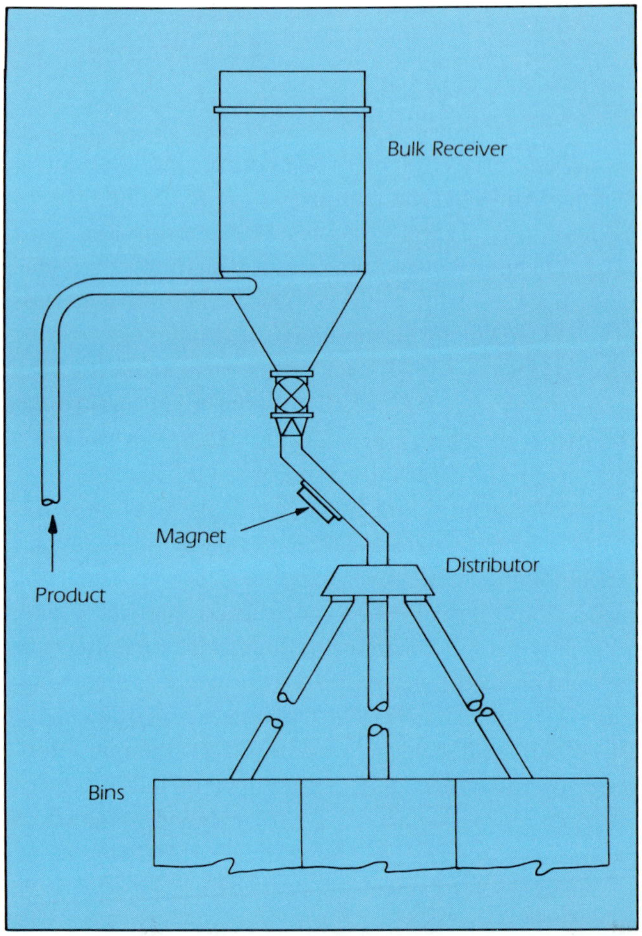

FIG. 11-2 As an alternative, magnets may be located after the bulk bins to monitor ingredients as they flow to processing.

with a specific contamination incident.

If mechanical unloading systems using belt, screw, or drag chain conveyors and bucket elevators are used, then magnets should be located in the spout feeding the bucket elevator. In the case of belt conveyors which feed the bucket elevators, magnets may be used in a suspended position over the belt.

Incoming Ingredients (Containerized). Magnets should also be used at the dump-in stations, where containerized ingredients are introduced into the system. Install these magnets at a point in the system where the feed is regulated *(Figure 11-3)*, as opposed to directly in the dump hopper where the rate of feed is uncontrolled.

Equipment Protection. Place magnets ahead of equipment which is subject to damage by metal. *(Figure 11-4)* Cutting knives, extruding dies, rolls, and grinders are just a few examples of equipment that should be protected.

As mentioned in the introduction, magnets should be used ahead of equipment containing explosive, dusty atmospheres to prevent explosions which could be caused by metal contaminants.

Equipment Contamination Detection. Some manufacturing equipment is more likely to malfunction and cause metal contamination because of its specific design characteristics. Magnets should be used at, or as close as possible to, the discharge of this equipment. *(Figure 11-5)* Such magnets not only offer protection, but also help identify equipment maintenance needs. They can head off real problems.

Packaging Point. Use magnets as close as possible to the point where the product is placed in the package. *(Figure 11-6)* Magnets at this point are very important to the product safety program. When magnets are used in conjunction with metal detectors at the packaging point, the result is a strong consumer protection effort.

SELECTION

The proper selection of magnets requires that we understand

a few magnetic principles, know something about the different types of magnets available and what they are designed to do, and know something about the capacity of a magnet, including the conditions which might affect its capacity rating.

Principles. Selecting the right magnet for the job is made simpler with a basic understanding of a few simple magnetic separation principles.

1. It is important to know that magnetic materials are attracted to a magnet with a force proportionate to their size. For this reason, large pieces of metal are pulled into the magnetic circuit with more authority than smaller pieces.

2. We should know that the strength of a magnet varies inversely with the square of the distance from the magnet. Simply stated, this means that if a magnet has a 40-ounce pulling force at a quarter-inch distance, it would pull only 25% as much, or 10 ounces, at twice that distance.

FIG. 11-3 Magnets should be used where incoming containerized food ingredients enter the process.

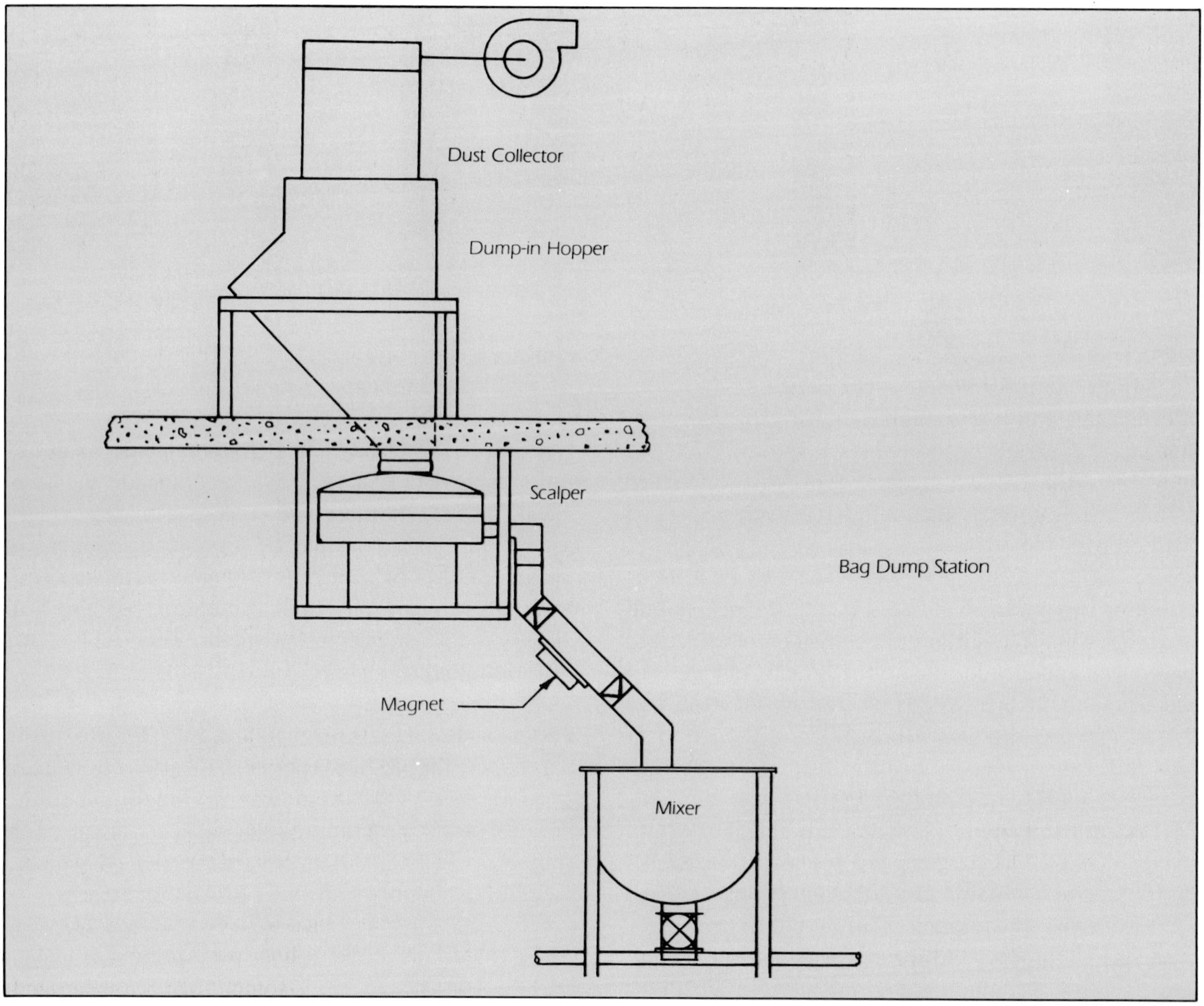

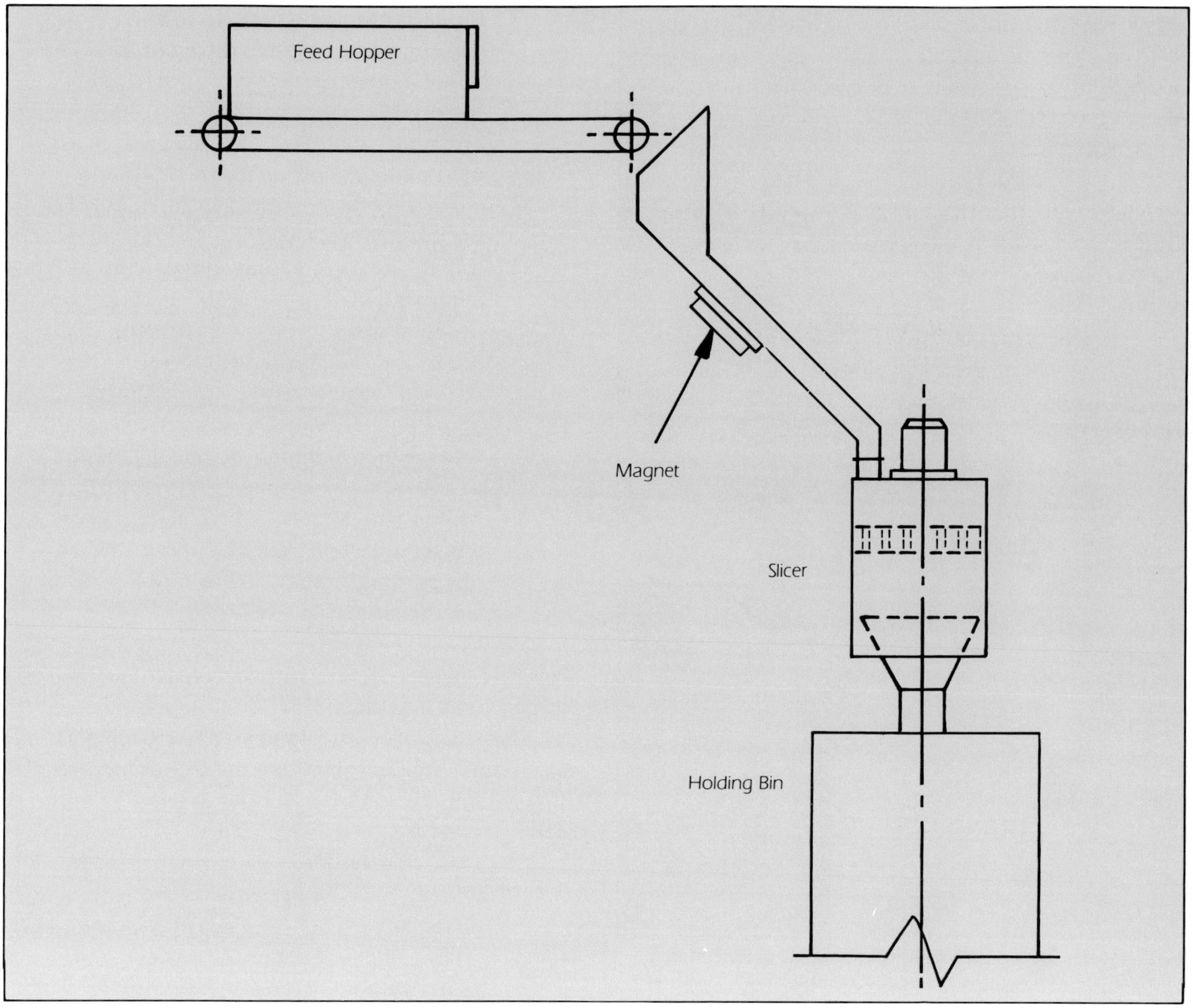

FIG. 11-4 Magnets can be used to protect equipment that is damaged easily by pieces of metal.

3. Magnets can be built with high or low gradients. Gradient is the change in magnetic strength between points at various distances perpendicular to the magnetic field. The greater the change in flux density, the greater the magnetic attraction.

   High-gradient fields are used when a magnet must have maximum holding power and when fine or weak magnetic materials must be attracted from short distances. High-gradient magnets are used by the steel industry for lifting operations.

   Low-gradient fields have a more gradual change in flux density at a given distance from the magnet. Low-gradient magnets establish a deeper magnetic field for entrapment. Most plate, pulley, and drum magnets are low gradient.

4. The unit of measure for the flux density is called gauss. Gauss is measured by an instrument which indicates the number of imaginary force lines per square centimeter or inch. Gauss is sometimes used to express

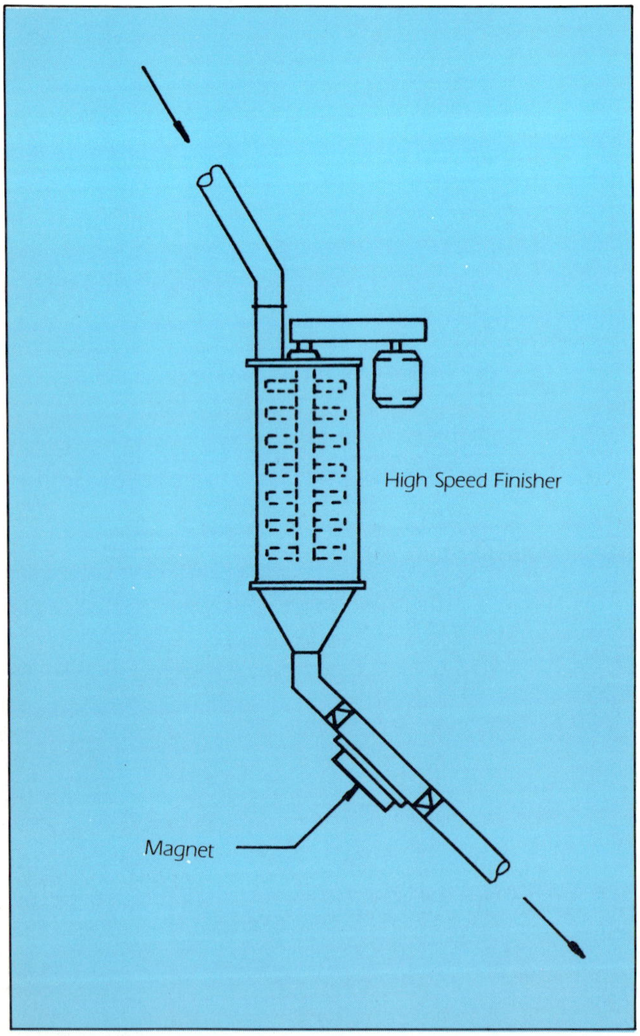

FIG. 11-5 Used after equipment which is prone to malfunction, magnets can prevent metal contamination of the food by the malfunctioning equipment.

the strength. A magnet having a flux density of 2,000 gauss would mean that there are 2,000 force lines per square centimeter.

5. It is important to know that you cannot insulate out a magnetic field. It can be baffled by a *like* magnetic field or shielded by steel or iron, but it cannot be insulated.

6. Most magnets are built from an alloy called alnico or from a pressed barium oxide ceramic material. Ceramics make the most permanent magnets.

7. Permanent magnets are subject to demagnetizing through different forms of abuse. Some magnets can be affected by forcing like poles together, disassembling assemblies that have been magnetized as a unit, and by allowing magnetic materials to contact the magnet castings in areas other than on the working face. Ceramic magnets usually are affected only by extreme temperatures. For every 100°F above room temperature, a 10% loss in strength can occur, but the magnet will recover when restored to room temperature. However, at approximately 900°F a magnet will totally demagnetize and must be remagnetized. Ceramic magnets can withstand -40°F without any effects. Alnico has no lower temperature limit and can operate in temperatures up to 750° to 850°F without a significant loss in strength.

Types of Magnets. Several different types of magnets are used in food manufacturing. Each type of magnet is designed to offer a particular feature or to handle some very specific problem. The plate magnetic is perhaps the most commonly used.

- Plate Magnets. The plate magnet is available in many sizes and strengths, and with many different capacity ratings. *(Figure 11-7)* The pulling and holding power varies significantly with the various strength magnets. This being the case, when removing fine ferrous contaminants from food products, it makes sense to start with a stronger magnet. This is important because, if you remember, the first principle of magnet selection states that ferrous contaminants are attracted to a magnet with a force proportionate to their size. The plate magnet is specifically designed for installation in the bottom of a chute or spout and should be used there whenever possible. It can also be used in a suspended position.

  The plate magnet is available with barrier strips *(Figure 11-8)*, single-taper steps

(Figure 11-9), and double-taper steps. *(Figure 11-10)* The barrier strip and taper step magnets are used to help retain fine ferrous and spherical contaminants. Contaminants trapped on the face of a plate magnet can be washed off by oncoming materials. The crevices created by the vertical walls of the barrier strip and the taper steps offer some protection for the trapped contaminants and make it extremely difficult for them to be washed free.

Plate magnet fields are considered to be low gradient, although in reality the gradient may be somewhere between high and low. Magnet design engineers constantly strive for a design that will have far-reaching capabilities and at the same time offer some real pulling and holding power.

FIG. 11-6 When used with metal detectors at the point of packaging, magnets help protect consumers. This magnet is suspended from a trolley located directly over the packaging scales.

- Hump Magnets. Hump magnets consist of two plate magnets in series. These are installed in transitions which change the direction of the product stream and thus slow it down. *(Figure 11-11)* This configuration was designed to deal with the more difficult separation jobs. Every time the product changes direction, a momentary delay occurs and the product is slowed. This increases the chance of trapping even some of the most difficult-to-trap contaminants; the faster the product goes, the lesser the magnet's ability to trap. Hump magnets should be used on free-flowing materials; sticky materials stand a greater chance of choking when their direction of movement is changed.

Hump magnets can be used in

FIG. 11-7 Plate magnets have a variety of applications throughout the food industry.

FIG. 11-8 Plate magnet with barrier strip

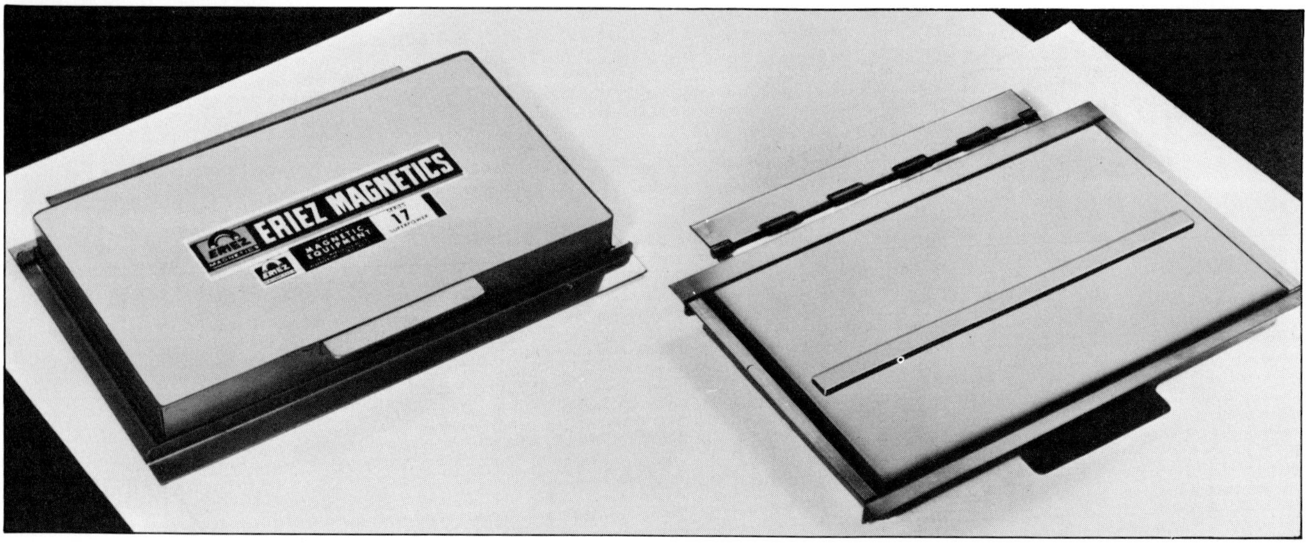

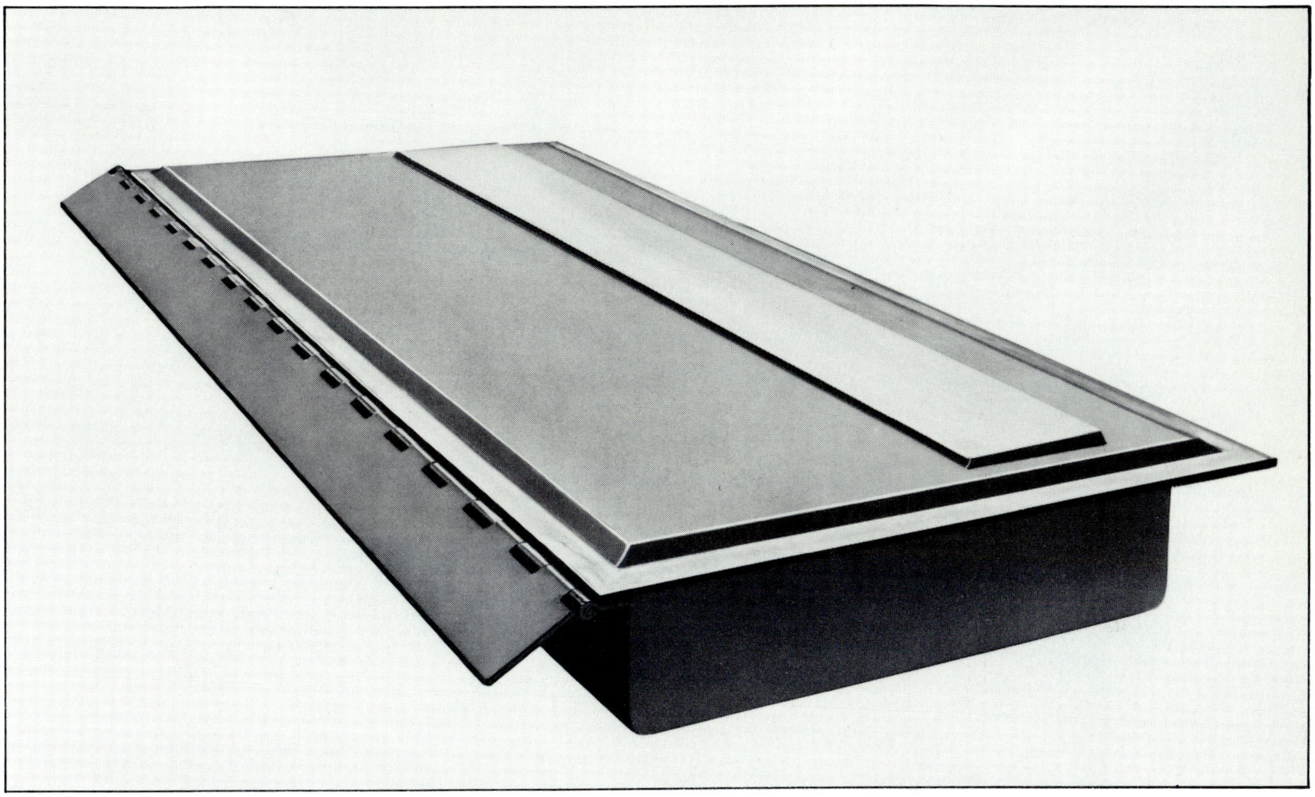

FIG. 11-9 Plate magnet with single-taper step

FIG. 11-10 Plate magnet with double-taper step

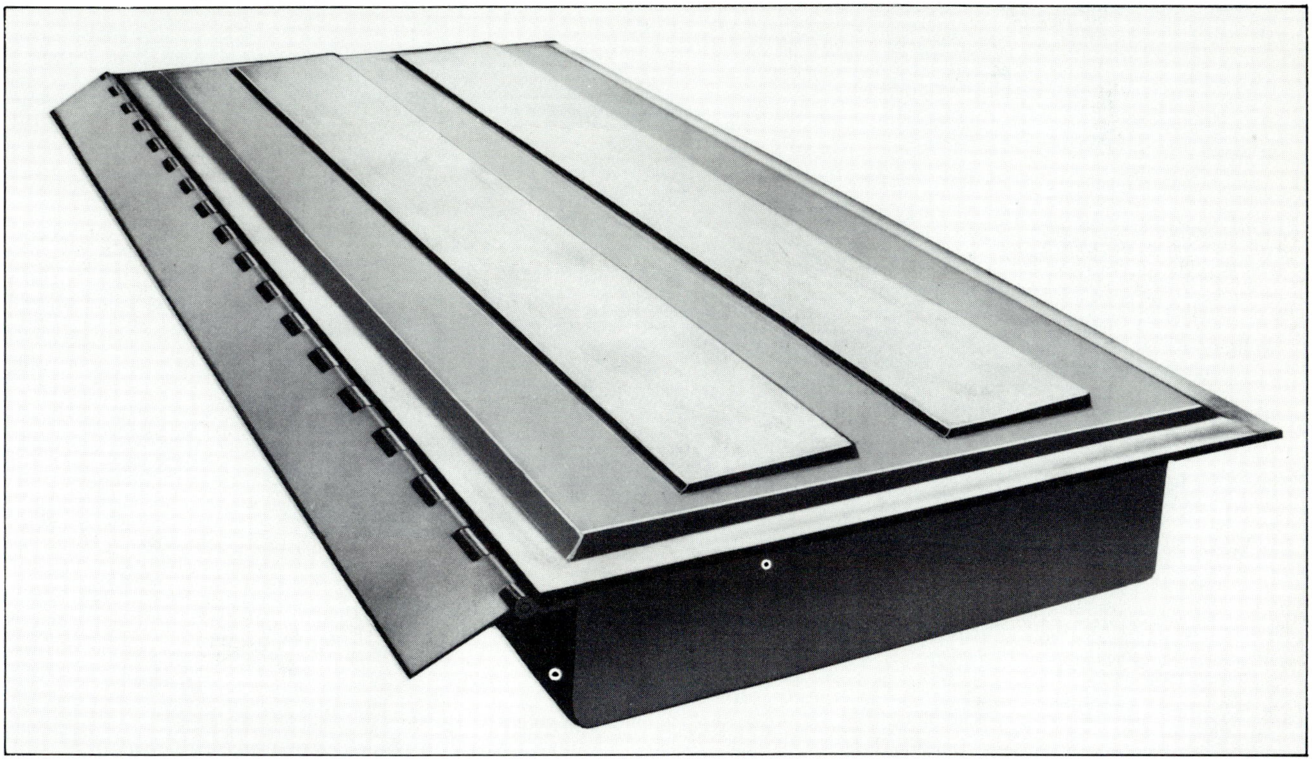

FIG. 11-11 Hump magnets should be used to solve difficult metal contamination separation problems.

FIG. 11-12 The round pipe separators shown include hinged magnet installations and a baffle to slow the product and direct it onto the magnet surface.

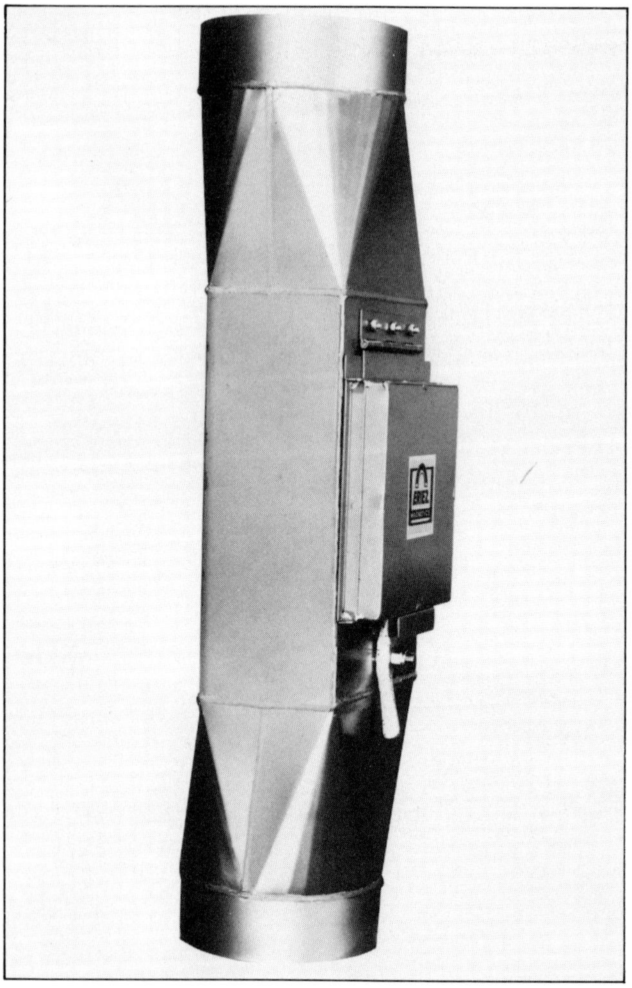

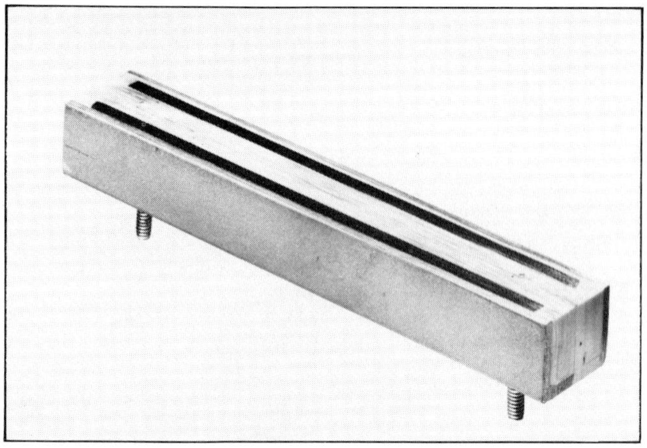

FIG. 11-13 Bar magnet

pneumatic lines as well as in gravity spouts, as long as the housing is designed to accommodate the suction or positive pressures of the system. However, the performance results of pneumatic line installations may leave something to be desired and should be demonstrated before this application is chosen.

- Round Pipe Separators. The round pipe separator is a plate magnet placed in a transition that has a baffle installed in it. *(Figure 11-12)* The baffle slows down the material to increase trapping efficiency.

- Bar Magnets. Bar magnets are high-gradient type magnets designed to handle fine ferrous contamination problems in shallow streams of product. *(Figure 11-13)* Bar magnets can be mounted under or over the flow of material. The material must be nonsticky and should not exceed a quarter to a half inch in depth. These conditions limit the choice of the bar magnet. Remember, however, that a high-gradient magnet produces a very strong field at a specific plateau; therefore, this magnet can be effective for some special types of problems.

- Grate Magnets. The grate magnet is specifically suited for fine and small ferrous

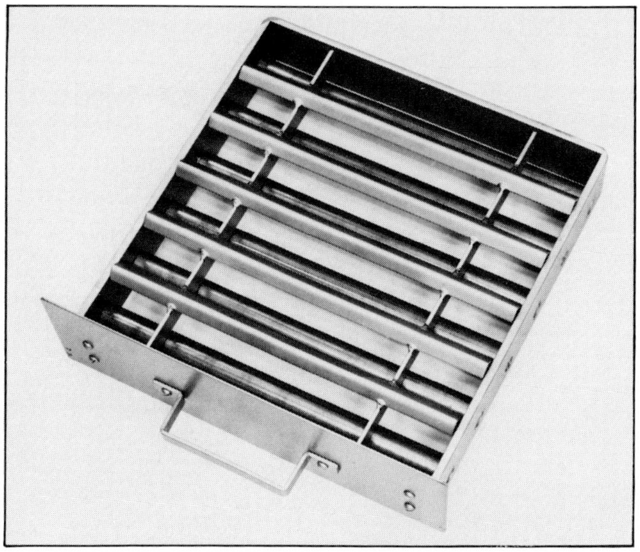

FIG. 11-14 Grate magnets are particularly well-suited for fine iron separation. The drawer grate magnet is pictured.

FIG. 11-16 Multiple-tier grate magnets are especially well-suited for difficult fine iron separation problems.

contamination problems. *(Figure 11-14)* One of the advantages of the grate magnet is that the trapped metal is washed to the underside of the round grate bar, where it is protected from the washing effect of the product. Grate magnets must be used on free-flowing materials. Material that is sticky, clusters easily, or lumps can possibly choke off at the grate openings.

Grate magnets are available with the round bars on 1-1/2 inch centers for fine ferrous contaminants, and on two-inch centers for small metal contamination. They can be made in the shape of a drawer for vertical installation in a transition. *(Figure 11-14)* A winged version of this magnet can be used in a sloping wall hopper. *(Figure 11-15)* Grate magnets also come in multiple layers for really difficult fine ferrous contamination problems. *(Figure 11-16)* Because the food must come in contact with the round grate bar for maximum separating effi-

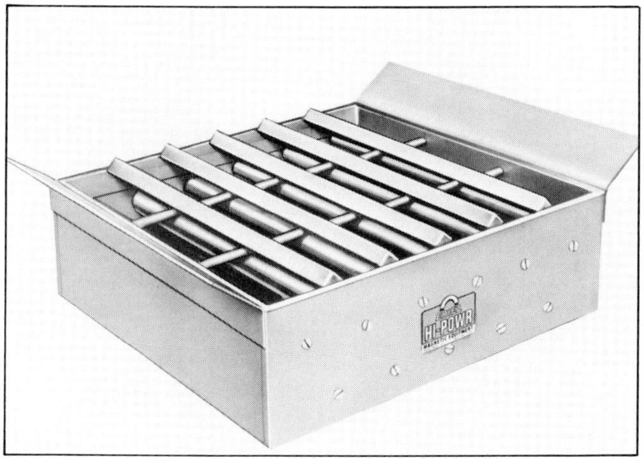

FIG. 11-15 The winged grate magnet is commonly used in sloped wall hoppers.

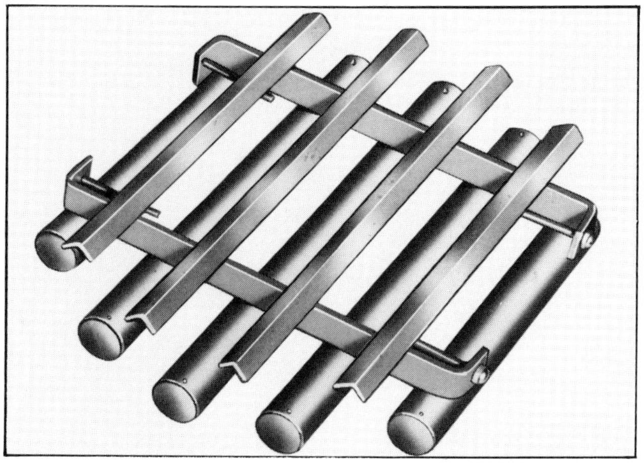

FIG. 11-17 The grill device directs the product onto the round grate bars.

FIG. 11-18 A pulley magnet can be used as a head pulley on a belt conveyor.

ciency, a grill-like grate is used to direct the flow of material onto the first layer of bars. *(Figure 11-17)* Subsequent grate bar layers are staggered to assure contact.

FIG. 11-19 Pulley magnets used as head pulleys on belt conveyors have self-cleaning features.

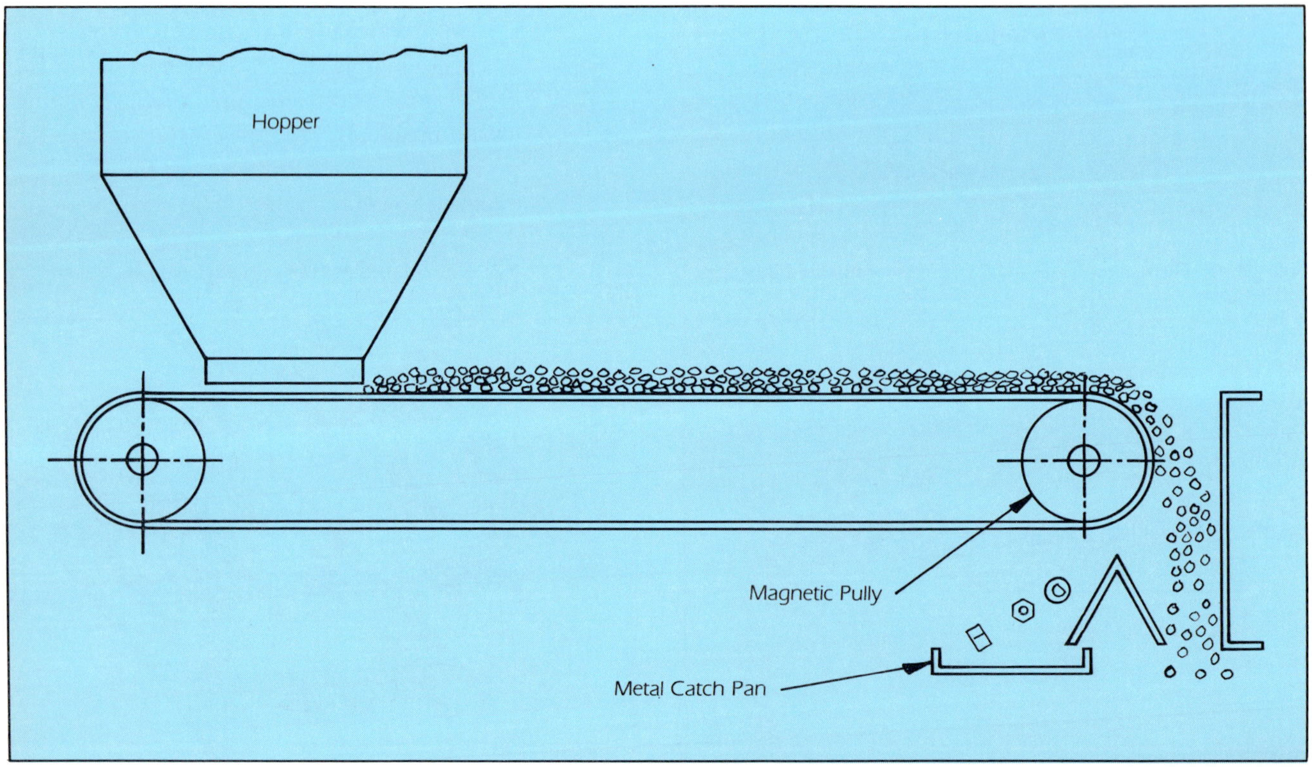

- Pulley Magnets. Pulley magnets are used as head pulleys on belt conveyors for the continuous automatic removal of tramp metal from product streams. *(Figure 11-18)* In effect, they are self-cleaning and truly test the principle that you cannot insulate out a magnetic field. The tramp metal contaminant first enters the magnetic field of the pulley which is reaching right through the belt. The contaminant is then held against the belt until it reaches the underside; finally it revolves out of the area of the field, and drops off into the tramp metal catch pan. *(Figure 11-19)*

- Drum Magnets. Drum magnets remove tramp metal from streams passing over them. *(Figure 11-20)* The drum magnet features a stationary permanent magnet installed inside a nonferrous stainless steel drum. The magnetic field penetrates the stainless steel drum and a contaminant

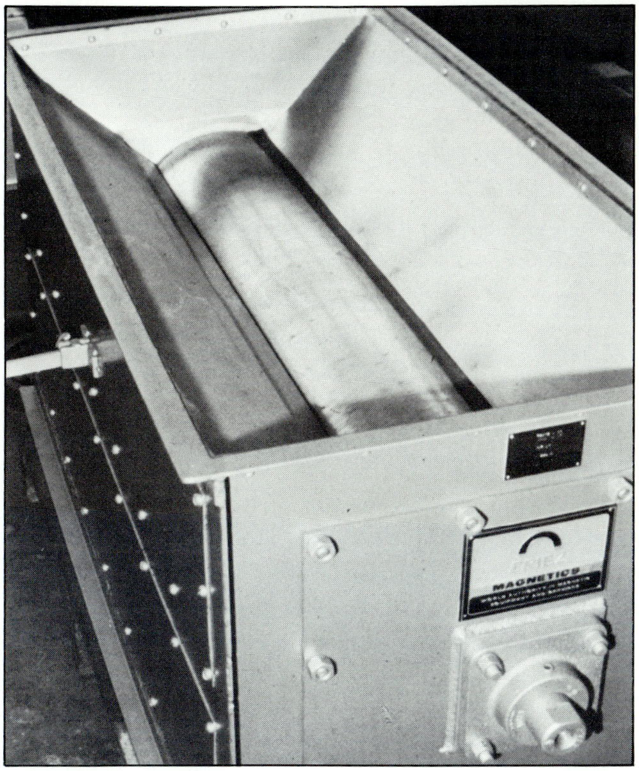

FIG. 11-20 The drum separator is used to separate tramp metal from raw commodity streams.

FIG. 11-21 Drum magnets are also self-cleaning.

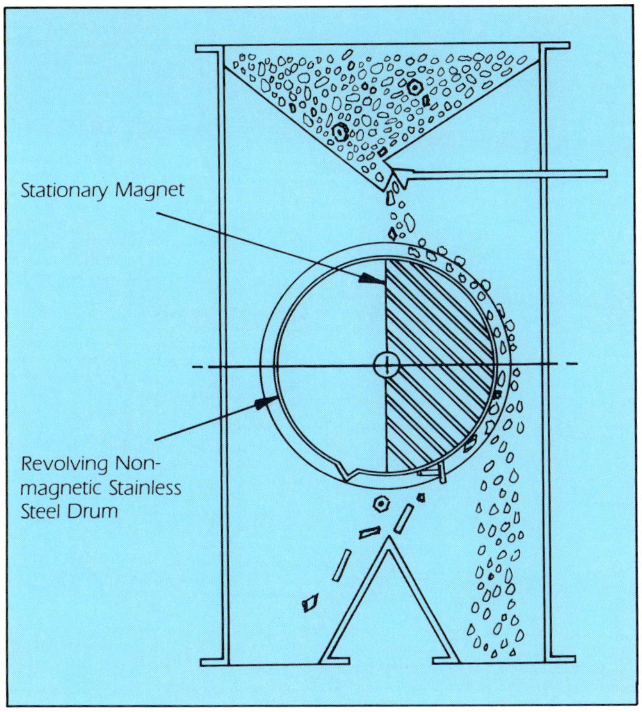

entering the field adheres to the revolving drum until it clears the field, at which time the contaminant drops into the reject chute. *(Figure 11-21)* Drum magnets would probably be best suited for use on a raw commodity stream such as corn or beans.

- Liquid Trap Magnets. Liquid trap magnets consist of a group of round magnetic bars installed vertically in a pipe fitting. Liquid or slurry materials contact the magnetic bars as they flow through. Metal contaminants are trapped and washed to the downstream side of the vertically suspended bars, where they receive some protection from the flowing liquids and cannot be readily washed off. Liquid traps are available with easy disassembly features. *(Figure 11-22)* They are also available in a sanitary version that features highly polished stainless steel surfaces.

Plate liquid trap magnets are also available. These units feature a standard plate magnet installed in a specially designed, highly polished stainless steel pipe fitting. This fitting is generally designed with

FIG. 11-22 Liquid trap magnet

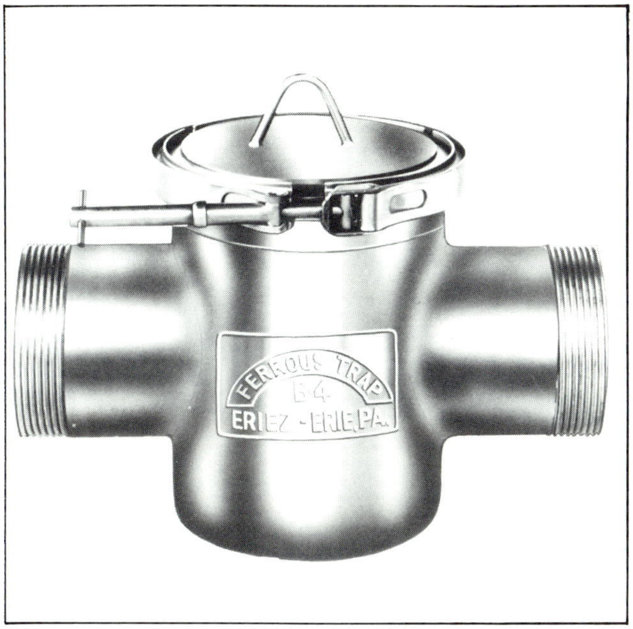

a weir to direct the liquid stream down onto the face of the magnet.

Magnet Capacity. There are several methods of rating the actual capacity of a given size magnet. These methods are all related to strength of the magnet as well as the physical size. Some manufacturers recommend magnet size according to the flow characteristics and depth of the product stream passing over the magnet. Other manufacturers state the number of cubic feet of material a certain magnet can handle per minute or per hour. Although any rating method leaves much to chance, those specifying cubic feet per minute or hour make it easier to relate to a given capacity.

Regardless of the method, capacity ratings have to be adjusted for the size of the contaminant as well as the velocity of the material as it flows over the magnet. Cubic-foot capacity ratings of plate magnets may need to be reduced by a factor of eight where the installation is at 60° and the problem is fine ferrous contaminants. Given these conditions, a magnet with a normal rating of 4,800 cubic feet per hour would only have a capacity of one-eighth of that, or 600 cubic feet per hour.

Pulley magnets can also be rated on a volume basis, using cubic feet per hour. However, adjustment is required for belt speed. If the belts run more slowly than recommended, the ratings must be factored. Slowing down the belt increases the depth of the product on the belt and increases the pulling burden that the magnetic field must overcome. The contaminant particle size also influences this rating—the smaller the contaminant, the greater the factor by which the capacity must be reduced.

Drum magnets can also be rated by the volume they can handle. Normally no factoring is required because this magnet is primarily used on tramp metal problems.

Grate magnets are usually sized to the opening of the transition or spout in which they are installed. The total area between all of the round grate bars must be 10% greater than the cross-section area of the inlet or outlet spout itself.

Bar magnets have very low capacity and must be used on very slow-moving products. The product stream depth should not exceed a quarter inch.

Liquid trap magnets are sized on a volumetric basis, using gallons per minute. Adjustments are normally made for viscosity.

Most magnet manufacturers will be glad to provide specific information on how to determine the proper size magnet for the specific job at hand.

INSTALLATION

Installing magnets is as important as selecting them. When magnets are installed with disregard for drawing instruction, the capacity, among other things, can be affected. For example, plate magnets installed in an incline spout more than two feet from a vertical juncture may experience a 10% capacity reduction for each foot beyond this point. At a point seven feet from the juncture a baffle should be used to slow the product down. *(Figure 11-23)*

Plate Magnets. Plate magnets should be installed at 45° or less whenever possible. The flatter the angle, the slower the velocity of product flow. When coupled with a shallow product stream, this installation is exceptionally efficient.

The maximum angle of a chute-mounted plate magnet should be 60°. If product flow characteristics require a steeper angle, consider using a baffle to slow

FIG. 11-23 Baffles may be required to slow the product velocity so that metal separation can occur.

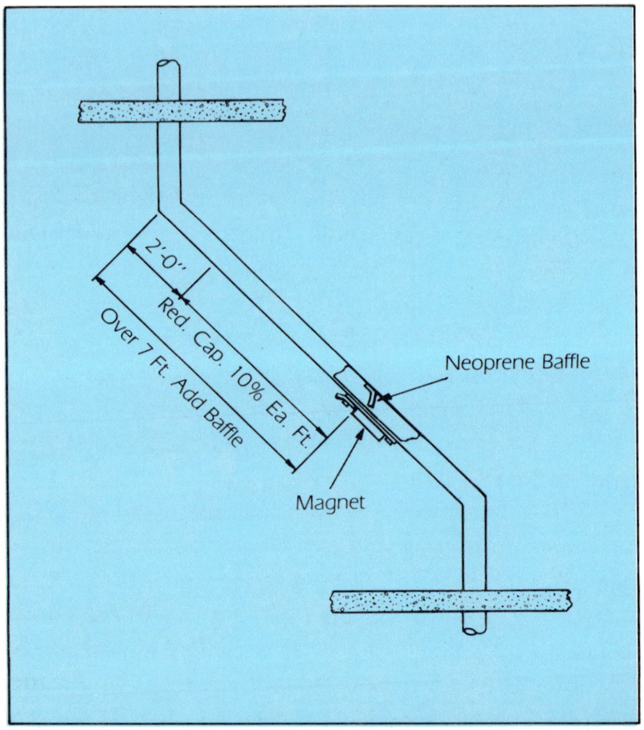

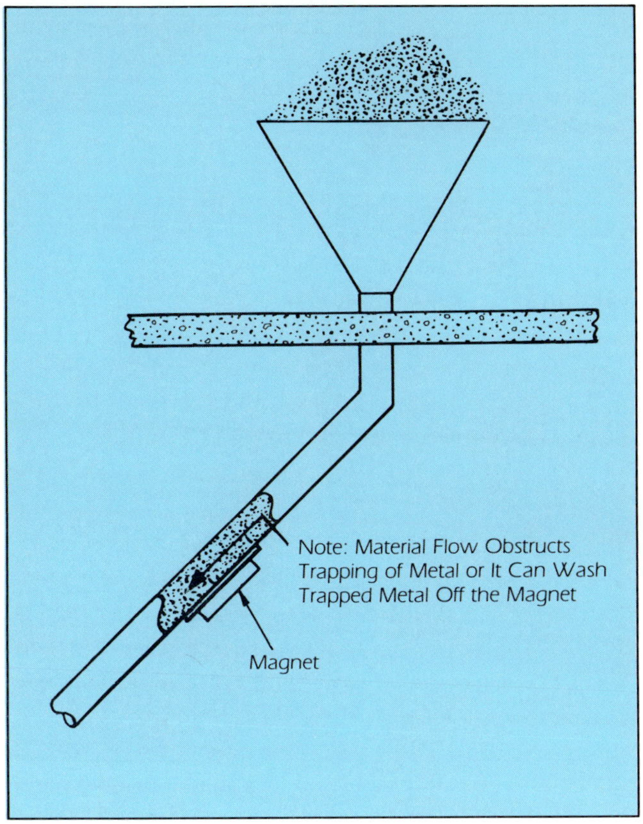

FIG. 11-24 Plate magnets should not be used in a choke-fed situation.

the product.

Plate magnets should not be used where the spout will be choke fed or 100% full. *(Figure 11-24)* Choke fed, the product becomes a pulling burden that many magnetic fields cannot overcome. This is particularly true when the contaminant is very small. Since a contaminant is attracted to a magnet with a force proportionate to its size, the magnetic field may not provide enough pulling force to overcome product resistance.

It is important that the magnet be accessible without tools. Small plate magnets should be installed with hinges and quick-release fasteners so they can be easily inspected and cleaned. *(Figure 11-25)*

Top access doors should be constructed when it is necessary to inspect and clean magnets on the run. A plastic plate can be placed over the magnet through the top access door to form a false bottom in the chute. *(Figure 11-26)* The plastic plate is secured in position by the long U-bolt which catches the top of the magnet transition at the base of the top door opening. The hinged magnet can be opened or lowered for cleaning, and meanwhile the product flows over the plastic false bottom.

Another installation for smaller magnets that provides access for cleaning on the run features two spout clamps that can be loosened by turning the wing nuts and pivoting the magnet transition 180° so that it is on top of the spout. *(Figure 11-27)* In the top position it can be opened for cleaning.

Plate magnets can also be installed on nonferrous rails so they can be pulled out for cleaning. *(Figure 11-28)* Since some of the larger magnets are quite heavy, it is important to make certain the slide rails are sized properly to accommodate the weight.

Plate magnets weighing over 20 pounds which are installed in the bottom of spouts or chutes should be fixed in place rather than hinged. Top access doors should be provided for cleaning. A-valves can be used in the spout just ahead of the magnet. This feature creates a bypass around the magnet which allows the product to be diverted while the magnet is being cleaned. Large hinged plate magnet installations often leak and are hard to secure.

Suspended Plate Magnets. Properly installed, suspended plate magnets can be effective in removing metal contaminants. The key to a successful suspended installation is remembering to apply a few of those principles we referred to earlier, particularly the inverse square law and the principle that a contaminant is attracted to the magnet with a force proportionate to its size. Of course, it is also important to choose the right magnet for the job. Some very strong low-gradient magnets can attract contaminants at a distance of five or six inches with a great deal of force. The design of the magnet field is crucial to the successful use of any magnet, but particularly when applied in the suspended position. *(Figure 11-29)*

Very large suspended magnets can be installed with beam dollies and pneumatic cylinders so that they can be moved from their normal surveillance position to a remote area for cleaning. *(Figure 11-30)* In this particular installation, a hinged nonmagnetic stainless steel plate is placed over the face of the magnet to help free contaminants that are held very firmly against it by the force field. The plate is hinged on the backside and a slight

FIG. 11-25 Install hinged magnets for easy cleaning.

FIG. 11-26 A top access door and a plastic plate forming a false bottom can allow a hinged magnet installation to be cleaned on the run.

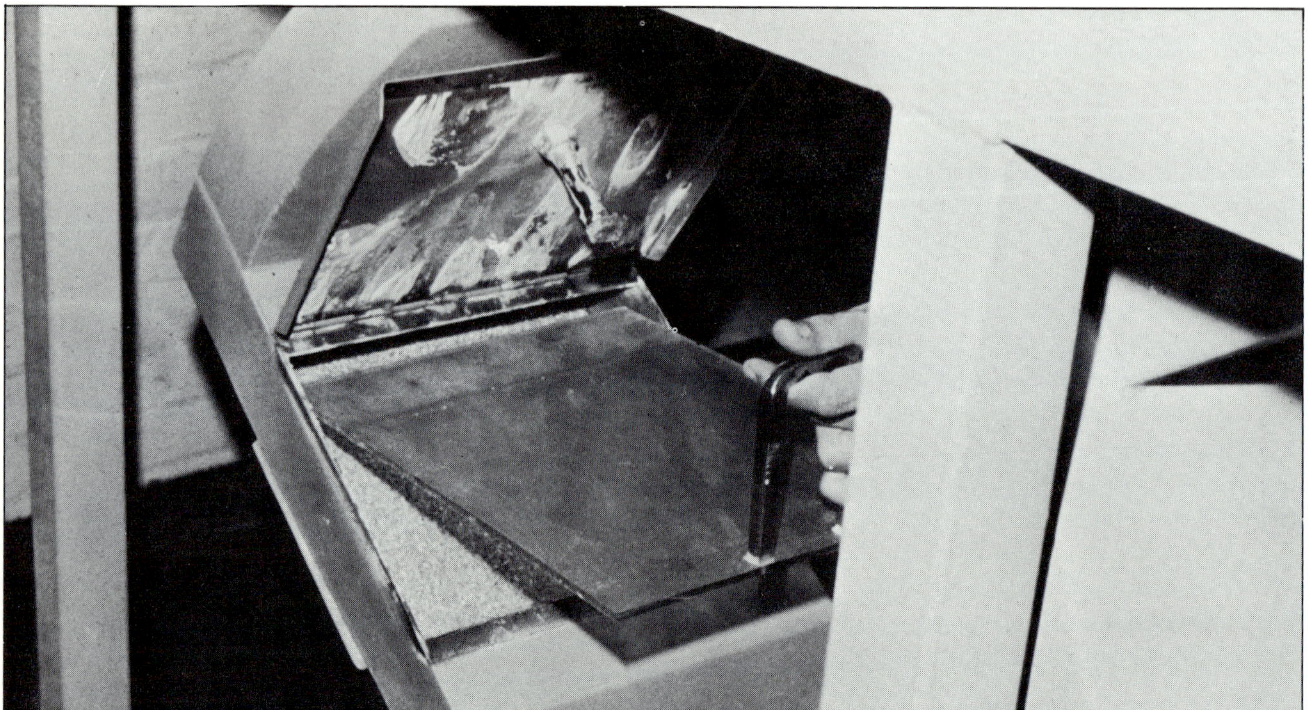

lever advantage is obtained when pressure is applied to the front ledge. *(Figure 11-31)*

Grate Magnets. Grate magnets should be installed so they can be easily removed for cleaning. The drawer type is particularly suitable for removal. Multiple layers should be installed on rails so that they can be removed a layer at a time. *(Figure 11-32)*

Liquid Traps. Liquid trap magnets should be installed with a bypass so they can be isolated by valving for cleaning. *(Figure 11-33)*

Hump Magnets. A hump magnet should be installed vertically in a gravity spouting application. *(Figure 11-34)* It should be so located that it can be easily serviced from floor level, if at all possible. Hump magnets can be installed in a vertical or horizontal position in a pneumatic conveying system.

Access to Magnets. As important as access into a magnet is access up to a magnet. Access up to the magnet means being able to get at it without having to crawl under or over equipment. Whenever possible, magnets should be installed at floor level. If magnets are located in the

FIG. 11-27 Two spout clamps and a hinged installation permit this magnet to be cleaned on the run.

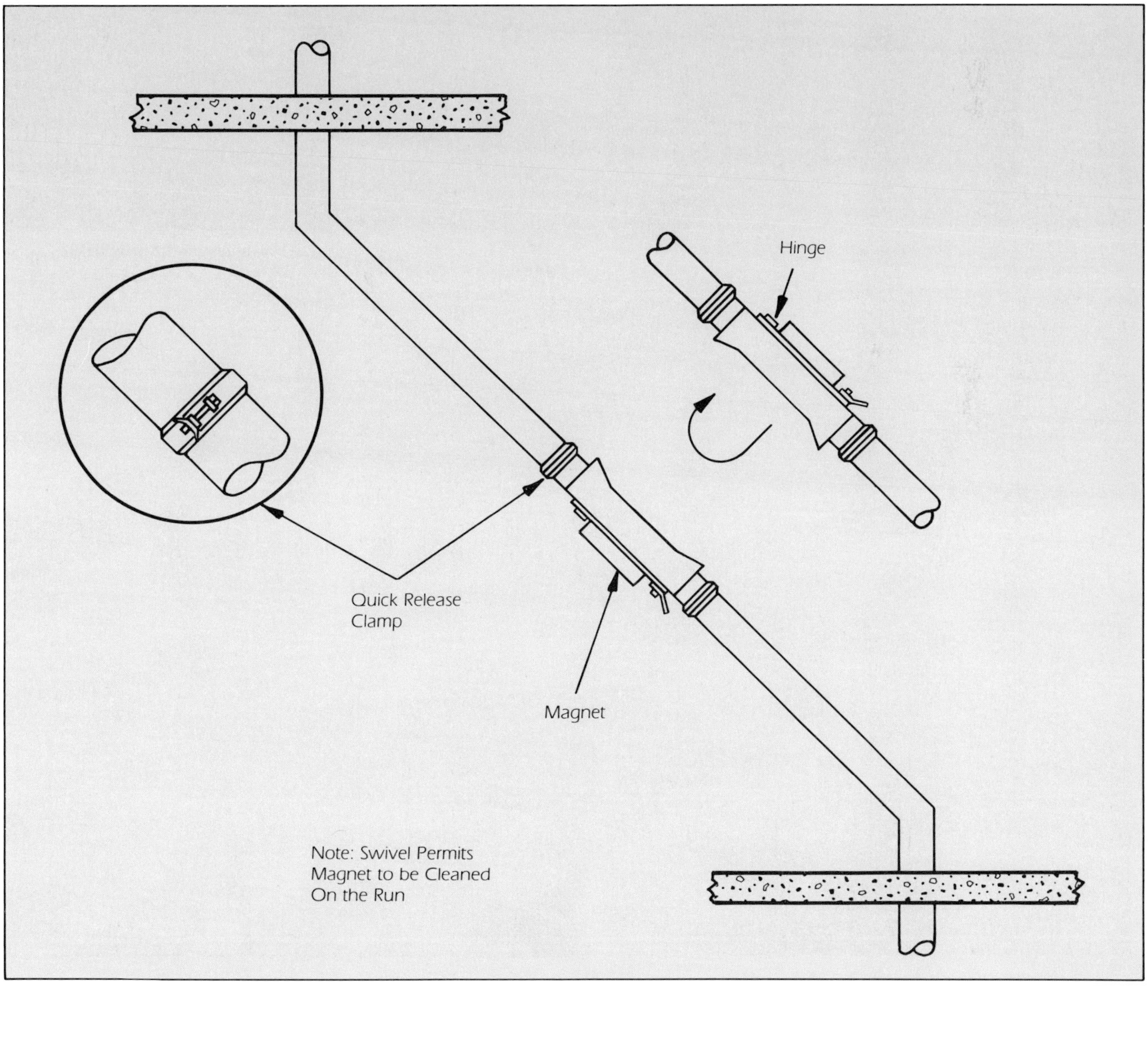

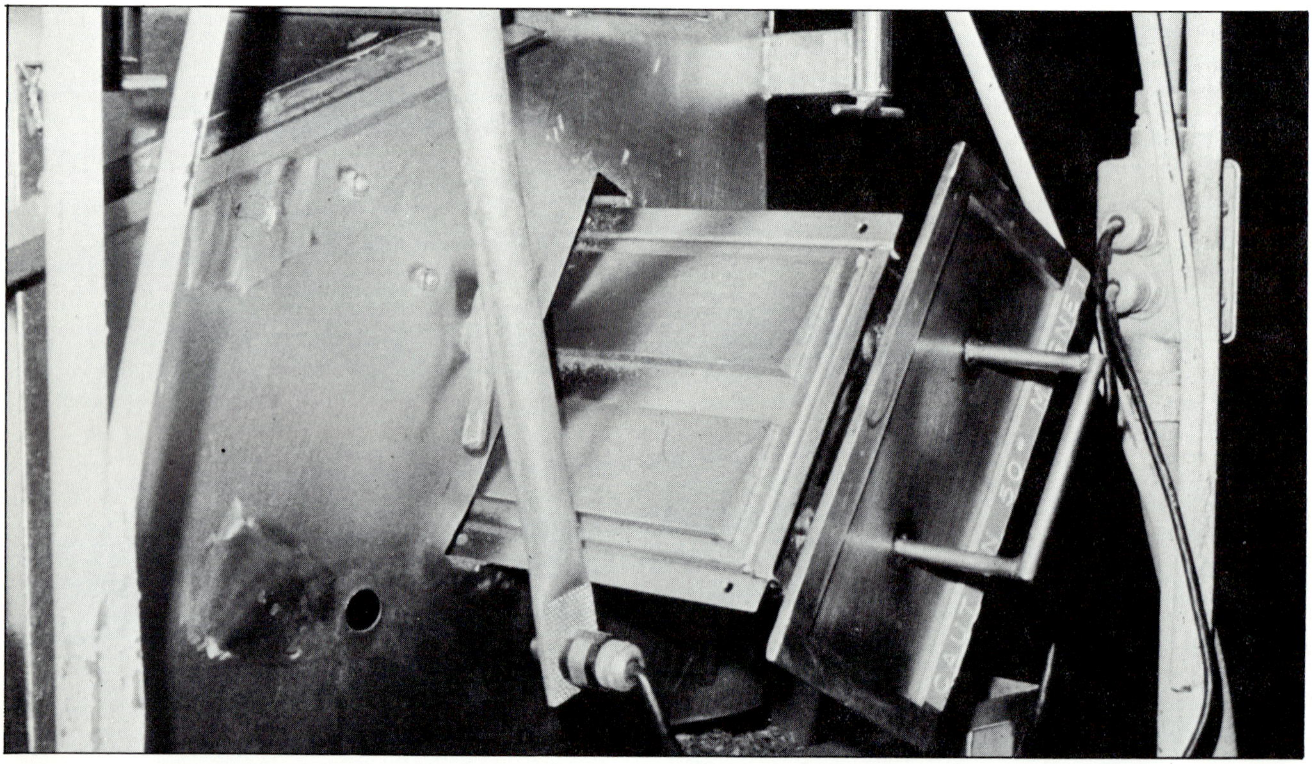

FIG. 11-28 Installed with nonmagnetic rails, this magnet can be easily removed for cleaning.

FIG. 11-29 This large suspended plate magnet is installed on a pneumatically operated dolly mechanism.

FIG. 11-30 An installation of this type allows the magnet to be easily removed for cleaning.

FIG. 11-31 The hinged face plate helps free trapped metals.

overhead, they should be arranged so there is adequate floor space to permit the use of mobile service platforms. In some cases, it may be necessary to install permanent service platforms.

INSPECTION AND CLEANING

If magnets are not inspected and cleaned on a regular schedule, there is little value in having them. A heavily contaminated magnet loses a great deal of its separating capability. *(Figure 11-35)* Trapped contaminants become magnetized and will trap other contaminants, but in the process the overall pulling and holding power of the magnet is greatly affected. Consequently, magnets should be cleaned regularly.

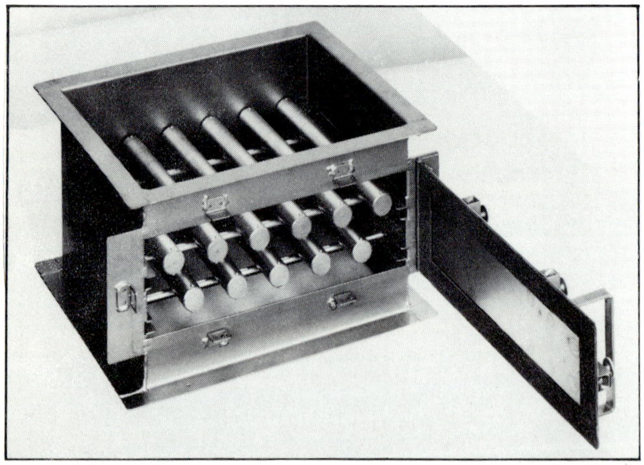

FIG. 11-32 These multiple-layer grates are installed on rails for easy removal of each layer individually.

FIG. 11-33 Liquid trap magnets can be installed with a bypass that allows for cleaning on the run.

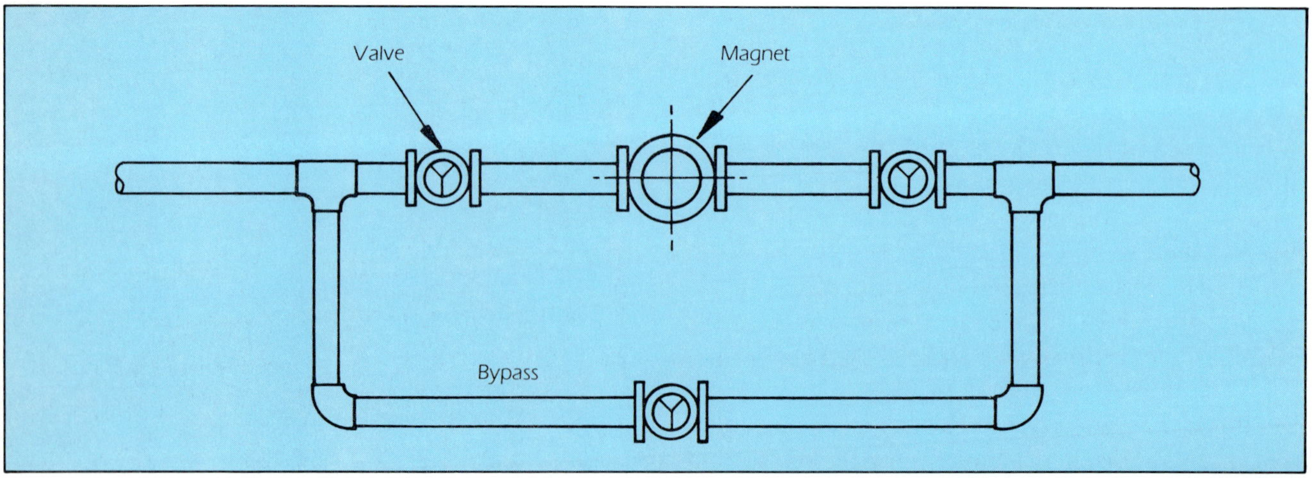

Fine metal contamination is difficult to remove from most magnets, particularly from grate magnets. A good sticky masking tape is one of the best ways to remove fine metal contamination from any type of magnet.

When metal contaminants are removed from the magnet, they should be logged in a log book. *(Figure 11-36)* The logging of material found on the magnet is very important to the product safety program. The log book can be helpful in establishing preventive maintenance schedules.

- Frequency. The frequency with which magnets are inspected and cleaned depends on two important factors: the food product and the location of the magnet.

- Food Product. If the food product is a raw commodity such as grain, the contamination level may be more suspect. Raw commodities have greater exposure to contamination in the fields and in the transporting process. Magnets used on raw commodities should be checked and cleaned often.

  As a rule, magnets used on processed ingredients do not need to be inspected as frequently. Experience should establish inspection and cleaning frequency.

- Location. Magnets located on the head end of a system where incoming ingredients or raw commodities are brought into the plant

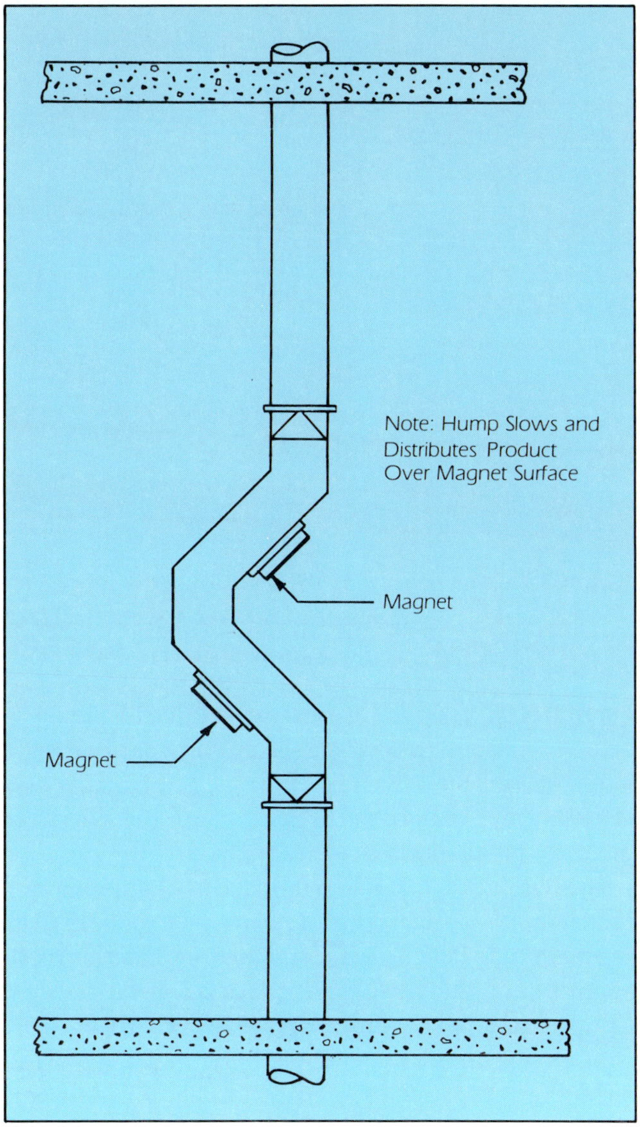

FIG. 11-34 Gravity hump magnet application

for the first time should be inspected after every lot is unloaded. In the case of raw commodities, it might be necessary to make very frequent checks.

Magnets protecting equipment such as knives, dies, and grinders should also be inspected frequently. Trapped contaminants should be removed so they are not washed from the magnet back into the equipment.

Log Book. Log trapped materials in the log book on an *exception* basis; that is, make an entry on the log sheet only when something has been found on the magnet. *(Figure 11-37)* Only in this way will you be able to detect a pattern which may be telling you something—that a particular piece of equipment is malfunctioning or coming apart for instance. Logging systems that require an entry each time a magnet is checked, regardless of findings, tend to mask a developing problem, as most entries will indicate that nothing was found.

Place trapped material in a small envelope and save them for a short time; they might be valuable in reconstructing a failure. They might also be valuable in determining the quantity of product involved in the contamination, and most important, the type of hazard the contaminant represents.

If there is a strong need to make certain that magnets are being inspected as scheduled, consider using a separate *check-off* sheet. *(Figure 11-38)* The check-off

FIG. 11-35 Contaminated magnets are less efficient.

sheet should include the magnet location, date of inspection, and initials of the inspector.

MAINTENANCE

There is very little that can go wrong with a magnet and therefore very little maintenance is required. However, some early-vintage magnets may require occasional testing. Through various forms of abuse some older magnets could lose all or part of their holding and pulling strength. Test kits available from magnet manufacturers will allow you to check both holding and pulling powers of plate magnets. The test results should be recorded and compared with previous test results or the original specifications. Magnets should be tested annually or when there is reason to believe they have lost some power.

Many of the misapplications or misconceptions associated with magnet installations can be avoided with proper problem identification and a little applied reasoning.

FIG. 11-36 Log book

METAL DETECTORS

A complete metal contamination prevention program for a food manufacturing facility must include the use of both magnets and metal detectors. This is because each device has its limits. Magnets cannot trap nonferrous materials; metal detectors have contaminant size and sensitivity limits. Whereas using both units as an integral part of the total program cannot guarantee that all metal will be removed, together the units can give reasonably good assurance from a technical standpoint that you are doing everything that can be done to prevent metal contamination.

To be effective, metal detectors must be applied with an understanding of how they work, what their detecting capabilities are, and with some knowledge of interferences that affect their detecting capabilities. Metal detectors cannot simply be put into a system and be expected to do a job. You must know a great deal, starting with what you can expect of the equipment itself. Metal detectors are most often used to detect metal in food which has been packaged. They can also be used to detect metal in a stream of ingredients or product back in the manufacturing process. Their application in the manufacturing process is not as wide because they monitor larger quantities of product at one time. Thus, if metal is detected and a reject occurs, a large amount of product might be lost. Since the search for the contaminant and refeeding are usually done by an operator, this can result in costly labor losses as well. Sometimes the use of a particular packaging material—such as aluminum foil, for example—requires that the metal detector be used in the manufacturing process. In situations like this there might be no alternative but to place the metal detector on a process stream—unless, of course, you are satisfied with a packaging line metal detector which can sense ferrous metals through nonferrous metallic foil packages. However, remember that it can only sense ferrous contaminants! In view of this, it is often more advisable to use a standard detector in the manufacturing process at a point as close to where it is packaged as possible.

HOW THEY WORK

Metal detectors are comprised of two major components: a search head and a control unit. *(Figure 11-39)* The search head consists of three coils wound around a forming block with an opening in the center. The opening is the search area through which the product passes, and is referred to as the aperture. The center coil, called the drive or transmitting coil, is connected to an oscillator, which produces an oscillating electromagnetic field in the aperture. This is

FIG. 11-37 Log sheet

MAGNET LOG SHEET

PLANT: ____________________

DEPT: ____________________

DATE: ____________________

NO.	DATE	MAGNET	SUPERVISOR	DESCRIPTION

FIG. 11-38 Check-off sheet

MAGNET CHECK SHEET

PLANT____________________ DEPARTMENT____________________

MAGNET LOCATION____________________ FREQUENCY CHECKED____________________

3165 GO-1

DATE	CHECKED BY	DATE	CHECKED BY	DATE	CHECKED BY	DATE	CHECKED BY	DATE	CHECKED BY

FIG. 11-39 This typical packaging line metal detector application shows the two main components, the search head and the control unit.

FIG. 11-40 The geometric center of a metal detector aperture is often the area where metal is the most difficult to sense.

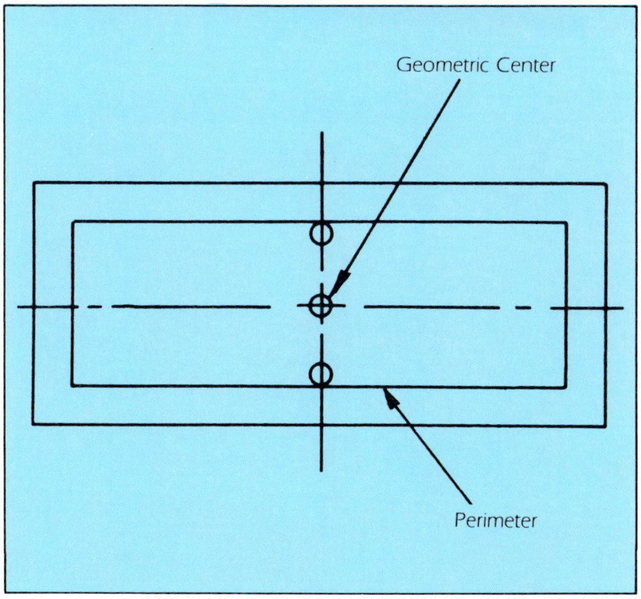

known as the search field. The field induces a signal in each of the two outer coils, called the search or receiving coils, which are equally spaced from the center or transmitting coil. The coils are electrically balanced and connected so that when the search field around both coils is undistorted, their combined output is zero. All metal objects have magnetic or electronic properties. When metal enters the search field, it disturbs the balanced field and produces a voltage signal from one of the receiving coils. This signal is processed electronically in the control unit to sound an alarm or automatically reject the product. The search head shape is such that the search field strength is fairly equal throughout the aperture. However, the field strength does not diminish with distance from the transmitting coil and is at its weakest in the geometric center of the aperture. It is strongest at the perimeter of the aperture. *(Figure 11-40)*

SEARCH HEAD SENSITIVITY

The maximum sensitivity of the search head is measured by the smallest object which will cause a signal. This is usually defined by the smallest diameter sphere that can be detected along the geometric centerline of the aperture. A sphere is chosen because it produces the same signal regardless of its orientation in the aperture. The following factors influence sensing capabilities: product effect, conductivity of the object, size, shape, orientation, and location of the object, size of the aperture, frequency of the electromagnetic field, background interference, and installation discrepancies. Depending on the influence of these factors and assuming a unit with an aperture whose smallest dimension is four inches (102 mm), the consistent rejection of a 3/64-inch (1 mm) diameter steel sphere is not too much to expect of an efficiently operating unit.

Product Effect. Some food products are very good conductors, particularly those with a high moisture level or salt content, and in themselves can cause a signal when passing through the search field. This is known as product effect. Some design changes can be made to minimize this problem but it will probably always be a factor. Adjustments can be made to most metal detectors in the field to compensate for the product effect. This adjustment may have a somewhat negative effect on the overall detecting ability by reducing sensitivity, particularly sensitivity to nonferrous metal conductors such as lead and nonmagnetic stainless steel.

Conductivity of the Object. Some metals are better conductors of electricity than others and will produce a stronger signal or be easier to detect as they pass through the aperture. The electrical signal produced by the passing metal varies in amplitude and phase according to the size and type of metal. These variations are shown in the vector diagram. *(Figure 11-41)* The diagram shows the relative amplitude obtained from a sphere of identical diameter, but of different metal. It is apparent from the length of the vectors that the search field is more sensitive to ferrous metals than to nonferrous ones and that it is least sensitive to nonmagnetic stainless steel.

Many metal detectors are equipped with phase discriminators which have adjustable detection envelopes. In order to be detected, a metal must produce a signal strong enough to pass through the periphery of the detection envelope. Superimposing the detection envelopes on the vector diagram shows how the discriminator can be used to increase or decrease sensitivity to different metals. *(Figure 11-41)*

The phase discriminator can also be used to compensate for product signals. Almost every product will produce a signal with a given amplitude and phase, just as different metals do. Rotating the detection envelope can make it less apt to produce a false signal. At the same time, however, the repositioned envelope will become more or less sensitive to other metals represented by the different vectors. Hence, it might not detect some of the very small, difficult-to-sense metals like stainless steel. Note also that rotating the envelope makes the unit more sensitive to vibration. *(Figure 11-41)*

Size/Shape/Orientation/Location. The size, shape, orientation, and location of an object as it passes through the search field have a great deal to do with its detection because the detector is highly sensitive to the surface area *seen* by the electromagnetic field. A thin wire is the most difficult to detect when passing through the aperture parallel to the electromagnetic field, as shown in position A. *(Figure 11-42)* It is the easiest to detect when passing through perpendicular to the field, as shown in position B.

When a particular problem involves small needle-like metals, it is extremely important to determine the worst position of orientation. As mentioned earlier, because of the variations of signal strength due to orientation, spherical shapes are used for standard sensitivity comparisons.

Size of the Aperture. The size of an object that can be sensed is linked to the size of the detector's aperture. As a very rough guide, most metal detectors can sense a 1-mm steel sphere in a four-inch high aperture, a 2-mm sphere in a six-inch high aperture, and so on. The longest dimension of the aperture does not affect the sensitivity very much unless it is exceptional, i.e., ten or more times greater than the smallest dimension.

FIG. 11-41 Vector diagrams show the amplitude and phase of the same size spheres of different metals.

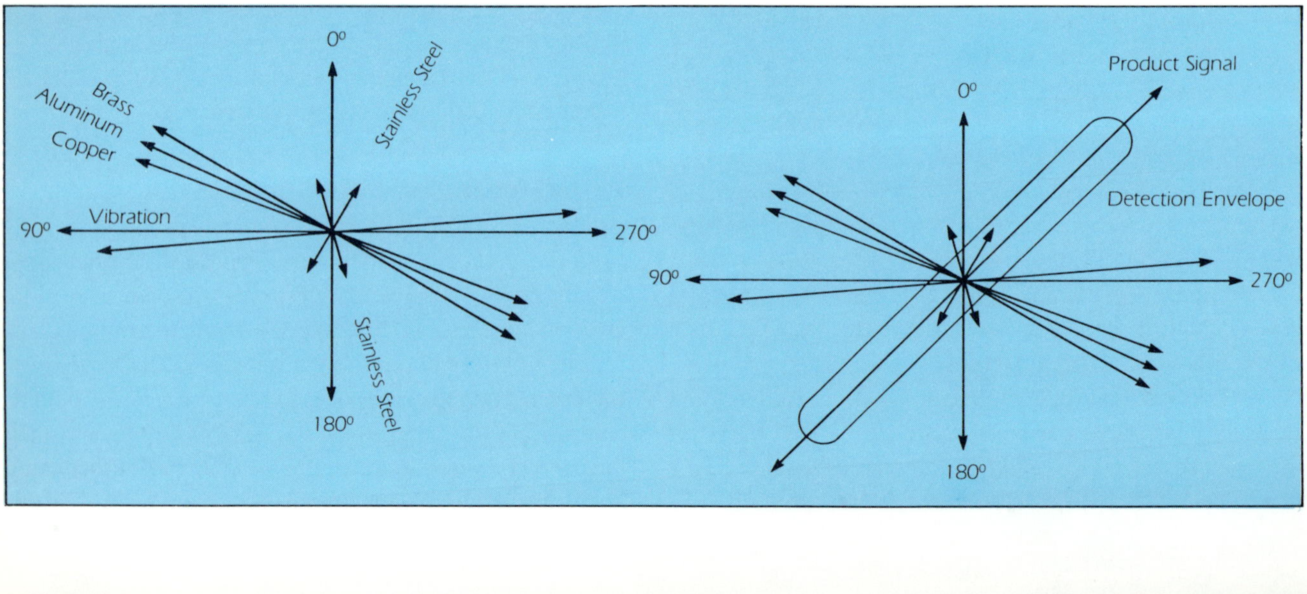

Frequency of the Electromagnetic Field. The choice of electromagnetic field frequency has a great influence on the strength of the product effect signal, particularly when dealing with highly conductive, moist products such as cheese, meat, and pickles. For a highly conductive product, a lower frequency greatly reduces the strength of the product-generated signal. *(Figure 11-43)*

Background Interference. In normal operation a metal detector must distinguish a wanted signal from a background of interference. These interferences can be mechanical noise, electrical noise within the circuitry, vibration, and product effect. Background noise has a significant effect on the threshold limit and detection level of the system. The threshold limit of a system is the level below which no signal will trigger the reject device. The amount that a metal signal exceeds the threshold is called the detection level. *(Figure 11-44)*

The threshold limit is adjustable and is set to ignore background interference. The lower the threshold is set, the more sensitive the detector is to background noise. The higher the threshold is set, the more immune it is to background noise; however, higher thresholds will also decrease sensitivity to very small metal objects.

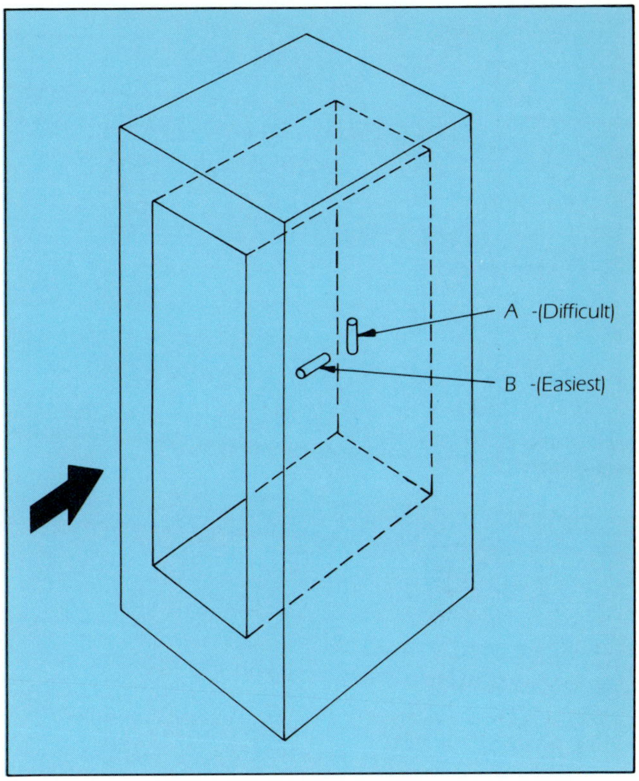

FIG. 11-42 The orientation of a metal object and its location within the search aperture influences its detection capability.

Installation. In planning the installation of metal detectors, certain factors must be considered to assure efficient operation of the system. These installation factors affect the threshold limit, and therefore the ultimate detection level.

First, the installation must be vibration free. Install the search head where there is no excessive floor vibration. Be sure that the equipment does not make contact with any adjacent equipment, including the conveyor that passes through the search field. Contact with this equipment will cause unwanted vibration.

Second, treat the search head with care. No metal of any kind should be located within a specified distance from the aperture. *(Figure 11-45)* The distance requirements may vary according to the detector manufacturer, so it is advisable to follow installation instructions carefully. Care should also be taken to prevent any secondary electromagnetic loops from being induced in the general area of the head. This can occur when metal shafts, pulleys, and bearings are placed too close to the search head.

Third, the conveyor belt that passes through the search field must be completely free from all metal on the top and bottom. Endless belt adhesives used to glue the splice should be free from any metallic properties.

REJECT DEVICES

Metal detectors reject contaminated products in several different ways. Any of the following rejection devices may be used. *(Figure 11-46)*

- Package kicker
- Package air jet
- Plough
- Package reject flapper
- Packaging marking device
- Two-way A-valve system (used primarily on bulk products)

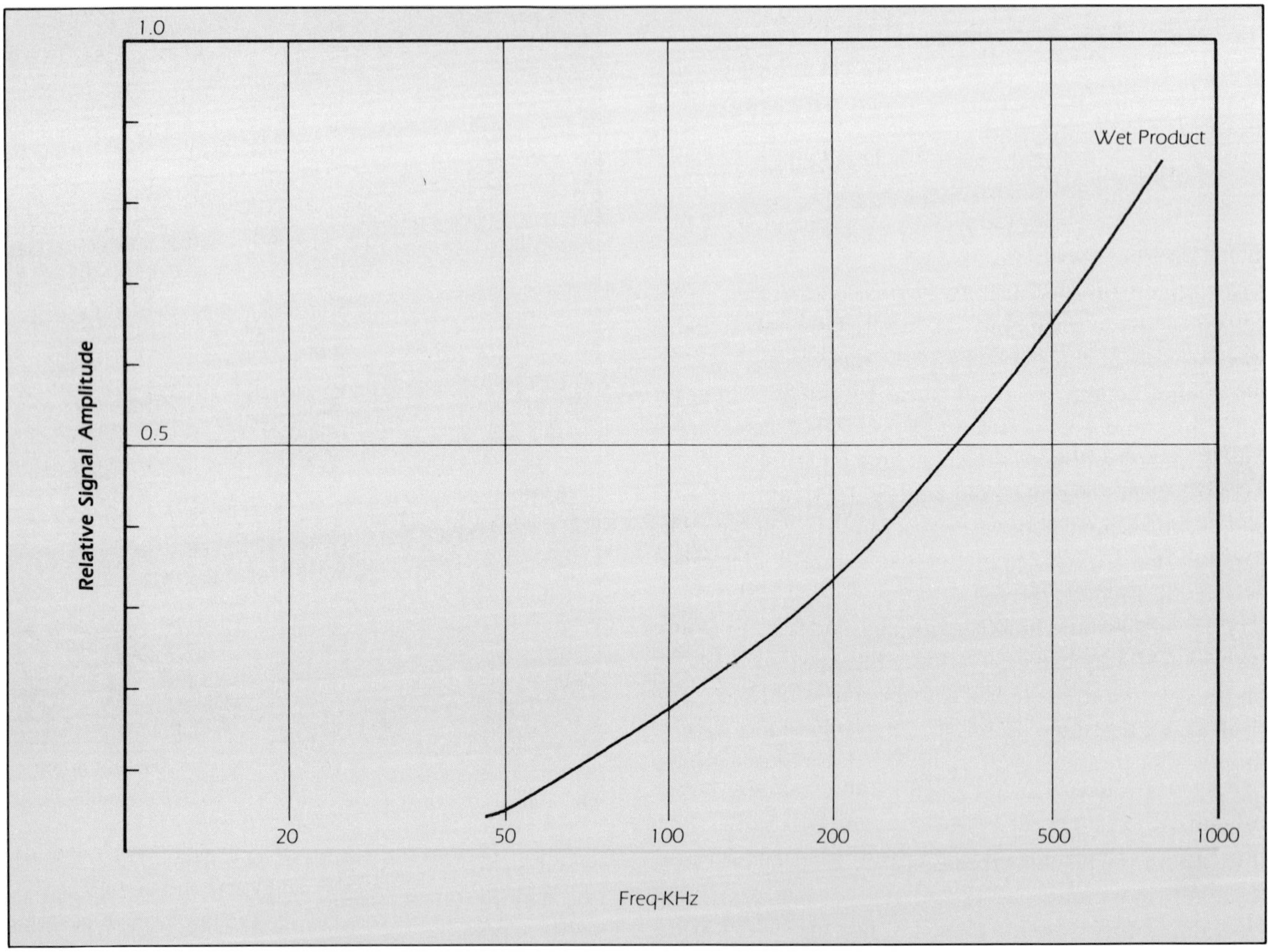

FIG. 11-43 Frequency variation curve

FIG. 11-44 Threshold limit

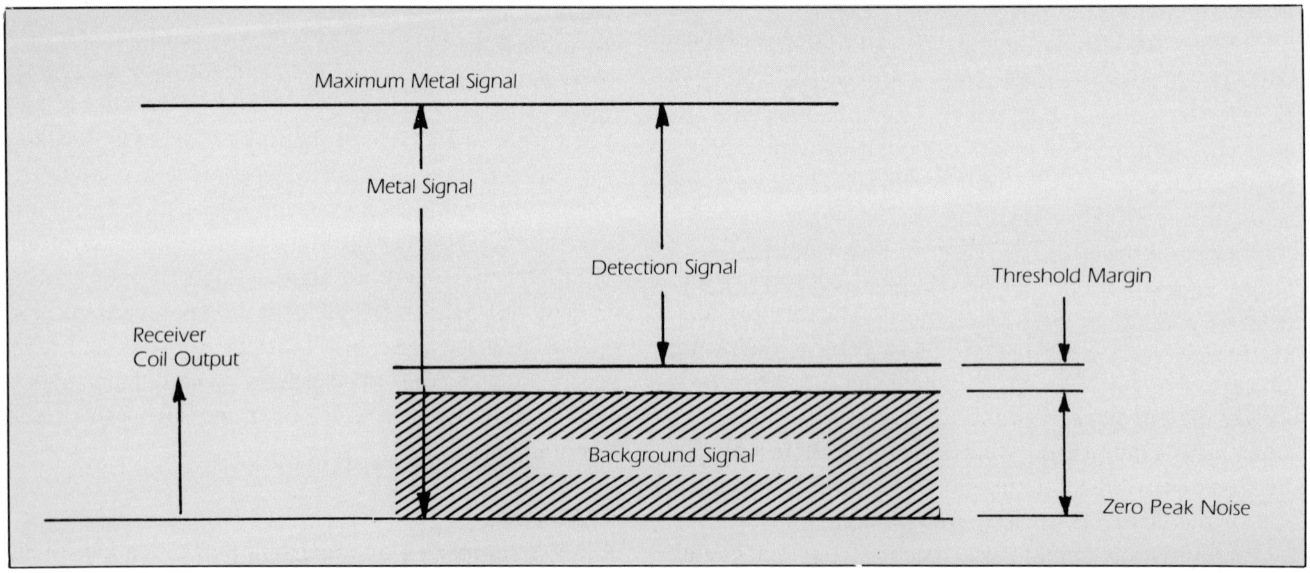

When choosing a rejection device, you will want to consider such important factors as whether the detected metal contaminant can be traced to the package it was found in, whether both the product and the product package can be recovered, and whether there will be production delays caused by stopping and starting packaging equipment.

When metal detectors have been installed in the manufacturing process to monitor a stream of product, a reject valve arrangement may be used to divert contaminated product. If the process allows the product stream to be interrupted, the reject signal could be used to shut down the system.

INSPECTION PROGRAMS

To keep metal detectors performing satisfactorily day in and day out, you will need to set up inspection programs to check detection capabilities of the equipment. Establish detection standards that call for consistent rejection of a given size metal sphere and require efficiency checks at regular intervals. Since the search field is sensitive to the location of the metal in the search area, make sure the test sphere passes through the area as close to geometric center as possible.

In the interest of identifying a particular equipment problem, metal detector rejects should be logged. The procedures can be identical to those used for magnet logging.

Satisfactory performance from metal detectors depends on careful selection and installation of reliable equipment and commitment to a good maintenance and inspection program. When metal detectors are combined with a good magnet application, metal contamination can be controlled.

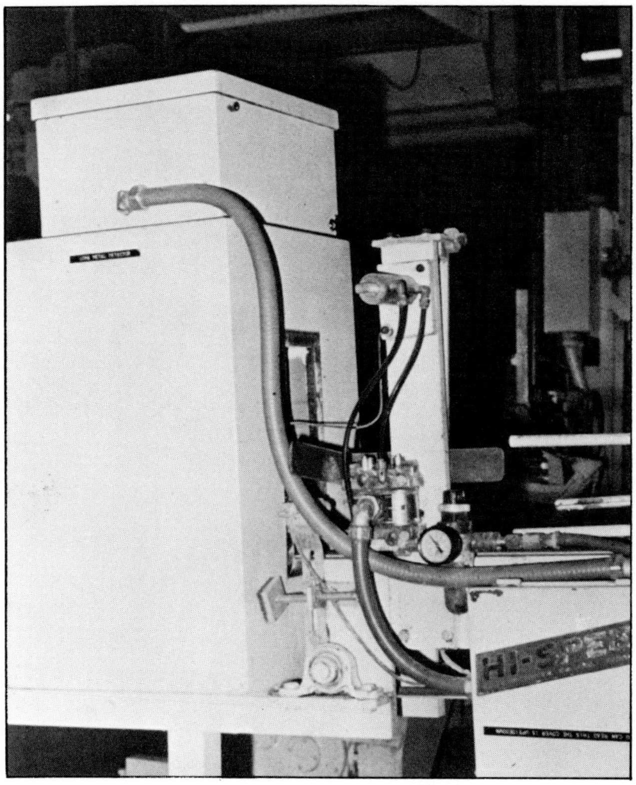

FIG. 11-45 Metal parts located close to the aperture can produce false signals.

SCALPING AND SIFTING

In the food manufacturing process, scalping and sifting are most frequently associated with grading and classifying. Emphasis on product safety and consumer protection shifts the function of scalping and sifting to insect removal and foreign material control. Even though many food processes involve ingredients or commodities that have undergone some form of initial processing or cleaning, there is always the possibility of recontamination from machinery failure or from human errors (such as articles falling from pockets). Recontamination could also occur from packaging materials or from the transporting vehicles. Whatever the source, the recontamination risk, although unlikely, is generally high enough to justify the addition of scalping or sifting equipment primarily for product safety purposes.

Scalping or sifting applications should include screening entire quantities of containerized ingredients and some bulk ingredients as they are introduced into the manufacturing system, as well as random sampling of bulk quantities of ingredients as they are brought into the plant and placed in storage. Scalping and sifting are often used to clean raw commodities such as grains and vegetable products. Used this way, equipment is viewed as an integral part of the manufacturing cleaning system, removing most or all of the foreign material. Since this involves many other operations (such as washing and grading) to complete the process, cleaning will not be included as part of this chapter. This chapter will discuss scalping and sifting to prevent recontamination.

SAMPLING

Random sampling is an important part of a good foreign

FIG. 11-46 These metal detector reject devices do not shut down the line for a rejection.

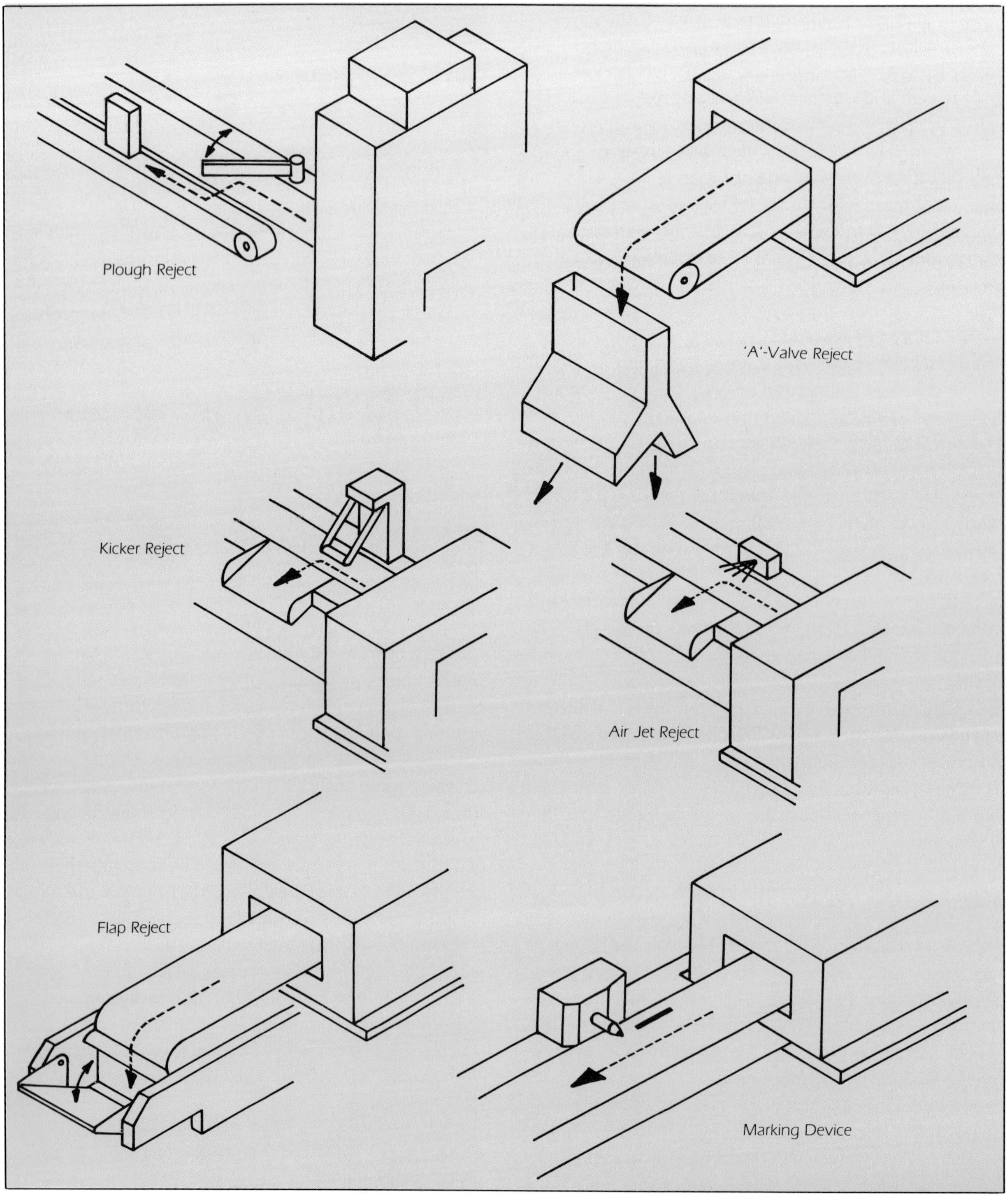

material and insect control program. It is most frequently used to sample incoming bulk ingredients. This is one means of qualifying the condition of ingredients brought into the plant each day, but it is by no means as acceptable as a program which calls for the complete sifting or scalping of all of the ingredients or finished food products. Random sampling can be accomplished by small automatic samplers that take a periodic or continuous sample of a bulk ingredient as it goes through the unloading cycle. The continuous sample is then screened. *(Figure 11-47)* Another method involves interrupting the full stream momentarily (for one or two seconds) to collect one large sample for a particular period in time. *(Figure 11-48)* This type of system uses timing equipment to produce a programmed sequence for taking full-stream samples. These samples are automatically diverted to a surge hopper, where they are fed into the scalping or sifting unit.

INSECT CONTROL

Scalping or sifting for insect control can be accomplished in two different ways. The first and most reliable method passes ingredients through the sifter screen and tails the insects over to reject. *(Figure 11-49)* The finer the ingredient particle size, the finer the mesh of the sifting screen must be, and the more efficient it is in removing all forms of insect life, including eggs. Stored-product insects and a few of the dermestidae are most threatening to grain-based food products and a number of dehydrated vegetable products. Dermestidae can be found around almost any food product. To remove adult stored-product insects, a minimum

FIG. 11-47 Continuous sampling of incoming ingredients

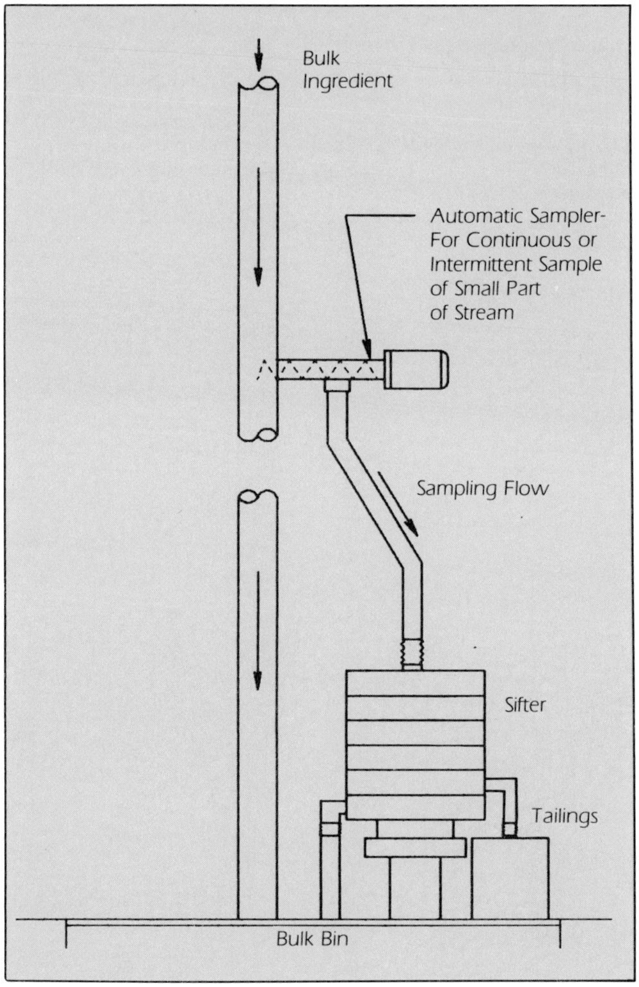

FIG. 11-48 Momentary sampling of a complete stream of incoming ingredients

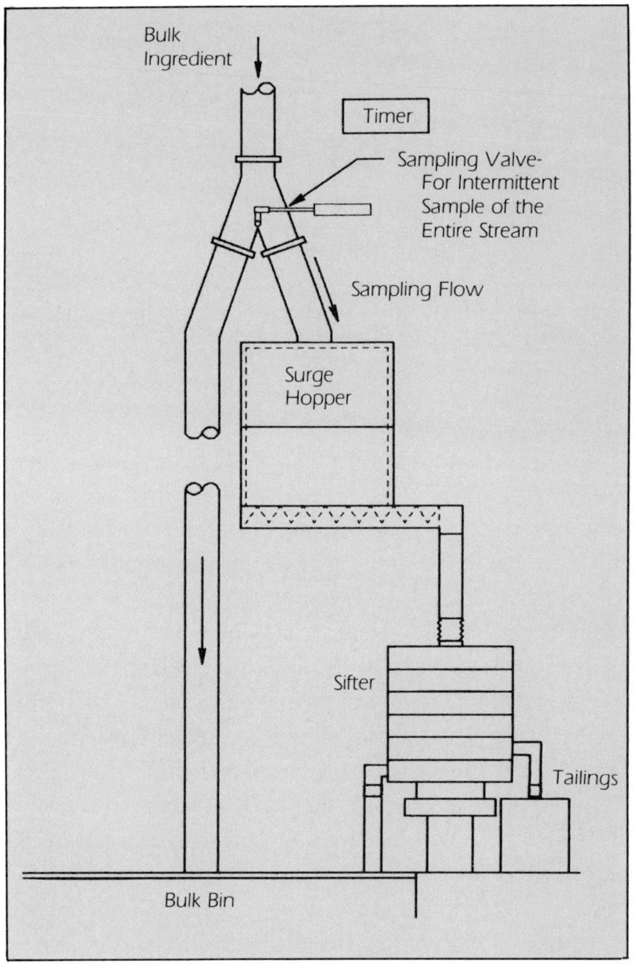

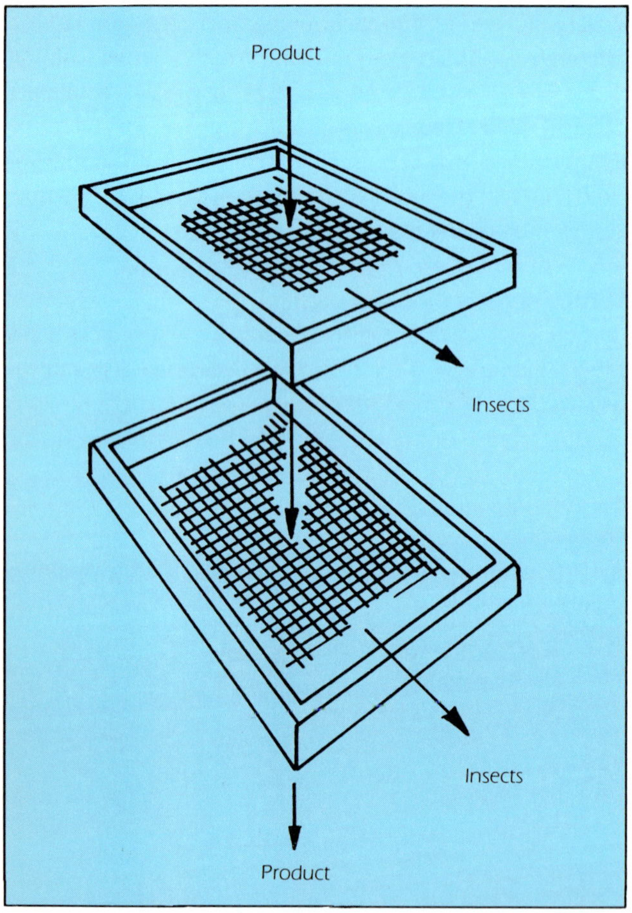

FIG. 11-49 The most reliable method of removing insects passes the product through the sifting screen and tails the insect over the screen.

30-mesh screen with a .016 mesh opening is needed. The 30-mesh will also remove most of the insects in the larva stage. Obviously, the finer the mesh, the better the opportunity to remove them all. A 10 XX screen is needed for egg removal. A 10 XX wire would have a mesh opening of .0053 inches.

In the second system of scalping or sifting for insect removal, the insect falls through the screen to reject and the food product tails over the screen. *(Figure 11-50)* This method is a long way from being completely effective, as insects can easily ride over the screen on top of the product. However, it is an excellent means of obtaining a measure of an insect problem in a product which has a large particle size, such as a pasta product. This particular method of insect removal should not be underestimated; it can be very helpful in identifying a serious insect problem.

PHYSICAL FOREIGN MATERIAL CONTROL

Designing a scalping or sifting system to remove physical foreign material is in many respects more difficult than designing a system to remove insects. Foreign material can be any size from minute particles to large objects. In addition, foreign material that is similar in size to the food product will be extremely difficult to remove.

Both bulk and containerized ingredients are subject to foreign material contamination. Since many of the contaminants are string, paper, and wood from packaging material, containerized ingredients are more likely to be contaminated. Whether the product is bulk or containerized, the scalping job can be done more efficiently if equipment is used for one specific ingredient so that the screen can be sized closely to its particle size. Since it is anticipated that most of the foreign material will be larger objects that come from packaging, most applications will be for a single screen scalping unit which allows the ingredients to pass through the opening and the foreign material to pass over the screen as rejected material. In cases

FIG. 11-50 Less reliable as a removal method, but worthwhile as a measurement tool, is passing the product over the screen and passing the insect through as a reject.

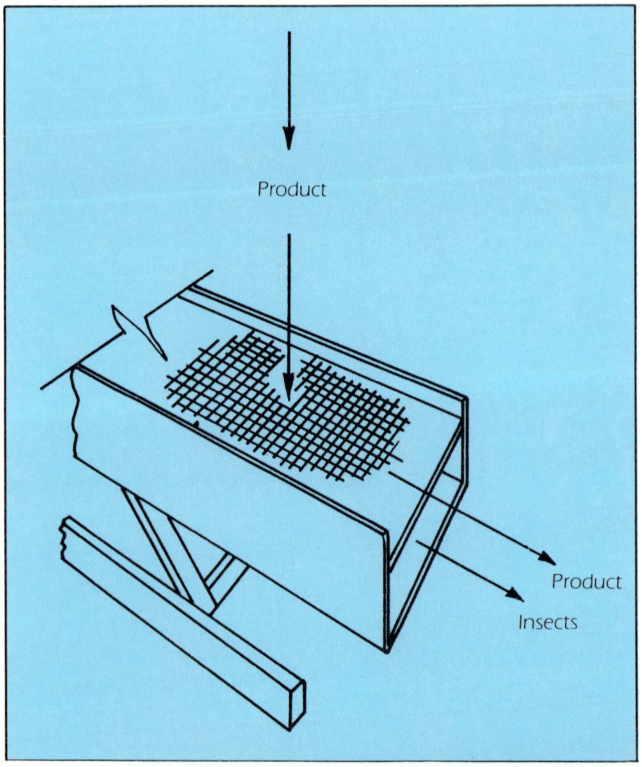

where the particle size of the food product is large, the reject process would be reversed so that the food product passes over the screen and the rejected material drops through.

In some instances, many different containerized ingredients are introduced into the system through a common dump-in using a single scalping unit. In such situations the scalper should have two screens and a spout valving arrangement. *(Figure 11-51)* With two screens, a wider variety of product particle sizes can be more efficiently screened. Several different flow options can be selected through the valving arrangements, depending on the anticipated problem. However, if some of the ingredients are sticky and would blind the lower screen, you will want to reconsider using a two-screen unit. An alternative scalping approach for dump-in systems that must accommodate a multiple range of ingredients (and where blinding might be a problem) would be a system which has two or more individual dump-in and scalper units, each designated to handle a certain group of ingredients of the same particle size.

EQUIPMENT

Scalping and sifting equipment comes in a variety of sizes and with a selection of rotary or reciprocal mechanical actions. Besides being of sufficient size to handle the rate

FIG. 11-51 Two-screen scalper unit

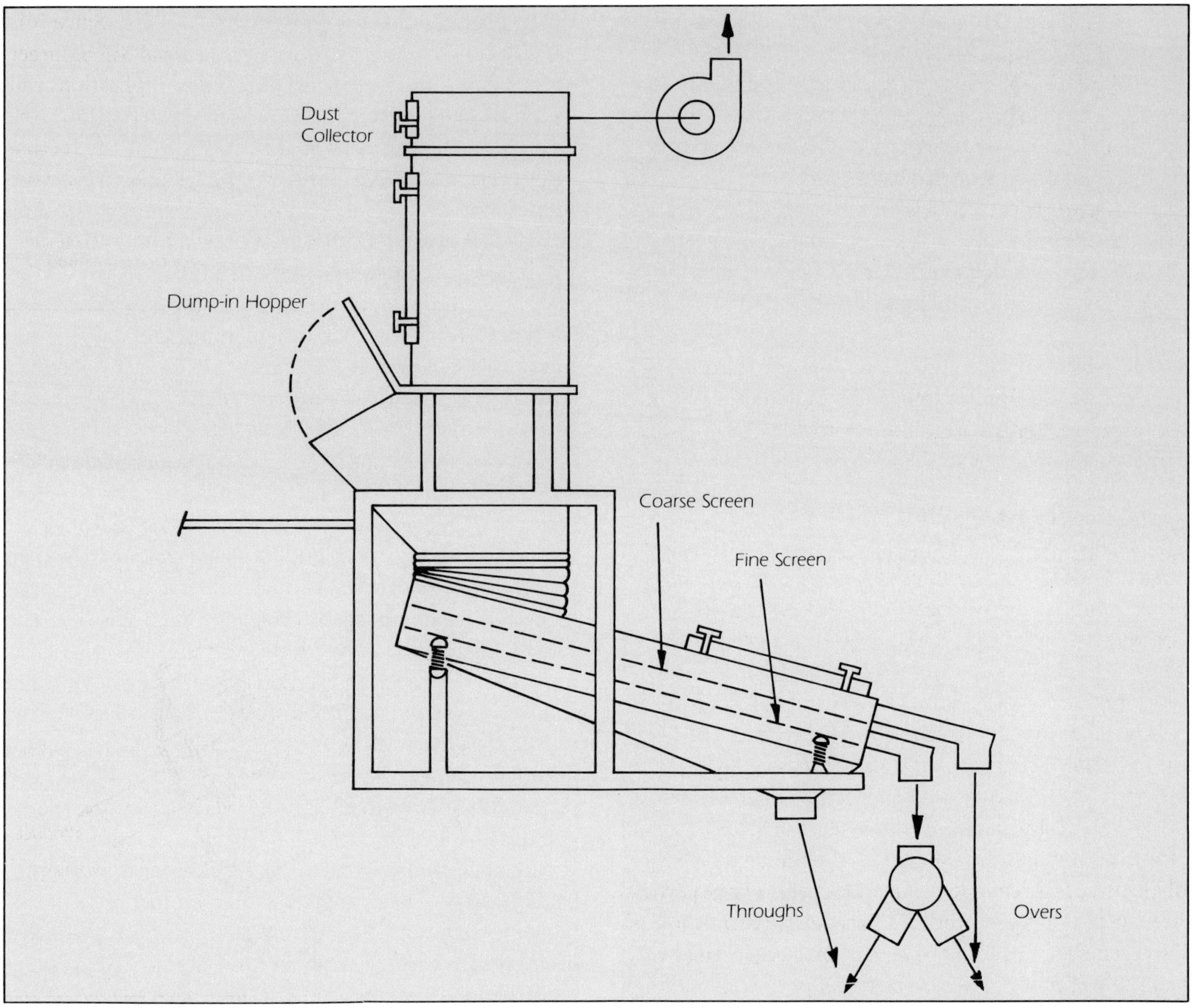

of flow, it must be appropriate for the job at hand. Not all scalpers and sifters are designed for foreign material control, despite the claims of the equipment manufacturers. It is important to know that the equipment is intended for foreign material removal, not grading and classifying. Some high-capacity scalping equipment uses mechanical agitators to improve the throughput. However, these same devices can force foreign objects through scalping screens; they were not designed for foreign material removal. Screen selection is an element of equipment choice that must not be overlooked. The mechanical actions and the abrasive characteristics can cause rapid deterioration of some screen materials. Deterioration can result in disintegration of the screen, and the screen material might end up in the product as a contaminant. How serious this type of contaminant is depends on the screen material.

Wire screen, which is the most durable and long lasting, can disintegrate into needle-like pieces that are potentially very harmful. If wire screens are used, they should be made of metal that has magnetic properties, such as mild steel. Where corrosion-resistant wire is needed, consider using zinc-plated or series 400 stainless steel wire. By using screen with magnetic properties, magnets can be added downstream to trap any wire which may have worked loose.

Many flour sifters use silk screens satisfactorily. Silk does not wear as well as some other materials, so it must be inspected and replaced more frequently. Silk is a contaminant, but it is not considered injurious.

Nylon and a number of synthetic screen materials are also being used satisfactorily. Many are more durable than silk. Synthetic materials stretch, causing efficiency problems, and they, too, wear out. The contamination is objectionable, although in most cases it should not be viewed as injurious. It could be harmful, however, depending on its consistency; sometimes synthetic material takes on a consistency similar to fish bones, particularly when exposed to high temperatures.

The value of scalping and sifting protection cannot be argued when it comes to dealing with recontamination in processed ingredients of finished food products. The equipment has done its job if it prevents even one serious injury. Product safety requires a complete foreign material protection program. Scalping and sifting is an integral part of the program.

FIG. 11-52 Cross section of a typical multi-pass aspirator, aspirators use air velocities to remove certain contaminants from various raw commodities while they are falling freely through the equipment.

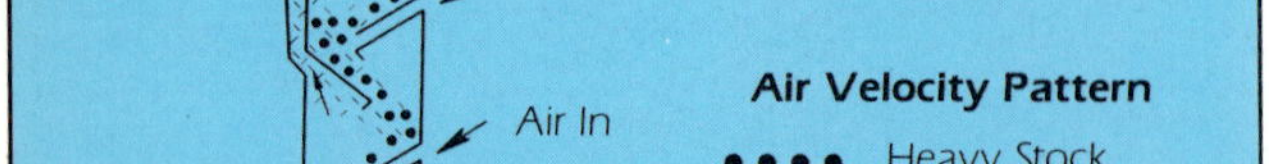

ASPIRATORS

Aspirators have been used for many years to remove infestation, weed seeds, and other contaminants from a variety of grains. Both single-pass and multiple-pass aspirators are commonplace. Separating efficiency levels vary with the equipment and with the products being processed. Aspirator equipment manufacturers suggest that high levels of separating efficiency can be easily obtained with multiple-pass aspirators in particular. These can operate in the 90% to 99.6% or better removal range.

Aspirator separation is based on the terminal velocity of a falling object. The force of gravity on an object is constant. However, the effect of air resistance on that object depends on its size, density, and shape. Theoretically, an object starts to fall at a slow rate and increases speed until it reaches maximum speed, called terminal velocity. The rate of fall is affected by resisting air movement. If air is moving in the opposite direction of the falling object, and if the velocity of this air equals the terminal velocity of the object, the object will float. If resisting air velocity exceeds terminal velocity, the object will rise. Where objects have different terminal velocities due to variance in size, shape, and density, heavier ones will fall while smaller, lighter, or broader ones will be lifted by the resisting air flow. Hence, small insects, insect fragments, and other lightweight contaminants can be effectively removed from processed grain streams when resistant air flows are applied to falling grain streams. These principles are illustrated. *(Figure 11-52)*

CHAPTER 12 FOOD LAWS and REGULATIONS

CHAPTER CONTENTS

INSPECTIONS

FEDERAL FOOD, DRUG, AND COSMETIC ACT, AS AMENDED

THE FEDERAL MEAT INSPECTION ACT AND THE FEDERAL POULTRY INSPECTION ACT

OTHER FEDERAL, STATE, AND LOCAL LAWS AND REGULATIONS

The food industry is quite heavily regulated in the United States. At the federal level, the Food and Drug Administration (F. D. A.) and the United States Department of Agriculture (U. S. D. A.) are the main agencies with jurisdiction. In addition, numerous state and local government agencies also have jurisdiction. Further, many states have cooperating agreements with the federal agencies whereby the state enforcing agencies may be asked to enforce federal food laws.

The Federal Food and Drug Administration obtains the power to regulate the food industry through the Federal Food, Drug, and Cosmetic Act, As Amended. The United States Department of Agriculture, which primarily regulates the meat and poultry industry, obtains its power through the Federal Meat Inspection Act and the Federal Poultry Inspection Act. Respective state and city statutes give jurisdiction to state and local regulatory agencies.

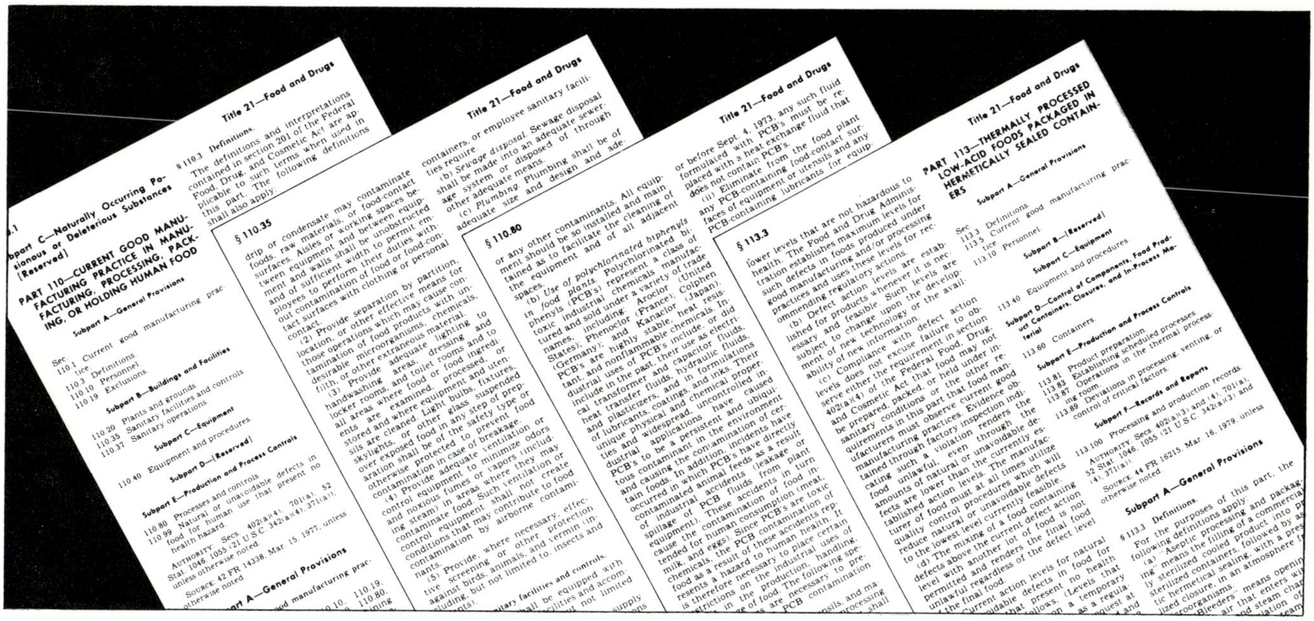

INSPECTIONS

The enforcement procedures used by the various regulatory agencies begin with food plant inspections by a field inspector. The U. S. D. A. usually assigns a full-time inspector to a meat and poultry plant, who grades meat and makes detailed sanitation inspections of the plant. The F. D. A. and many state and local agencies make unannounced inspections of selected plants. Upon arrival, the inspector presents his or her credentials to the plant management and usually states the reason for the inspection. The inspector then proceeds with the inspection, which may take just a few hours or as long as one or more full-time days, depending on the type and size of the facility as well as the type of inspection. Sometimes an inspector just samples products or confirms interstate movements of food products instead of making a detailed general inspection which takes considerably more time.

After a general sanitation inspection, the inspector normally offers comments regarding the food manufacturing practices he or she has observed. Depending on the results of the total inspection, a plant might receive praise for a satisfactory sanitation level, a written reprimand or—if conditions are extremely insanitary—demands for product seizure, court injunction, or both. Monetary penalties can also be imposed. Regulatory agencies represent the law and they have the means to enforce it through the judicial system.

In addition to inspecting food manufacturing facilities, regulatory agencies go about enforcement in a number of other ways. Inspectors select products from retail stores or from storage warehouses and examine them for product safety. An examination may include checking for bacteria, mold, and chemical contamination, or for label compliance or weight, or simply for physical contamination. Regulatory agencies also respond to consumer complaints received by their offices. The response may be a simple request for an explanation or it may result in a full-scale inspection of a plant.

The food industry as a whole is very supportive of food laws and regulations and is appreciative of the intent and purpose of regulation, which is to assure people everywhere of safe food products. In many instances, the industry has taken an active part in the formation of meaningful and helpful legislation. This cooperating effort benefits the industry, the regulatory agencies, and most importantly, the consumer.

FEDERAL FOOD, DRUG, AND COSMETIC ACT, AS AMENDED

The Federal Food, Drug and Cosmetic Act, As Amended includes all amendments through the most current publishing date. The Act, which appears in the United States Code under Title 21, is very complete. Chapter IV, *Food,* establishes definitions and standards for foods and food packaging, adulterated and misbranded foods, tolerances for poisonous ingredients in food, and for food additives. In Chapter II, entitled *Prohibited Acts And Penalties,* the prohibited acts, injunction proceedings, penalties, seizures, and hearing procedures are explained. Chapter VII, *General Administrative Provisions,* establishes the authority to conduct examinations and investigations of food products and food manufacturing plants, as well as the authority to promulgate regulations for efficient enforcement. Copies of the Act can be obtained from the U. S. Government Printing Office and should be part of the plant library.

THE UMBRELLA GOOD MANUFACTURING PRACTICE (GMP)

Chapter VII, Section 701(A), of the Federal Food, Drug and Cosmetic Act As Amended establishes a group of the regulations under the heading *Current Good Manufacturing Practice in Manufacturing, Processing, Packing, Or Holding Human Food.* These are commonly referred to as the Umbrella GMP. Like other regulations established in this manner, they have the force and effect of law. The Umbrella GMP is very detailed, specifying standards for all food manufacturing functions including processing, packing, and storing of raw materials, processed ingredients, packaging materials, and finished food products. Because the F. D. A. customarily does not issue standards or approve products or equipment, the Umbrella GMP is the closest thing to regulatory standards for building and operating a food manufacturing facility. A complete excerpt from a recent issue follows. However, regulations are subject to frequent revision so you may want to obtain a more recent copy.

Subpart C—Naturally Occurring Posionous or Deleterious Substances [Reserved]

* PART 110—CURRENT GOOD MANUFACTURING PRACTICE IN MANUFACTURING, PROCESSING, PACKING, OR HOLDING HUMAN FOOD

Subpart A—General Provisions

Sec.
110.1 Current good manufacturing practice.
110.3 Definitions.
110.10 Personnel.
110.19 Exclusions.

Subpart B—Buildings and Facilities

110.20 Plants and grounds.
110.35 Sanitary facilities and controls.
110.37 Sanitary operations.

Subpart C—Equipment

110.40 Equipment and procedures.

Subpart D—[Reserved]

Subpart E—Production and Process Controls

110.80 Processes and controls.
110.99 Natural or unavoidable defects in food for human use that present no health hazard.

AUTHORITY: Secs. 402(a)(4), 701(a), 52 Stat. 1046, 1055 (21 U.S.C. 342(a)(4), 371(a)), unless otherwise noted.

SOURCE: 42 FR 14338, Mar. 15, 1977, unless otherwise noted.

Subpart A—General Provisions

§ 110.1 Current good manufacturing practice.

The criteria in §§ 110.10, 110.19, 110.20, 110.35, 110.37, 110.40, 110.80, and 110.99 shall apply in determining whether the facilities, methods, practices, and controls used in the manufacture, processing, packing, or holding of food are in conformance with or are operated or administered in conformity with good manufacturing practices to assure that food for human consumption is safe and has been prepared, packed, and held under sanitary conditions.

§ 110.3 Definitions.

The definitions and interpretations contained in section 201 of the Federal Food, Drug, and Cosmetic Act are applicable to such terms when used in this part. The following definitions shall also apply:

(a) "Adequate" means that which is needed to accomplish the intended purpose in keeping with good public health practice.

(b) "Plant" means the building or buildings or parts thereof, used for or in connection with the manufacturing, processing, packaging, labeling, or holding of human food.

(c) "Sanitize" means adequate treatment of surfaces by a process that is effective in destroying vegetative cells of pathogenic bacteria and in substantially reducing other microorganisms. Such treatment shall not adversely affect the product and shall be safe for the consumer.

§ 110.10 Personnel.

The plant management shall take all reasonable measures and precautions to assure the following:

(a) *Disease control.* No person affected by disease in a communicable form, or while a carrier of such disease, or while affected with boils, sores, infected wounds, or other abnormal sources of microbiological contamination, shall work in a food plant in any capacity in which there is a reasonable possibility of food or food ingredients becoming contaminated by such person, or of disease being transmitted by such person to other individuals.

(b) *Cleanliness.* All persons, while working in direct contact with food preparation, food ingredients, or surfaces coming into contact therewith shall:

(1) Wear clean outer garments, maintain a high degree of personal cleanliness, and conform to hygienic practices while on duty, to the extent necessary to prevent contamination of food products.

(2) Wash their hands thoroughly (and sanitize if necessary to prevent contamination by undesirable microorganism) in an adequate hand-washing facility before starting work, after each absence from the work station

and at any other time when the hands may have become soiled or contaminated.

(3) Remove all insecure jewelry and, during periods where food is manipulated by hand, remove from hands any jewelry that cannot be adequately sanitized.

(4) If gloves are used in food handling, maintain them in an intact, clean, and sanitary condition. Such gloves should be of an impermeable material except where their usage would be inappropriate or incompatible with the work involved.

(5) Wear hair nets, headbands, caps, or other effective hair restraints.

(6) Not store clothing or other personal belongings, eat food or drink beverages, or use tobacco in any form in areas where food or food ingredients are exposed or in areas used for washing equipment or utensils.

(7) Take any other necessary precautions to prevent contamination of foods with microorganisms or foreign substances including, but not limited to, perspiration, hair, cosmetics, tobacco, chemicals, and medicants.

(c) *Education and training.* Personnel responsible for identifying sanitation failures or food contamination should have a background of education or experience, or a combination thereof, to provide a level of competency necessary for production of clean and safe food. Food handlers and supervisors should receive appropriate training in proper food-handling techniques and food-protection principles and should be cognizant of the danger of poor personal hygiene and insanitary practices.

(d) *Supervision.* Responsibility for assuring compliance by all personnel with all requirements of this Part 110 shall be clearly assigned to competent supervisory personnel.

§ 110.19 Exclusions.

The following operations are excluded from coverage under these general regulations; however, the Commissioner will issue special regulations when he believes it necessary to cover these excluded operations: Establishments engaged solely in the harvesting, storage, or distribution of one or more raw agricultural commodities, as defined in section 201(r) of the act, which are ordinarily cleaned, prepared, treated or otherwise processed before being marketed to the consuming public.

Subpart B—Buildings and Facilities

§ 110.20 Plants and grounds.

(a) *Grounds.* The grounds about a food plant under the control of the operator shall be free from conditions which may result in the contamination of food including, but not limited to, the following:

(1) Improperly stored equipment, litter, waste, refuse, and uncut weeds or grass within the immediate vicinity of the plant buildings or structures that may constitute an attractant, breeding place, or harborage for rodents, insects, and other pests.

(2) Excessively dusty roads, yards, or parking lots that may constitute a source of contamination in areas where food is exposed.

(3) Inadequately drained areas that may contribute contamination to food products through seepage or foot-borne filth and by providing a breeding place for insects or microorganisms.

If the plant grounds are bordered by grounds not under the operator's control of the kind described in paragraphs (a) (1) through (3) of this section, care must be exercised in the plant by inspection, extermination, or other means to effect exclusion of pests, dirt, and other filth that may be a source of food contamination.

(b) *Plant construction and design.* Plant buildings and structures shall be suitable in size, construction, and design to facilitate maintenance and sanitary operations for food-processing purposes. The plant and facilities shall:

(1) Provide sufficient space for such placement of equipment and storage of materials as is necessary for sanitary operations and production of safe food. Floors, walls, and ceilings in the plant shall be of such construction as to be adequately cleanable and shall be kept clean and in good repair. Fixtures, ducts, and pipes shall not be so suspended over working areas that

drip or condensate may contaminate foods, raw materials, or food-contact surfaces. Aisles or working spaces between equipment and between equipment and walls shall be unobstructed and of sufficient width to permit employees to perform their duties without contamination of food or food-contact surfaces with clothing or personal contact.

(2) Provide separation by partition, location, or other effective means for those operations which may cause contamination of food products with undesirable microorganisms, chemicals, filth, or other extraneous material.

(3) Provide adequate lighting to handwashing areas, dressing and locker rooms, and toilet rooms and to all areas where food or food ingredients are examined, processed, or stored and where equipment and utensils are cleaned. Light bulbs, fixtures, skylights, or other glass suspended over exposed food in any step of preparation shall be of the safety type or otherwise protected to prevent food contamination in case of breakage.

(4) Provide adequate ventilation or control equipment to minimize odors and noxious fumes or vapors (including steam) in areas where they may contaminate food. Such ventilation or control equipment shall not create conditions that may contribute to food contamination by airborne contaminants.

(5) Provide, where necessary, effective screening or other protection against birds, animals, and vermin (including, but not limited to, insects and rodents).

§ 110.35 Sanitary facilities and controls.

Each plant shall be equipped with adequate sanitary facilities and accommodations including, but not limited to, the following:

(a) *Water supply.* The water supply shall be sufficient for the operations intended and shall be derived from an adequate source. Any water that contacts foods or food-contact surfaces shall be safe and of adequate sanitary quality. Running water at a suitable temperature and under pressure as needed shall be provided in all areas where the processing of food, the cleaning of equipment, utensils, or containers, or employee sanitary facilities require.

(b) *Sewage disposal.* Sewage disposal shall be made into an adequate sewerage system or disposed of through other adequate means.

(c) *Plumbing.* Plumbing shall be of adequate size and design and adequately installed and maintained to:

(1) Carry sufficient quantities of water to required locations throughout the plant.

(2) Properly convey sewage and liquid disposable waste from the plant.

(3) Not constitute a source of contamination to foods, food products or ingredients, water supplies, equipment, or utensils or create an insanitary condition.

(4) Provide adequate floor drainage in all areas where floors are subject to flooding-type cleaning or where normal operations release or discharge water or other liquid waste on the floor.

(d) *Toilet facilities.* Each plant shall provide its employees with adequate toilet and associated hand-washing facilities within the plant. Toilet rooms shall be furnished with toilet tissue. The facilities shall be maintained in a sanitary condition and kept in good repair at all times. Doors to toilet rooms shall be self-closing and shall not open directly into areas where food is exposed to airborne contamination, except where alternate means have been taken to prevent such contamination (such as double doors, positive air-flow systems, etc.). Signs shall be posted directing employees to wash their hands with cleaning soap or detergents after using toilet.

(e) *Hand-washing facilities.* Adequate and convenient facilities for hand washing and, where appropriate, hand sanitizing shall be provided at each location in the plant where good sanitary practices require employees to wash or sanitize and dry their hands. Such facilities shall be furnished with running water at a suitable temperature for hand washing, effective hand-cleaning and sanitizing preparations, sanitary towel service or suitable drying devices, and, where appropriate, easily cleanable waste receptacles.

(f) *Rubbish and offal disposal.* Rubbish and any offal shall be so conveyed, stored, and disposed of as to minimize the development of odor, prevent waste from becoming an attractant and harborage or breeding place for vermin, and prevent contamination of food, food-contact surfaces, ground surfaces, and water supplies.

§ 110.37 Sanitary operations.

(a) *General maintenance.* Buildings, fixtures, and other physical facilities of the plant shall be kept in good repair and shall be maintained in a sanitary condition. Cleaning operations shall be conducted in such a manner as to minimize the danger of contamination of food and food-contact surfaces. Detergents, sanitizers, and other supplies employed in cleaning and sanitizing procedures shall be free of significant microbiological contamination and shall be safe and effective for their intended uses. Only such toxic materials as are required to maintain sanitary conditions, for use in laboratory testing procedures, for plant and equipment maintenance and operation, or in manufacturing or processing operations shall be used or stored in the plant. These materials shall be identified and used only in such manner and under conditions as will be safe for their intended uses.

(b) *Animal and vermin control.* No animals or birds, other than those essential as raw material, shall be allowed in any area of a food plant. Effective measures shall be taken to exclude pests from the processing areas and to protect against the contamination of foods in or on the premises by animals, birds, and vermin (including, but not limited to, rodents and insects). The use of insecticides or rodenticides is permitted only under such precautions and restrictions as will prevent the contamination of food or packaging materials with illegal residues.

(c) *Sanitation of equipment and utensils.* All utensils and product-contact surfaces of equipment shall be cleaned as frequently as necessary to prevent contamination of food and food products. Nonproduct-contact surfaces of equipment used in the operation of food plants should be cleaned as frequently as necessary to minimize accumulation of dust, dirt, food particles, and other debris. Single-service articles (such as utensils intended for one-time use, paper cups, paper towels, etc.) should be stored in appropriate containers and handled, dispensed, used, and disposed of in a manner that prevents contamination of food or food-contact surfaces. Where necessary to prevent the introduction of undesirable microbiological organisms into food products, all utensils and product-contact surfaces of equipment used in the plant shall be cleaned and sanitized prior to such use and following any interruption during which such utensils and contact surface may have become contaminated. Where such equipment and utensils are used in a continuous production operation, the contact surfaces of such equipment and utensils shall be cleaned and sanitized on a predetermined schedule using adequate methods for cleaning and sanitizing. Sanitizing agents shall be effective and safe under conditions of use. Any facility, procedure, machine, or device may be acceptable for cleaning and sanitizing equipment and utensils if it is established that such facility, procedure, machine, or device will routinely render equipment and utensils clean and provide adequate sanitizing treatment.

(d) *Storage and handling of cleaned portable equipment and utensils.* Cleaned and sanitized portable equipment and utensils with product-contact surfaces should be stored in such a location and manner that product-contact surfaces are protected from splash, dust, and other contamination.

Subpart C—Equipment

§ 110.40 Equipment and procedures.

(a) *General.* All plant equipment and utensils should be (1) suitable for their intended use, (2) so designed and of such material and workmanship as to be adequately cleanable, and (3) properly maintained. The design, construction, and use of such equipment and utensils shall preclude the adulteration of food with lubricants, fuel, metal fragments, contaminated water, or any other contaminants. All equipment should be so installed and maintained as to facilitate the cleaning of the equipment and of all adjacent spaces.

(b) *Use of polychlorinated biphenyls in food plants.* Polychlorinated biphenyls (PCB's) represent a class of toxic industrial chemicals manufactured and sold under a variety of trade names, including: Aroclor (United States); Phenoclor (France); Colphen (Germany); and Kanaclor (Japan). PCB's are highly stable, heat resistant, and nonflammable chemicals. Industrial uses of PCB's include, or did include in the past, their use as electrical transformer and capacitor fluids, heat transfer fluids, hydraulic fluids, and plasticizers, and in formulations of lubricants, coatings, and inks. Their unique physical and chemical properties and widespread, uncontrolled industrial applications have caused PCB's to be a persistent and ubiquitous contaminant in the environment and causing the contamination of certain foods. In addition, incidents have occurred in which PCB's have directly contaminated animal feeds as a result of industrial accidents (leakage or spillage of PCB fluids from plant equipment). These accidents in turn cause the contamination of food intended for human consumption (meat, milk, and eggs). Since PCB's are toxic chemicals, the PCB contamination of food as a result of these accidents represents a hazard to human health. It is therefore necessary to place certain restrictions on the industrial uses of PCB's in the production, handling, and storage of food. The following special provisions are necessary to preclude accidental PCB contamination of food:

(1) New equipment, utensils, and machinery for handling or processing food in or around a food plant shall not contain PCB's.

(2) On or before September 4, 1973, the management of food plants shall:

(i) Have the heat exchange fluid used in existing equipment or machinery for handling or processing food sampled and tested to determine whether it contains PCB's, or verify the absence of PCB's in such formulations by other appropriate means. On or before Sept. 4, 1973, any such fluid formulated with PCB's must be replaced with a heat exchange fluid that does not contain PCB's.

(ii) Eliminate from the food plant any PCB-containing food-contact surfaces of equipment or utensils and any PCB-containing lubricants for equipment or machinery that is used for handling or processing food.

(iii) Eliminate from the food plant any other PCB-containing materials wherever there is a reasonable expectation that such materials could cause food to become contaminated with PCB's either as a result of normal use or as a result of accident, breakage, or other mishap.

(iv) The toxicity and other characteristics of fluids selected as PCB replacements must be adequately determined so that the least potentially hazardous replacement is used. In making this determination with respect to a given fluid, consideration should be given to (*a*) its toxicity; (*b*) the maximum quantity that could be spilled onto a given quantity of food before it would be noticed, taking into account its color and odor; (*c*) possible signaling devices in the equipment to indicate a loss of fluid, etc.; and (*d*) its environmental stability and tendency to survive and be concentrated through the food chain. The judgment as to whether a replacement fluid is sufficiently nonhazardous is to be made on an individual installation and operation basis.

(3) For the purposes of this section, the provisions do not apply to electrical transformers and condensers containing PCB's in sealed containers.

Subpart D—[Reserved]

Subpart E—Production and Process Controls

§ 110.80 Processes and controls.

All operations in the receiving, inspecting, transporting, packaging, segregating, preparing, processing, and storing of food shall be conducted in accord with adequate sanitation principles. Overall sanitation of the plant shall be under the supervision of an individual assigned responsibility for this function. All reasonable precautions, including the following, shall be taken to assure that production procedures do not contribute contamination such as filth, harmful chemicals, undesirable microorganisms, or any other objectionable material to the processed product:

(a) Raw material and ingredients shall be inspected and segregated as necessary to assure that they are clean, wholesome, and fit for processing into human food and shall be stored under conditions that will protect against contamination and minimize deterioration. Raw materials shall be washed or cleaned as required to remove soil or other contamination. Water used for washing, rinsing, or conveying of food products shall be of adequate quality, and water shall not be reused for washing, rinsing, or conveying products in a manner that may result in contamination of food products.

(b) Containers and carriers of raw ingredients should be inspected on receipt to assure that their condition has not contributed to the contamination or deterioration of the products.

(c) When ice is used in contact with food products, it shall be made from potable water and shall be used only if it has been manufactured in accordance with adequate standards and stored, transported, and handled in a sanitary manner.

(d) Food-processing areas and equipment used for processing human food should not be used to process nonhuman food-grade animal feed or inedible products unless there is no reasonable possibility for the contamination of the human food.

(e) Processing equipment shall be maintained in a sanitary condition through frequent cleaning including sanitization where indicated. Insofar as necessary, equipment shall be taken apart for thorough cleaning.

(f) All food processing, including packaging and storage, should be conducted under such conditions and controls as are necessary to minimize the potential for undesirable bacterial or other microbiological growth, toxin formation, or deterioration or contamination of the processed product or ingredients. This may require careful monitoring of such physical factors as time, temperature, humidity, pressure, flow-rate and such processing operations as freezing, dehydration, heat processing, and refrigeration to assure that mechanical breakdowns, time delays, temperature fluctuations, and other factors do not contribute to the decomposition or contamination of the processed products.

(g) Chemical, microbiological, or extraneous-material testing procedures shall be utilized where necessary to identify sanitation failures or food contamination, and all foods and ingredients that have become contaminated shall be rejected or treated or processed to eliminate the contamination where this may be properly accomplished.

(h) Packaging processes and materials shall not transmit contaminants or objectionable substances to the products, shall conform to any applicable food additive regulation (Parts 170 through 189 of this chapter), and should provide adequate protection from contamination.

(i) Meaningful coding of products sold or otherwise distributed from a manufacturing, processing, packing, or repacking activity should be utilized to enable positive lot identification to facilitate, where necessary, the segregation of specific food lots that may have become contaminated or otherwise unfit for their intended use. Records should be retained for a period of time that exceeds the shelf life of the product, except that they need not be retained more than 2 years.

(j) Storage and transportation of finished products should be under such conditions as will prevent contamination, including development of pathogenic or toxigenic microorganisms, and will protect against undesirable deterioration of the product and the container.

§ 110.99 Natural or unavoidable defects in food for human use that present no health hazard.

(a) Some foods, even when produced under current good manufacturing and/or processing practices, contain natural or unavoidable defects at lower levels that are not hazardous to health. The Food and Drug Administration establishes maximum levels for such defects in foods produced under good manufacturing and/or processing practices and uses these levels for recommending regulatory actions.

(b) Defect action levels are established for products whenever it is necessary and feasible. Such levels are subject to change upon the development of new technology or the availability of new information.

(c) Compliance with defect action levels does not excuse failure to observe either the requirement in section 402(a)(4) of the Federal Food, Drug, and Cosmetic Act that food may not be prepared, packed, or held under insanitary conditions or the other requirements in this part that food manufacturers must observe current good manufacturing practices. Evidence obtained through factory inspection indicating such a violation renders the food unlawful, even through the amounts of natural or unavoidable defects are lower than the currently established action levels. The manufacturer of food must at all times utilize quality control procedures which will reduce natural or unavoidable defects to the lowest level currently feasible.

(d) The mixing of a food containing defects above the current defect action level with another lot of food is not permitted and renders the final food unlawful regardless of the defect level of the final food.

(e) Current action levels for natural and unavoidable defects in food for human use that present no health hazard are as follows. (Levels that have been adopted on a temporary basis prior to publication as a regulation may be obtained upon request at the Office of Public Affairs, Food and Drug Administration, Room 15B-42, Parklawn Building, 5600 Fishers Lane, Rockville, MD 20857.)

[42 FR 14338, Mar. 15, 1977, as amended at 46 FR 8459, Jan. 27, 1981]

PART 113—THERMALLY PROCESSED LOW-ACID FOODS PACKAGED IN HERMETICALLY SEALED CONTAINERS

Subpart A—General Provisions

Sec.
113.3 Definitions.
113.5 Current good manufacturing practice.
113.10 Personnel.

Subpart B—[Reserved]

Subpart C—Equipment

113.40 Equipment and procedures.

Subpart D—Control of Components, Food Product Containers, Closures, and In-Process Material

113.60 Containers.

Subpart E—Production and Process Controls

113.81 Product preparation.
113.83 Establishing scheduled processes.
113.87 Operations in the thermal processing room.
113.89 Deviations in processing, venting, or control of critical factors.

Subpart F—Records and Reports

113.100 Processing and production records.

AUTHORITY: Secs. 402(a)(3) and (4), 701(a), 52 Stat. 1046, 1055 (21 U.S.C. 342(a)(3) and (4), 371(a)).

SOURCE: 44 FR 16215, Mar. 16, 1979, unless otherwise noted.

Subpart A—General Provisions

§ 113.3 Definitions.

For the purposes of this part, the following definitions apply:

(a) "Aseptic processing and packaging" means the filling of a commercially sterilized cooled product into presterilized containers, followed by aseptic hermetical sealing, with a presterilized closure, in an atmosphere free of microorganisms.

**Part 110, Current Good Manufacturing Practice in Manufacturing, Processing, Packing, or Holding Human Food.* Code of Federal Regulations, Food and Drugs, Title 21, Subchapter B—Food for Human Consumption (April 1, 1983), Washington, D. C.: U. S. Government Printing Office. Pages 82-88.

SPECIFIC GMPS

In addition to the Umbrella GMP, a number of specific *Good Manufacturing Practices* have been promulgated for certain segments of the food industry. A few of them are:

- Cacao products and confectionery
- Smoked and smoke-flavored fish
- Frozen raw breaded shrimp
- Processing and bottling of bottled drinking water

More of the specific industry GMPs are planned. Like the Umbrella GMP, they are subject to constant updating so you will want to make certain that you have a copy of the most current issue.

THE FEDERAL MEAT INSPECTION ACT AND THE FEDERAL POULTRY INSPECTION ACT

The Federal Meat Inspection Act and the Federal Poultry Inspection Act also appear in the United States Code under Title 21. They, too, are very thorough, establishing definitions and standards for meat, poultry, and meat and poultry products, as well as for the packaging of these products. Further, these acts provide for the inspection of animals about to be slaughtered and entered into the various processing and rendering systems for use in commerce, including post-mortem examination. The acts authorize inspection of processing or manufacturing facilities and give an inspector the authority to refuse a system start-up if conditions are insanitary.

The U. S. D. A. takes an active role in the design of a new meat or poultry processing plant or in the remodeling of an existing facility by approving (or disapproving) certain design recommendations. The U. S. D. A. has also established an approval process for building materials including those used in fabricating, processing, and packaging equipment. It is generally just a matter of submitting pertinent data and samples of a particular product for U. S. D. A. examination. The U. S. D. A., in turn, may issue a letter stipulating where and how the material may be used.

The U. S. D. A. also issues a number of documents on the subject of meat and poultry plant sanitation as a service to the industry, making readily available information on various materials that have been previously approved for use in the food plant.

OTHER FEDERAL, STATE, AND LOCAL LAWS AND REGULATIONS

In addition to those mentioned, many other directly and indirectly related federal, state, and local laws and regulations supported by law affect the production of human foods. Many states have their own statutes that govern food production and storage. Local city health authorities also maintain jurisdiction over food plants within their boundaries. Virtually all of these food laws include compliance inspection programs. In many instances, compliance problems are associated with sanitary design failures. In most cases, these failures will not be cited as violations of specific written standards or instructions, but instead will be cited as a violation of other regulatory compliance documents such as Section 402 A-4 of the Federal Food, Drug, and Cosmetic Act, As Amended. This Act states: *A food shall be deemed to be adulterated if it has been prepared, packed, or held under insanitary conditions whereby it may have become contaminated with filth, or whereby it may have been rendered injurious to health.*

In some rare instances, state and local laws or regulations might conflict, but usually where there are differences, it is simply a matter of one law being more stringent than another. In such cases, your design must meet the most stringent requirement. To resolve actual conflict, confront each regulatory office for clarification and direction. And by all means, be sure your plant library includes current copies of the available state and local food laws and regulations.

REFERENCES

BOOKS

American Society of Heating, Refrigerating and Air Conditioning Engineers, Inc., *ASHRAE Handbook and Product Directory—1980.* New York: American Society of Heating, Refrigerating and Air Conditioning Engineers, Inc., 1980.

Association of Food Industry Sanitarians, Inc. in cooperation with the National Canners Association. *Sanitation for the Food Preservation Industries.* New York: McGraw-Hill, 1952.

Cotton, Richard T. *Pests of Stored Grain and Grain Products.* Minneapolis, Minn.: Burgess Publishing Company, 1963.

Eshbach, Ovin W. *Handbook of Engineering Fundamentals.* 2nd ed. New York: John Wiley and Sons, 1953.

Jones, F. D. and Erik Obery. *Machinery Handbook,* 15th ed. New York: Industrial Press, 1955.

Parker, Milton E. *Food Plant Sanitation.* New York: McGraw-Hill, 1948.

BOOKLETS AND BROCHURES

Architects and Tile Contractors Chemical Construction Materials Guide. Mertztown, Pa.: Atlas Minerals and Chemicals Inc.

Federal Food and Drug Administration. *Current Good Manufacturing Practice in Manufacturing, Processing, Packing, or Holding of Human Food.* Reprinted Washington, D. C.: G. P. O., 1977.

Gilbert, Don. *Industrial Insectology.* Jonesboro, Ark.: 4-D Enterprises, 1980.

Kice Metal Products Aspiration Brochure. Wichita, Kans.: Kice Metal Products Co. Inc.

Klenzade Handbook Committee. *Klenzade Dairy Sanitation Handbook.* 4th ed. Beloit, Wis.: Klenzade Products, Incorporated, 1960.

Nixon, Harold E. and Revis L. Stephenson. *Design of Industrial Vacuum Cleaning Systems.* New York: Clarkson Industries, Inc., 1974.

U. S. Department of Health, Education and Welfare, Consumer Protection and Environmental Health Service, Environmental Control Administration. *Control of Domestic Rats and Mice* by Bayard F. Bjornson, Kent S. Littig and Harry D. Pratt. Washington, D. C.: G. P. O., 1969.

INDEX

Access platforms. *See* Catwalks
Air exchange. *See also* Ventilation
  equipment, 81-86
  in mold control, 4
Aspirators, 266

Bacteria
  carried by cockroaches, 210
  growth of, 2
Beetles, 210-211
Bins, *See* Storage bins
Building design
  ceilings, 67-68
  design, 38
  employee facilities, 38, 101-104
  landscaping, 38-39
  materials, 38
  security lighting, 39-41
  size and kind of structure, 38
Building layout, preliminary considerations
  cleaning systems, 19
  flow, 16-18
  layout, 18
  lighting, 19
  manufacturing environment, 19
  pests, 20
  pollution control, 19
  product safety, 20
  stories, number of, 14
  waste handling, 20
Building materials and equipment inspection, 164-165

Catwalks
  construction details, 33-34
  materials, 2
  superstructure, 33
Ceilings
  concrete, 67
  gypsum, 68
  maintenance, 172-173
  metal, 67
  plaster, 68
Cleaning systems
  central systems, 190-195
  dry cleaning, 200-202
  process pipe systems, 187-190
  recirculating, 186-187, 190
  room, 185-186
  spray, 187-188
  wet cleaning, 183-184
Cockroaches, 210-212
Concrete floors. *See* Floors
Contamination of food
  foreign material control, 233-266
  insect, 5-7
  microbe, 2-4
  physical/chemical material, 9, 33
  rodent, 7-9
  sand blasting floors, 48
Construction features
  assembly detail, 126-129
  fabrication detail, 118-126
  paint finishes, 129-130
Construction, contractor responsibilities in, 164
Construction progress inspections, 165-168
Conveying systems
  belt, 133-134
  bucket, 136-138
  elevators, 138-139
  pneumatic, 139-143

screw, 134-136

Decks. *See* Catwalks
Doors
air curtain, 72-75
docks, 75-76
strip, 72
trackwell, 69-72
truck, 69
Drainage systems
insulation, 97-100
interceptors, 93
openings, 96
pipe arrangement, 93-95
plant, 91
sanitary, 89
supports, 97
Dust collection systems, 151-154

Electric light insect traps, 213-216
Electrical design requirements
insect-sensitive, 156
microbe-sensitive, 156
Electrical equipment
control panels, 157
glass, 159
motors, 158
push-button stations, 159
switch gear and control centers, 157
transformers, 156-157
Electrical hardware
conduit, 160
conduit support, 161-162
ductwork/wireways, 161
junction boxes, 160
switch boxes, 160-161
Electrical systems
design requirements, 156
insect control, 213-216
Employee facilities
basins and showers, 102
cafeterias, 104
locker rooms, 101
lockers, 101-102
toilet rooms, 103-104
Equipment cleaning rooms, 185-186
Equipment construction materials
metals, 114-115
other materials, 116-118
plated metals, 115-116
Equipment design
accessibility, 106-107
bearings, 108
compressed air, 111-113
covers, 110
drives, drive motors, 108
food contact surfaces, 106
framing and supports, 108
hoods, 110
installation, 107
kettles, 110
regulations, 272-273
seals, 108-110
steam cooking, 110-111
Equipment inspection on building site, 164-165
Equipment layout
arrangement, 23
floor trenches, 29
machinery loft, 25
overhead and ceilings, 25-29
preliminary considerations, 21-25
service lines, 23
space requirements, 21-22
Equipment maintenance
approved lubricants, 181
cleanout, 178-180
contamination, 175-177
importance of, 174
temporary repairs, 178
Equipment supports, 29-32

Fabrication
detail, 118-126
on job site, 168
Facility design. *See* Design of building
Federal Food, Drug, and Cosmetic Act, 267-268
Floors

bituminous/asphalt, 54
concrete, 47-52, 55
crushed rock, 54
fiberglass, 52-54
maintenance, 170
metal tiles, 52
resilient materials, 52
service machinery, 25-28
warehouse, 55
wood, 52
Flow
departmental, 17
process and packaging, 18
sheets, 16-18
Food poisoning
botulism, 3-4
clostridium perfringens, 3-4
salmonellosis, 3-4
sources of, 4
staphylococcus, 3-4
Foundations and framing
concrete framing, 46
pest considerations, 41
plated surfaces, 45
steel framing, 41-45
Fumigation, 226-229

Grating for catwalks, 33

HACCP (Hazard Analysis and Critical Control points), 11
Hair
sanitary standards, 11
Heat sterilization insect control, 219-226
Hygiene, personnel practices
construction personnel, 164
hair, sanitary standards, 11
rules of, 10

Impact insect control system, 232
Insect-sensitive foods, 5
Insects
common, 210-212
control of, 6-7, 263-264
control systems of, 213-232
development of, 5
in floors, 46, 51
in walls, 58
Inspections
of building materials and equipment, 164-165
meat and poultry laws, 276
plant, 274
Insulation, 68

Landscaping, 38-39
Layout. *See also* Building design
preliminary, 18
Lighting
design, 76
importance of, 19
maintenance, 173
security, 39-41
types, 77-79

Magnets
location, 234-235
inspection/cleaning, 252-256
installation, 246-252
maintenance, 254
selection, 235-246
Maintenance
building, 170-174
cleanup after, 178-180
during operations, 180
equipment, 174-177
lubricants, 181
materials, 181
preventive, 170
shop and storage area, 180
temporary, 178
Meat and poultry processing regulations, 276
Metal detectors
how they work, 255-257
inspection programs, 261
reject devices, 259-261

search-head sensitivity, 257-259
Microbe
control, 4
-sensitive electrical design, 156
-sensitivity, 2
in wall materials, 61

Paint, finishes, 129-130
Pests
in ceilings, 67
control of, 6-9
control systems, 213-232
floor construction, consideration of, 47
insect development, 5
insect-sensitive electrical design, 156
rodent infestations and control, 7-9
Plant
controls, 19-20
flow of facility, 16-18
location, 14
size and design, 14-16
Plumbing system, 86-87
Pollution
control planning, 19
from floor construction, 49
from roof water, 65
Process controls
engineering, 130-132
regulations, 273-275
Poultry and meat processing regulations, 276
Product risks
foreign material, 9
insects, 5-7
microbes, 2-4
rodents, 7-9

Rodent
control with ultra-sonics, 232
factors in plant location, 14
infestation and control, 7-9
-proofing dock doors, 69, 72
in walls, 61
warehouse floors, and 55

Roofs
construction, 63-65
maintenance, 173

Sanitation
construction specifications, 34-35
for food safety, 2
in floor material selection, 47-49
in plant location, 14
laws and regulations, 269-275
lighting in, 19
programs, 11
Scalping and sifting
application, 261
equipment, 265-266
foreign material control, 264-265
insect control, 263
sampling, 261-263
Security, lighting, 39-41
Sensitivities, product
bacteria, 2
food poisoning, 2
insect, 5
microbe, 2, 4
mold, 4
Site, factors in selecting, 14
Spray cleaning
recirculation, 187-188
Spray/misting insect control system, 216-219
Storage, liquid and transfer
tanks, 148-150
transfer systems, 150-151
Storage bins, bulk and use
rounds, 143-146
use/holding, 146-147

Ventilation
air changes, 79-80
equipment, 81-86
filters, 80
need for, 79
odor control, 81
plumbing, 86-87

Walls
  concrete, concrete block 56-58
  finishes of; 58-59
  construction details, 63
  maintenance, 170-172
  other wall materials, 59-63
Waste
  disposal systems, 203-207
  plant location consideration, 14
  salvage, 207
Water
  backflow, 88-89
  drainage systems, 89
  industrial wastes, 93
  insulation, 97
  interceptors, 93
  openings, wall and floor, 96
  pipe lines, 89
  pipe supports, 97
  piping, 93-95
  potable/nonpotable, 87-88
  reuse of, 88
  storage, 89
Wet cleaning systems
  central cleaning systems, 191-196
  clean-in-place, 186-190
  clean-out-of-place, 185-186
  individual hot water hose station, 195-199
Windows
  construction, 68-69
  maintenance, 173